INTEGER PROGRAMMING

Theory and Practice

The Operations Research Series

Series Editor: A. Ravi Ravindran
Dept. of Industrial & Manufacturing Engineering
The Pennsylvania State University, USA

Operations Research: A Practical Approach
Michael W. Carter and Camille C. Price

Operations Research Calculations Handbook
Dennis Blumenfeld

Integer Programming: Theory and Practice
John K. Karlof

Forthcoming Titles

Applied Nonlinear Optimization Modeling
Janos D. Pinter

Operations Research and Management Science Handbook
A. Ravi Ravindran

INTEGER PROGRAMMING
Theory and Practice

Edited by John K. Karlof

CRC Press
Taylor & Francis Group
Boca Raton London New York

CRC Press is an imprint of the
Taylor & Francis Group, an **informa** business
A TAYLOR & FRANCIS BOOK

CRC Press
Taylor & Francis Group
6000 Broken Sound Parkway NW, Suite 300
Boca Raton, FL 33487-2742

First issued in paperback 2019

© 2006 by Taylor & Francis Group, LLC
CRC Press is an imprint of Taylor & Francis Group, an Informa business

No claim to original U.S. Government works

ISBN-13: 978-0-8493-1914-3 (hbk)
ISBN-13: 978-0-367-39211-6 (pbk)
Library of Congress Card Number 2005041912

Library of Congress Cataloging-in-Publication Data

Integer programming : theory and practice / edited by John K. Karlof.
    p. cm.
  ISBN 0-8493-1914-5 (alk. paper)
  1. Integer programming. I. Karlof, John K.

T57.74.I547 2005
519.7'7--dc22   2005041912

Visit the Taylor & Francis Web site at
http://www.taylorandfrancis.com

and the CRC Press Web site at
http://www.crcpress.com

Preface

Integer programming is a rich and fertile field of applications and theory. This book contains a varied selection of both. I have purposely included applications and theory that are usually not found in contributed books in the hope that the book will appeal to a wide variety of readers. Each of the chapters was invited and refereed. I want to thank the contributors as well as the referees, who took great care in reviewing each submitted chapter.

The Boolean optimization problem (BOOP) is based on logical expressions in prepositional first-order logic, with a profit associated with variables having a true (or false) value subject to these variables making a logical expression true (or false). BOOP represents a large class of binary optimization models that include weighted versions of set covering, graph stability, set partitioning, and maximum-satisfiability problems. In Chapter 1, Lars Hvattum, Arne Løkketangen, and Fred Glover describe new constructive and iterative search methods for solving the BOOP.

Duan Li, Xiaoling Sun, and Jun Wang report recent developments in Chapter 2 on convergent Lagrangian techniques that use objective level-cut and domain-cut methods to solve separable nonlinear integer-programming problems. The optimal solution to the Lagrangian-relaxation problem does not necessarily solve the original problem, even for linear or convex integer-programming problems. This circumstance is the duality gap. The idea of the new cutting methods is based on the observation that the duality gap can be eliminated by reshaping the perturbation function. Thus, the optimal solution can be exposed to the convex hull of the revised perturbation function, which guarantees the convergence of the Lagrangian method on the revised domain.

Robert Nauss discusses the generalized assignment problem (GAP) in Chapter 3. The GAP is concerned with assigning m agents to M tasks so that the assignment costs are minimized, each task is assigned to exactly one agent, and resource limitations for each agent are enforced. The GAP may be formulated as a binary-integer linear-programming problem. The problem can be very difficult to solve with as few as 35 agents and tasks. A special-purpose branch-and-bound algorithm that utilizes a number of tools such as Lagrangian relaxation, subgradient optimization, lifted cover cuts, logical feasibility testing, and specialized feasible-solution generators is presented.

In Chapter 4, Ted Ralphs and Matthew Galati discuss the use of decomposition methods to obtain bounds on the optimal value of solutions to integer linear-programming problems. Let P be the convex hull of feasible solutions. Most bounding procedures are based on the generation of a polyhedral approximation to P. The most common approximation is the continuous approximation. Traditional dynamic procedures for augmenting the continuous approximation fall generally into cutting-plane methods and methods that dynamically generate extreme points to improve the approximation. Ralphs and Galati describe the principle of decomposition and its application in the traditional setting. They extend the traditional framework to show how the cutting-plane method can be integrated with either the Dantzig-Wolfe method or the Lagrangian method to generate improved bounds. They introduce a new concept called structured separation and show how it can be used in a decomposition framework. Software implementation is also discussed.

Chapter 5 contains models and solution algorithms for the rescheduling of airlines that result from the temporary closure of airports. Shangyao Yan and Chung-Gee Lin consider the operations of a multiple fleet with one-stop and nonstop flights when a single airport is temporally closed, most often because of weather. A basic model is first constructed as a multiple time-space network, from which several strategic network models are developed for scheduling. These network models are formulated as pure network-flow problems, network-flow problems with side constraints, or multiple-commodity network-flow problems. The first are solved by use of the network-simplex method, and the others are solved by application of a Lagrangian relaxation-based algorithm. The models are shown to be useful in actual operations by tests on the operations of a major airline.

Chapter 6 and Chapter 7 deal with the determination of an optimal mix of self-owned and chartered vessels of different types that are needed to transport a product. Chapter 6 considers transportation between a single source and a single destination, and Chapter 7 considers multiple sources to various destinations. Hanif Sherali and Salem Al-Yakoob develop integer-programming models to determine an optimal fleet mix and schedule. The new models are solved by application of an optimization package and are compared to an *ad hoc* scheduling procedure that simulates how schedules are generated by a major petroleum corporation.

Chapter 8 presents an application of integer programming that involves the capture, storage, and transmission of large quantities of data collected during a variety of possible testing scenarios that might involve military ground vehicles, cars, medical applications, large equipment, missiles, or aircraft. The particular application that this chapter focuses on is testing military aircraft. A large amount of data relating to parameters such as speed, altitude, mechanical stress, and pressure is collected. Typically, several hundred or possibly thousands of parameters are continuously sampled during the flight, with a subset of these transmitted to a ground station. The parameters to be transmitted are multiplexed into a data structure called a *data cycle map*, a sequence of digital words. Data cycle maps are constructed subject to certain regulations. One of the most constraining features of the data cycle map construction process is that each parameter must appear periodically within the map. In Chapter 8, David Panton, Maria John, and Andrew Mason show how a set-packing integer-programming model may be used to find data cycle map constructions that are feasible and efficient.

Govind Daruka and Udatta Palekar consider in Chapter 9 the problem of determining the assortment of products that must be carried by the stores in a retail chain to maximize profit. They develop an integer linear-programming model to solve this problem. The model considers sales forecasts and constrains the assortments on the basis of available space, desired inventory turns, advertising restrictions, and other product-specific restrictions. The resulting model is solved by use of a column-generation approach. The model and algorithm were implemented for a large retail chain and have been successfully used for several years.

Chapter 10 contains an overview of noncommercial software tools for the solution of mixed-integer linear programs (MILP). Jeff Linderoth and Ted Ralphs first review solution methodologies for MILP, and then present an overview of the available software, including detailed descriptions of eight publicly available software packages. Each package is categorized as a black-box solver, a callable library, a solver framework, or some combination of these. The distinguishing features of all eight packages are fully described. The chapter concludes with case studies that illustrate the use of two of the solver frameworks to develop custom solvers for specific problem classes and with benchmarking of the six black-box solvers.

The Editor

John Karlof grew up in upstate New York and graduated from The State University of New York, Oswego, in 1968 with a B.A. in mathematics. He received M.A. (1970) and Ph.D. (1973) degrees from the University of Colorado, Boulder, in mathematics. He attended The State University of New York, Stony Brook, in 1981 and received an M.S. in operations research. Dr. Karlof was professor of mathematics at the University of Nebraska at Omaha from 1974 to 1984 and is currently professor of mathematics and graduate coordinator at the University of North Carolina, Wilmington, where he has been since 1984. His research interests include coding theory, job scheduling, and mathematical programming. He has published several papers and directed masters theses in these areas. In his spare time, Dr. Karlof enjoys racing and cruising on his 30-foot sailboat, *Epsilon,* in coastal Carolina.

Contributors

Govind P. Daruka Department of Mechanical and Industrial Engineering, University of Illinois, Urbana, Illinois

Matthew V. Galati Analytical Solutions — Operations R&D, SAS Institute, Cary, North Carolina

Fred Glover Leeds School of Business, University of Colorado, Boulder, Colorado

Lars M. Hvattum Molde University College, Molde, Norway

Maria John Center for Industrial and Applied Mathematics, University of South Australia, Mawson Lakes, South Australia

Duan Li Department of Systems Engineering and Engineering Management, The Chinese University of Hong Kong, Hong Kong, People's Republic of China

Chung-Gee Lin Department of Business Mathematics, Soochow University, Taipei, Taiwan

Jeffrey T. Linderoth Department of Industrial and Systems Engineering, Lehigh University, Bethlehem, Pennsylvania

Arne Løkketangen Molde University College, Molde, Norway

Andrew Mason Department of Engineering Science, University of Auckland, Auckland, New Zealand

Robert M. Nauss College of Business Administration, University of Missouri, St. Louis, Missouri

Udatta S. Palekar Department of Mechanical and Industrial Engineering, University of Illinois, Urbana, Illinois

David Panton Center for Industrial and Applied Mathematics, University of South Australia, Mawson Lakes, South Australia

Ted K. Ralphs Department of Industrial and Systems Engineering, Lehigh University, Bethlehem, Pennsylvania

Hanif D. Sherali Department of Industrial and Systems Engineering, Virginia Polytechnic Institute and State University, Blacksburg, Virginia

Xiaoling Sun Department of Mathematics, Shanghai University, Shanghai, People's Republic of China

Jun Wang Department of Systems Engineering and Engineering Management, The Chinese University of Hong Kong, Hong Kong, People's Republic of China

Salem M. Al-Yakoob Department of Mathematics and Computer Science, Kuwait University, Kuwait City, Kuwait

Shangyao Yan Department of Civil Engineering, National Central University, Chungli, Taiwan

Referees

Kurt Brethauer Operations and Decision Technologies Department, Kelley School of Business, Indiana University, Bloomington, Indiana, U.S.

Yaw Chang Department of Mathematics and Statistics, University of North Carolina Wilmington, U.S.

Marco Dorigo Iridia, Universite' Libre de Bruxelles, Belgium

John Forrest IBM, T. J. Watson Research Center, Yorktown Heights, New York, U.S.

Lou Hafer School of Computing Science, Simon Fraser University, Burnaby, British Columbia, Canada

Tabitha James Department of Business Information Technology, Virginia Polytechnic Institute and State University, Blacksburg, Virginia, U.S.

Charles Jones

Andrew Lim   Department of Industrial Engineering and Engineering Management, Hong Kong University of Science and Technology, Hong Kong

Tassos Perakis   Department of Naval Architecture and Marine Engineering, University of Michigan, Ann Arbor, Michigan, U.S.

David Ronen   College of Business Administration, University of Missouri — St. Louis, U.S.

Matthew Tenhuisen   Department of Mathematics and Statistics, University of North Carolina Wilmington, U.S.

Dusan Teodorovic   Department of Civil and Environmental Engineering, Virginia Polytechnic Institute and State University, Blacksburg, Virginia, U.S.

Joris van de Klundert   Department of Mathematics, Maastricht University, Maastricht, The Netherlands

Mutsunori Yagiura   Department of Applied Mathematics and Physics, Kyoto University, Kyoto, Japan

Contents

1

New Heuristics and Adaptive Memory Procedures for Boolean Optimization Problems

Lars M. Hvattum, Arne Løkketangen, and Fred Glover

CONTENTS

1.1   Introduction

The Boolean Optimization Problem (BOOP) represents a large class of binary optimization models, including weighted versions of Set Covering, Graph Stability, Set Partitioning, and Maximum Satisfiability problems. These problems are all NP-hard, and exact (provably convergent) optimization methods encounter severe performance difficulties in these particular applications, being dominated by heuristic search methods even for moderately sized problem instances.

Previous heuristic work on this problem is mainly by Davoine, Hammer, and Vizvári [2], employing a greedy heuristic based on pseudo-boolean functions. Hvattum, Løkketangen, and Glover [11] describe simple iterative heuristic methods for solving BOOP, starting from random initial solutions. Although equipped with no long-term mechanism apart from a random restart procedure, they obtain very good results compared to the work by Davoine, Hammer, and Vizvári, and also by an even greater margin when compared to CPLEX and XPRESS/MP on the larger problems.

The remainder of this chapter is organized as follows. Section 1.2 provides BOOP problem formulations and details of previous work. Section 1.3 describes new local search mechanisms, designed to diversify the search, while Section 1.4 describes our new constructive methods. In Section 1.5 we address the Weighted Maximum Satisfiability problem (W-MAX_SAT), and show how to transform it into a BOOP formulation framework. Computational results are given in Section 1.6, followed by the conclusions in Section 1.7.

1.2   Problem Formulation and Search Basics

1.2.1   Problem Formulation

The Boolean Optimization Problem (BOOP), first formulated in Davoine, Hammer, and Vizvári [2], is based on logical expressions in prepositional, first-order logic, with an extra cost (or profit) associated with the variables having a *true* (or *false*) value. One formulation can be (assuming maximization)

$$Max\, z = \sum_{i=1}^{N} (c_i | x_i = true/false)$$

such that

$$\Phi(x) = \Phi(x_1, \ldots, x_N) = \begin{cases} true \\ false \end{cases}$$

where $\Phi(x)$ is the logical expression, and N the number of variables. The solution to this problem is the set of truth value assignments to the x_i variables that yields the highest objective function value z, while satisfying the logical expression. The logical expression can in general be arbitrary, but we restrict ourselves to formulations in *conjunctive normal form, CNF.* (The *disjunctive normal form* can be obtained by a simple transformation.) Informally, a BOOP can be regarded as a *satisfiability problem* (SAT) with an objective function added on. For more information on SAT, see, for example, Cook [1], and Du, Gu, and Pardalos [4].

Applying simple transformations described in Hvattum, Løkketangen, and Glover [11], we get the following model by splitting each x_i into its *true* and *false* component y_i and $y_{i\#}$:

$$Max\, z = \sum_{i=1}^{N} c_i y_i$$

s.t.

$$Dy \geq 1$$
$$y_i + y_{i\#} = 1$$

where D is the 0-1 matrix obtained by substituting the y's for the x_i's. The last constraint is handled implicitly in the search heuristics we introduce.

1.2.2   Local Search Basics

To better understand the mechanisms described in this chapter, some background from previous work is helpful. Further details can be found in Hvattum, Løkketangen, and Glover [11], whereas an introduction to tabu search can be found in, e.g., Glover and Laguna [10]. The basic strategy of the earlier work includes the following features:

- The *starting solution* (or starting point) is based on a random assignment to the variables. This solution may be primally infeasible, and hence the search must be able to move in infeasible space.
- A *move* is the flip of a variable by assigning the opposite value (i.e., change $1 \to 0$ or $0 \to 1$).
- The *search neighborhood* is the full set of possible flips, with a *neighborhood size* of N, the number of variables.
- *Move evaluation* is based on both the change in objective function value, and the change in amount of infeasibility.
- The *move selection* is greedy (i.e., take the best move according to the move evaluation).
- Simple randomized *tabu tenure* and a *new best* aspiration criterion are used.
- A random restart is applied after a certain number of moves, to *diversify* the search.
- The *stopping criterion* is a simple time limit.

The manner in which we incorporate these features, and add new ones to our current method, is sketched in the following sections.

1.2.3 Move Evaluation Function

The move evaluation function, F_{Mi}, has two components. The first is the change in objective function value. The cost coefficients, c_i, are initially normalized to lie in the range $(0,1)$. This means that the change in objective function value per move, Δz_i, is in the range $(-1, +1)$.

The second component is the change in the number of violated clauses (or constraint rows), for the flipping of each variable. This number, ΔV_i will usually be a small positive or negative integer. For a different way to handle infeasible solutions, see Løkketangen and Glover [12].

These two components are combined to balance the considerations of obtaining solutions that are feasible and that have a good objective function value. The relative emphasis between the two components is changed dynamically to focus the search in the vicinity of the feasibility boundary, using the following move evaluation function:

$$F_{Mi} = \Delta V_i + w * \Delta z_i$$

The value of w, the adaptive component, is initially set to 1. It is adjusted after each move so that:

- If the current solution is feasible: $\quad\quad w = w + \Delta w_{inc}$
- If the current solution is not feasible, and $w > 1$: $\quad w = w - \Delta w_{dec}$

The effect of this adaptation is to induce a strategic oscillation around the feasibility boundary. A different approach appears in Glover and Kochenberger [9], where the oscillation is coupled with the use of a critical event memory, forcing the search into new areas.

1.3 Local Search Improvements

The simple local search described in Hvattum, Løkketangen, and Glover [11] relies on a sophisticated adaptive move evaluation scheme for achieving the type of balance between feasibility and objective function quality previously described. From their computational results, however, it is evident that for the larger test cases a better form of diversification than random restart is needed to be able to explore larger parts of the search space.

The extra mechanisms come at a cost. There is a tradeoff between the gains provided by improved search guidance or diversification, and the associated computational effort to perform the extra calculations and to maintain the auxiliary data structures. In the current setting, the additional mechanisms reduce the number of search iterations done in a given amount of computational time.

We have implemented two processes for diversification: *Adaptive Clause Weighting*, and *Probabilistic Move Acceptance*.

1.3.1 Adaptive Clause Weights

In the basic local search scheme, all violated clauses (i.e., constraint rows) contribute the same amount to the move evaluation function, F_{Mi}, as described in Section 1.2.3. However, some of the clauses will be more difficult to satisfy than others, and should be given more emphasis. We achieve this by attaching a separate weight, CW, to each clause. Previous work on adaptive clause weights can be found in Løkketangen and Glover [13].

All clauses start with $CW = 1$. The weight is updated only after iterations where a clause becomes violated, at which point the weight of the newly violated clause is incremented by a small amount, ΔCW. To prevent clause weights from growing prohibitively large, they are renormalized by dividing all the clause weights by a constant CW_{DIV}, whenever one weight becomes greater than some CW_{LIM}.

Such a procedure constitutes a long-time learning approach. The move evaluation function drives the search out of the feasible region to seek solutions with high objective function quality in nearby infeasible space. Having adaptive clause weights helps the search to better adapt to the infeasibility border of the search space, thus enabling the search to cross back over the border to find different, and better, feasible solutions. As shown in Section 1.6.1, the tradeoff between the extra time taken to update the weights, and the resulting improved search guidance pays the greatest dividends for the larger problems.

1.3.2 Probabilistic Move Acceptance

In every iteration the search method generates a list that identifies a subset of possible moves to execute, and the *best* move from this list is selected. Usually this *best* equates with best *move evaluation value*. But the move evaluation function is rather myopic, only looking at the local neighborhood, and we modify it by using recency and frequency measures as proposed in tabu search. (See, e.g., Glover and Laguna [10] and Gendreau [6]).

In a sorted list of possible moves, the presumably *best* moves will be at the front of the list, but not necessarily in strict order. A simplified variant of this principle from Glover [7] is also employed in GRASP, where the chosen move is randomly selected among the top half of the moves (see Feo and Resende [5]).

We use this approach by selecting randomly from the top of the list, but in a way biased toward the moves having the highest evaluations. This is called *Probabilistic Move Acceptance*, *PMA*, as described in Løkketangen and Glover [12]. The selection method is as follows:

PMA:

1. Select a move acceptance probability, p.
2. In each iteration sort the admissible moves according to the move evaluation function.
3. Reject moves with probability $(1 - p)$ until a move is accepted.
4. Execute the selected move in the normal way.
5. If not finished, go to 2 and start the next iteration.

This can also be viewed as using randomness to diversify the search (as a substitute for deterministic use of memory structures), but in a guided way.

In our local search setting, using PMA generally yields worse results than the deterministic approach of always taking the best nontabu move. This implies that the move evaluation function is good, and that rejecting the top moves deteriorates the search.

The inclusion of PMA yields better results for the largest problems (with up to 1000 variables and 10,000 clauses). This indicates that the PMA introduces some necessary diversification that the basic mechanisms lack.

1.4 Constructive Methods

Constructive methods in the literature are mainly used for creating good, or feasible, starting solutions for subsequent local search heuristics. We show how proper use of adaptive memory structures derived from tabu search can be used to create iterated constructive learning heuristics. These generate a series of solutions, where the constructive guidance is modified by the outcome of the previous searches. Our ideas are based on principles for exploiting adaptive memory to enhance multi-start methods given in Glover [8]. A general discussion of multi-start methods can be found in Martí [14].

In each iteration of our constructive method we start with all the variables unassigned, and then greedily assign one variable to each step, based on an evaluation of the available assignments, until all variables have received a value.

We focus in particular on implementing the principles embodied in the *PAM* (Persistent Attractiveness Measure) and *MCV* (Marginal Conditional Validity) concepts. As is customary, our methods also incorporate a short local search after each constructive phase.

1.4.1 PAM–Persistent Attractiveness Measure

The *Persistent Attractiveness Measure*, PAM, is a measure of how often a specific value assignment to a variable is considered attractive, without actually being selected for inclusion in the solution. It is reasonable to assume that early assignments in the construction phase have more impact than later assignments, and that good variable assignments are usually given good evaluations. The attractiveness should also increase for a variable assignment that is repeatedly ranked high without being chosen.

If we index the assignment steps with s, and the individual rankings in a step with r, we would like the PAM to have the following properties:

PAM(r, s) is decreasing for increasing s (earlier steps are more important)

PAM(r, s) is decreasing for increasing r (higher rank is better)

We only calculate PAM for the top ranked moves.

The PAM evaluator can be formulated as $E(s, r) = E'(s) + E''(r)$, where, for BOOP, we set

$$E'(s) = as^* - as$$

and

$$E''(r) = br^* - br$$

where $s^* = N$ (number of variables) and r is a parameter. The constants a and b are determined experimentally as subsequently described.

The PAM-values corresponding to a given assignment are summed over all the constructive steps to yield an overall measure of attractiveness for each possible assignment.

PAM values for several consecutive constructive runs can be accumulated in a measure of attractiveness, e.g., by exponential smoothing:

New PAM $=$ (Last PAM $+$ Last Accumulated Total PAM)$/2$

We thus expand the move evaluation indicated earlier to become:

$$F(y_{i(\#)}) = \Delta V_{i(\#)} + w^*(z_{i(\#)} + PAM_{i(\#)})$$

The values for the PAM-measure are limited to an interval $[0, k]$, with k chosen to match w in some way, again as specified later. The parameter w is updated as in the local search (see Section 1.2.3), but only after the completion of each constructive run.

1.4.2 MCV–Marginal Conditional Validity

The choices made at the beginning of a constructive search are based on less information than later choices, and are thus more likely to be *bad*. When later choices are made, the problem has been reduced by the earlier choices, and better choices can be made (but in the context of the earlier ones). Later decisions are thus likely to make earlier decisions look better. We call this the *Marginal Conditional Validity* principle.

After the constructive phase we analyze the completed solution to find assignments that should have been different. There are two cases that can be used as a foundation:

1. A variable is *true*, but there are unsatisfied rows where the negated variable is present.

2. A variable is *false*, but the negated variable is present only in rows that would be satisfied even if the variable had been flipped.

In the first case the opposite value assignment to the variable would possibly satisfy more rows, while in the second case we would get an increase in the objective function value, without violating any new constraints. Each of the assignments recognized in this manner is enforced in the start of the next constructive run with a given probability p.

1.4.3   A Comparison with GRASP

The *Greedy Randomized Adaptive Search Procedure*, or GRASP, is a well known, memoryless, constructive heuristic relying heavily on randomization (see Resende, Pitsoulis, and Pardalos, 1997). A constructive run can be followed by a short greedy local search.

We have adapted and implemented GRASP to work for BOOP, for comparison purposes. We use the same basic objective function value as before, but without any adaptive memory or learning. The only parameter required for GRASP is the proportion of moves to be considered for execution in each constructive assignment, called α.

GRASP:

1. Start with all variables unassigned, rate all possible assignments.
2. Select an assignment randomly among those who are within $\alpha\%$ of best evaluation.
3. When all variables are assigned, possibly do a local search.
4. Go to 1, if not finished.

We use time as a stopping criterion, and try three versions of local search (LS): No LS, complete LS, or "steepest ascent" LS.

We also tried to augment GRASP with learning capabilities by introducing the adaptive component w in the evaluation function, as for our other constructive approach (see Section 1.4.1). Computational results are in Section 1.6.4.

1.5   Weighted Maximum Satisfiability

To support the claim that BOOP can represent many different problem classes, this section outlines how Weighted Maximum Satisfiability problems (W-MAX_SAT) can be easily transformed to BOOP. Section 1.6.5 gives computational results for this case, without any effort to specialize our procedure to handle the special structure of this problem.

A W-MAX_SAT instance can informally be regarded as an unsatisfiable instance of a SAT problem that in addition has *weights* on the clauses (rows). The objective is then to find a truth assignment that maximizes the sum of the weights on the satisfied clauses. This is similar to BOOP, except that weights are attached to the clauses rather than the variables. A W-MAX_SAT instance can be transformed to BOOP by adding a new variable to each clause to carry information about weights. The clause weights are transformed to objective function value coefficients for the new variables in the corresponding clauses, while the original n variables will have objective function value coefficients of 0.

Thus, if the W-MAX_SAT has n variables and m clauses, the BOOP will have $(n + m)$ variables and m clauses. The number of clauses (rows), m, is often large compared to the number of variables, n, giving a BOOP encoding for W-MAX_SAT having many more variables. (In the test instances used in Section 1.6.5 n is 100 and m is 800 to 900, giving 900 to 1000 variables for the BOOP encoding, compared to 100 for W-MAX_SAT.)

As we can see in the computational results Section 1.6.5, our BOOP code compares favorably to the GRASP heuristic on the same problem instances (Resende, Pitsoulis, and Pardalos [15]), and is only slightly worse than the special purpose method of Shang [17] in spite of the fact that no specialization is used in our procedure.

1.6   Computational Results

This section reports the final parameter settings applied to each of the different methods or mechanisms during testing, as well as overall computational results. Section 1.6.6 attempts to compare all the different methods and mechanisms in a meaningful way.

The same BOOP test cases as used in the previous work (Hvattum, Løkketangen, and Glover [11] and Davoine, Hammer, and Vizvári [2]) are used for testing. The test-set consists of 5485 instances, ranging in size from 50 to 1000 variables, and 200 to 10000 clauses (rows). Results are reported as the average of solution values relative to results obtained by Davoine, Hammer, and Vizvári using CPLEX 6.0.

The testing of W-MAX_SAT is based on modifying the unsatisfiable *jnh**, as used in Resende, Pitsoulis, and Pardalos [15]. These all have 100 variables and 800 to 900 clauses. For preliminary testing to fix parameter values, we selected the same three test cases as in Hvattum, Løkketangen, and Glover [11].

1.6.1   Effect of Adaptive Clause Weights

The first addition to the mechanisms for BOOP described in Hvattum, Løkketangen, and Glover [11], is the inclusion of adaptive clause weights (see Section 1.3.1). Preliminary testing showed that the results were not very sensitive to the values of CW_{LIM} (the maximum weight value) or CW_{DIV} (the renormalization factor). For our final testing we used $CW_{LIM} = 4.0$ and $CW_{DIV} = 2.0$. The best value for ΔCW was chosen to be 0.003, based on preliminary testing. The actual value is not sensitive, but it should be much smaller than 1.

Computational results are shown in Table 1.1 The results using Adaptive Clause Weights (ACW) are compared to the results from Hvattum, Løkketangen, and Glover [11] (TS), with computational time of 5 and 60 seconds.

TABLE 1.1

Adaptive Clause Weights

	TS 5	TS 60	ACW 5	ACW 60
Classes 1–22	100.001	100.001	100.002	100.002
Classes 23–49	101.214	101.215	101.212	101.214
Classes 50–54	106.305	106.982	107.628	107.866
Classes 55–63	102.463	102.465	102.462	102.464
Classes 1–63	101.373	101.427	101.477	101.497

As can be seen, the overall results show an improvement for both the 5 second and 60 second cutoff. For classes 55 to 63 the results are slightly inferior to those of our earlier approach.

1.6.2   Effect of Probabilistic Move Acceptance

The important parameter for PMA is the probability of move acceptance, p (see Section 1.3.2). In Table 1.2 and Table 1.3 are shown the results for a selected test case for various values of p without and with adaptive clause weights (ACW). As is indicated in the tables, a fairly large value should be chosen for p. In our subsequent test we use the value 0.9. Overall computational results are shown in Table 1.4. The use of PMA gives in general slightly inferior results, except for the largest problems. This is as expected, as the search guidance (through the move evaluation value) should be better for smaller problems. The PMA also introduces a certain amount of diversification that is helpful for the larger problems.

1.6.3   PAM and MCV

Preliminary testing gave the following values for the *PAM (Persistent Attractiveness Measure)* and *MCV (Marginal Conditional Validity)* parameters, whose

TABLE 1.2

PMA without ACW

p	Obj. value	Time
0.1	142796	3.86
0.2	143138	3.47
0.3	143255	7.05
0.4	143315	5.27
0.5	143356	4.17
0.6	143367	4.02
0.7	143372	1.98
0.8	143372	1.20
0.9	143372	1.59
1.0	143372	1.33

TABLE 1.3

PMA with ACW

p	Obj. value	Time
0.1	142845	5.29
0.2	143148	6.35
0.3	143246	6.18
0.4	143323	5.51
0.5	143357	6.11
0.6	143365	4.62
0.7	143369	2.27
0.8	143372	1.96
0.9	143372	1.27
1.0	143372	0.82

role is sketched in Section 1.4.1 and Section 1.4.2:

$$a = 2$$
$$b = 3$$
$$r^* = 4$$

The PAM-value of each variable assignment is scaled to lie between 0 and 0.3 before it is used in the move evaluation function as specified in Section 1.4.1.

Figure 1.1 shows results for the given test case for various values of p, the probability that determines when to apply the MCV principle. For this test case the best results are when the MCV principle ($p = 0$) is not applied. The results with $p = 0.4$ give best results when applying MCV, and this value is used in the computational testing.

Table 1.5 shows the computational results for our constructive learning heuristic applying both PAM and MCV. The column PAM/MCV-NO LS gives the results when no local search was applied after each constructive run. PAM/MCV-STEEP indicates that a steepest ascent local search was applied after each construction, and PAM/MCV-TS 500 indicates that a tabu search limited to 500 iterations, as described in Section 1.3.1, was used for improvement. All the runs were for 60 seconds. This new constructive method, even without the local search, performs much better than the basic GRASP approach (see Section 1.6.4). The constructive approach without local search (LS)

TABLE 1.4

Results for PMA

	Tabu Search		PMA w.o. ACW		PMA w. ACW	
	TS 5	**TS 60**	**PMA 5**	**PMA 60**	**PMA 5**	**PMA 60**
Classes 1–22	100.001	100.001	99.998	100.000	99.996	99.998
Classes 23–49	101.214	101.215	101.205	101.213	101.205	101.211
Classes 50–54	106.305	106.982	105.787	106.168	107.438	107.778
Classes 55–63	102.463	102.465	102.446	102.463	102.450	102.461
Classes 1–63	101.373	101.427	101.324	101.361	101.455	101.487

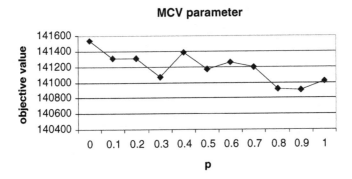

FIGURE 1.1
Values for p for MCV.

also beats the results in Davoine, Hammer, and Vizvári [2] on small instances, and beats, with the addition of a short LS to the constructive approach, these results on all the instances.

It seems that when combining the constructive learning heuristic with the TS from Section 1.3.1, most of the benefit comes from the TS. However, the method PAM/MCV-TS 500 was the only method that finds the optimum of all the small instances (classes 1 to 22, 5280 instances). In fact, all the optima were found within 2 seconds. This seems to reflect the trend we have observed for our constructive heuristics, that they are more effective for the smaller problem instances and do not often contribute improved results for the largest problem instances.

1.6.4   Comparison with GRASP

Results for the GRASP heuristic outlined in Section 1.4.3 are shown in Table 1.6, allowing for 5 or 60 seconds search time. A value of $\alpha = 0.5$ was used. The column GRASP–NO LS shows the results when no local search is applied after the constructive phase, and GRASP–CLS shows the results when a complete, recursive, local search is applied. GRASP–STEEP shows the results when steepest, ascent is used.

These results indicate that GRASP is better than Davoine, Hammer, and Vizvári [2] on small instances, but does not scale well for the larger problems.

TABLE 1.5

Results for PAM and MCV

	PAM/MCV-NO LS	PAM/MCV-STEEP	PAM/MCV-TS 500
Classes 1–22	99.359	99.780	100.002
Classes 23–49	99.571	100.074	101.205
Classes 50–54	97.202	98.545	106.778
Classes 55–63	99.581	99.942	102.448
Classes 1–63	99.310	99.831	101.405

TABLE 1.6

Results for GRASP

	GRASP-NO LS		GRASP-CLS		GRASP-STEEP	
	5 sec	60 sec	5 sec	60 sec	5 sec	60 sec
Classes 1–22	97.483	98.413	99.387	99.680	99.389	99.662
Classes 23–49	95.400	96.149	97.216	97.983	98.326	98.826
Classes 50–54	86.554	89.748	90.164	91.857	93.844	95.969
Classes 55–63	95.647	96.550	97.114	97.950	97.302	97.988
Classes 1–63	95.461	96.489	97.400	98.085	98.195	98.772

When we apply our adaptive component w, in order to balance the importance of feasibility vs. the objective function value, GRASP functions much better. Table 1.7 shows the results using the adaptive component, w, and complete local search. The same values are used for $\Delta w_{inc}(= 0.20)$ and $\Delta w_{dec}(= 0.15)$ as for the TS. The results are now better than Davoine, Hammer, and Vizvári [2], except on classes 50 to 54. This shows that a modified GRASP can compete with other heuristics on small and medium sized instances, while other mechanisms may be needed for the larger ones. The recent work on marrying GRASP with path relinking offers promise in this regard. (See Resende and Ribeiro [16].)

1.6.5 Results for Weighted Maximum Satisfiability

We use the encoding of W-MAX_SAT in the BOOP framework outlined in Section 1.5. Our problem instances are from Resende, Pitsoulis, and Pardalos [15], based on the unsatisfiable "jnh" instances from second DIMACS Implementation Challenge. These test cases have 100 variables, and 800 to 900 clauses (rows). Our BOOP encoding of these problems thus has 900 to 1000 variables and 800 to 900 rows, being somewhat inflated compared to the original encoding.

Computational results are shown in Table 1.8. The settings for ACW 60 (see Section 1.6.1), without any changes, are used. GRASP* shows the results reported in Resende, Pitsoulis, and Pardalos [15]. The column DML shows the results for DML, a Lagrange-based method specially tailored to the problem (Shang [17]).

TABLE 1.7

Results for GRASP with learning

	GRASP w. Learning
Classes 1–22	99.972
Classes 23–49	100.717
Classes 50–54	96.587
Classes 55–63	101.901
Classes 1–63	100.298

TABLE 1.8

Results for W-MAX_SAT

Problem	Optimal	GRASP*	DML	ACW 60
jnh1	420925	−188	0	0
jnh4	420830	−215	−41	−85
jnh5	420742	−254	0	−116
jnh6	420826	−11	0	−15
jnh7	420925	0	0	0
jnh8	420463	−578	0	0
jnh9	420592	−514	−7	−327
jnh10	420840	−275	0	0
jnh11	420753	−111	0	−250
jnh12	420925	−188	0	0
jnh13	420816	−283	0	0
jnh14	420824	−314	0	−172
jnh15	420719	−359	0	−52
jnh16	420919	−68	0	−5
jnh17	420925	−118	0	0
jnh18	420795	−423	0	−207
jnh19	420759	−436	0	0
jnh201	394238	0	0	0
jnh202	394170	−187	0	−126
jnh203	394199	−310	0	−137
jnh205	394238	−14	0	0
jnh207	394238	−137	0	−9
jnh208	394159	−172	0	−162
jnh209	394238	−207	0	0
jnh210	394238	0	0	0
jnh211	393979	−240	0	0
jnh212	394238	−195	0	0
jnh214	394163	−462	0	0
jnh215	394150	−292	0	−199
jnh216	394226	−197	0	0
jnh217	394238	−6	0	0
jnh218	394238	−139	0	0
jnh219	394156	−436	0	−103
jnh220	394238	−185	0	−33
jnh301	444854	−184	0	0
jnh302	444459	−211	−338	0
jnh303	444503	−259	−143	−414
jnh304	444533	−319	0	−570
jnh305	444112	−609	−194	−299
jnh306	444838	−180	0	0
jnh307	444314	−155	0	−685
jnh308	444724	−502	0	−699
jnh309	444578	−229	0	0
jnh310	444391	−109	0	0
Average	415914	−233	−16	−106

Our computational results compare favorably to those of the GRASP heuristic on the same problem instances. Our outcomes are only slightly worse than those of the special purpose DML method of Shang [17], although we are undertaking to solve the much larger transformed problem and make no use of any specialization.

1.6.6 Performance Profiles

It is always very difficult to compare different methods based on tables of computational results, unless one method is best on all the tests. We therefore also compare our methods using the ideas given in Dolan and Moré [3]. Based on the time used to find the best solution, we can construct a performance profile as follows.

Let P be the set of problem instances, S be the set of solvers, and n_p be the number of problems. Define $t_{p,s}$ to be the time used by solver s to solve problem p. Let

$$r_{p,s} = \frac{t_{p,s}}{\min\{t_{p,s*}|s^* \in S\}}$$

be the ratio between the performances of solver s to the *best* solver on the problem p. If a solver fails to solve a problem, then set $r_{p,s} = r_M$, where $r_M \geq r_{p,s}$ for all p and s.

A measure of the performance of a solver can be given by

$$\rho_s(\tau) = \frac{1}{n_p}size\{p \in P|r_{p,s} \leq \tau\}$$

where $\rho_s(\tau)$ is the probability that for solver s, the ratio of performance $r_{p,s}$ is within a factor τ of the best ratio. A plot of $\rho_s(\tau)$ for the different solvers will give interesting characteristics of the solvers. Please note that $\rho_s(1)$ gives the proportion of problems where s is winning over the other solvers.

For many of our problem instances the reported solution time is *very* small, and the solvers report 0. All these instances are removed from this comparison. This is not necessarily a drawback, as the remaining problems' instances presumably are the most interesting ones.

Figure 1.2 shows performance profiles for the following six methods:

- **ACW 60** – Tabu Search with adaptive clause weights
- **TS 60** – Tabu Search without adaptive clause weights
- **PMA 60 TS** – Tabu Search without adaptive clause weights, but with PMA
- **CON ACW** – Constructive Search, followed by TS
- **CON LS** – Constructive Search, followed by Steepest Ascent
- **CON** – Constructive Search – No LS

The allotted solution times are 60 seconds per problem instance.

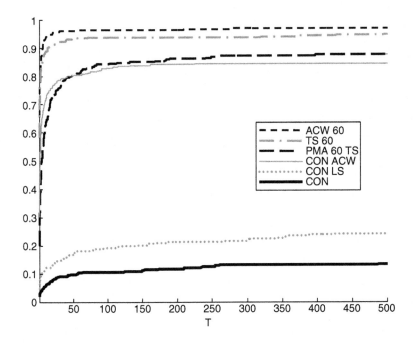

FIGURE 1.2
Performance profiles.

Of the original 5485 problem instances, 352 were left after removing those where at least one of the solvers reported a solution time of 0 seconds. Of these remaining problems there are 299 where not all the solvers find the same solution value.

As can be seen, the method ACW 60 is the best. It is of interest to note that when solution time is small (less than a factor 40 from the best solver on each particular instance), is that CON ACW is better than PMA 60 TS, while for longer solution times PMA 60 TS is better.

1.7 Conclusions

We have shown the value of certain types of adaptive memory to improve the performance of heuristics, both iterative and constructive. Our results compare very favorably to previous published results, and are significantly better than those obtained by exact solvers (XPRESS and CPLEX).

For BOOP, we have achieved the best results by using a tabu search based heuristic, augmented by an adaptive move evaluation function, and adaptive clause weights. Very good results are also obtained for constructive heuristics augmented by the learning schemes of *PAM* (*Persistent Attractiveness Measure*) and *MCV* (*Marginal Conditional Validity*).

We also show that our approach can be applied to Weighted Maximum Satisfiability problems by transforming them into (larger) BOOP problems, and that without specialization to the W-MAX_SAT setting we obtain results comparable to those of the better specialized methods from the literature.

References

1. Cook, S. A., The complexity of theorem-proving procedures, in *Proc. Third ACM Symp. on Theory of Computing*, 1971, 151.
2. Davoine, T., Hammer P. L., and Vizvári, B., A heuristic for Boolean optimization problems, *J of Heuristics*, 9, 229, 2003.
3. Dolan, E. D. and Moré J. J., Benchmarking optimization software with performance profiles, preprint ANL/MCS-P861-1200, Mathematics and Computer Science Division, Argonne National Laboratory, 2001.
4. Du, D., Gu, J., and Pardalos, P., Eds., *Satisfiability Problem: Theory and Applications*, DIMACS Series in Discrete Mathematics and Theoretical Computer Science, Vol. 35, AMS, 1997.
5. Feo, T. A. and Resende M. G. C., A probabilistic heuristic for a computationally difficult set covering problem, *Operations Research Letters*, 8, 67, 1989.
6. Gendreau, M., An introduction to tabu Search, in *Handbook of Metaheuristics*, Glover, F. and Kochenberger, G., Eds., Kluwer Academic Publishers, 2003, 37.
7. Glover, F., Tabu search — part I, *ORSA Journal on Computing*, 1, 190, 1989.
8. Glover, F., Multi-start and strategic oscillation methods — principles to exploit adaptive memory, in *OR Computing Tools for Modeling, Optimization and Simulation: Interfaces in Computer Science and Operations Research*, Laguna, M. and González-Velarde, J. L., 2000, 1.
9. Glover, F. and Kochenberger, G., Critical event tabu search for multidimensional knapsack problems, in *Meta Heuristics: Theory and Applications*, Osman, I. H. and Kelly, J. P., Eds., Kluwer Academic Publishers, 1996, 407.
10. Glover, F. and Laguna, M., *Tabu Search*, Kluwer Academic Publishers, 1997.
11. Hvattum, L. M., Løkketangen, A., and Glover, F., Adaptive memory search for Boolean optimization problems, *Discrete Applied Mathematics*, 142, 99, 2004.
12. Løkketangen, A. and Glover F., Probabilistic move selection in tabu Search for 0/1 mixed integer programming problems, in *Metaheuristics: Theory and Applications*, Eds.: Osman, I. H. and Kelly, J. P., Eds., Kluwer Academic Publishers, 1996, 467.
13. Løkketangen, A. and Glover, F., Surrogate constraint analysis — new heuristics and learning schemes for satisfiability problems, in *Satisfiability Problem: Theory and Applications*, Du, D., Gu, J., and Pardalos, P. M., Eds., DIMACS Series in Discrete Mathematics and Theoretical Computer Science, Vol. 35, AMS, 1997, pp. 537–572.
14. Martí, R., Multi-start methods, in *Handbook of Metaheuristics*, Glover, F. and Kochenberger, G., Eds., Kluwer Academic Publishers, 2003, 355.
15. Resende, M. G. C., Pitsoulis, L. S., and Pardalos, P. M., Approximate solution of weighted MAX-SAT problems using GRASP, in *Satisfiability Problem: Theory and Applications*, Du, D., Gu, J., and Pardalos, P. M., Eds., DIMACS Series in

Discrete Mathematics and Theoretical Computer Science, Vol. 35, AMS, 1997, 393.

16. Resende, M. G. C. and Ribeiro, C. C., GRASP with path-relinking: recent advances and applications, submitted to *Metaheuristics: Progress as Real Problem Solvers*, Ibaraki, T., Nonobe, K., and Yagiura, M., Eds., Springer, 2005.

17. Shang, Y., Global search methods for solving nonlinear optimization problems, Ph.D. Thesis, Dept. of Computer Science, University of Illinois, Urbana-Champaign, 1997.

2

Convergent Lagrangian Methods for Separable Nonlinear Integer Programming: Objective Level Cut and Domain Cut Methods

Duan Li, Xiaoling Sun, and Jun Wang

CONTENTS

2.1 Introduction

We consider the following general class of separable integer programming problems:

$$(P) \quad \min \ f(x) = \sum_{j=1}^{n} f_j(x_j)$$

$$\text{s.t.} \ g_i(x) = \sum_{j=1}^{n} g_{ij}(x_j) \leq b_i, \ i = 1, \ldots, m, \quad (2.1)$$

$$x \in X = X_1 \times X_2 \times \ldots \times X_n,$$

where all f_js are integer-valued functions, all g_{ij}s are real-valued functions and all X_js are finite integer sets in $\mathbb{R}$.

Problem (P) has a wide variety of applications, including resource allocation problems and nonlinear knapsack problems. In particular, manufacturing capacity planning, stratified sampling, production planning, and network reliability are special cases of (P) (see [2][10][22] and the references therein).

The solution methods for problem (P) and its special cases can be classified into three major categories, dynamic programming ([4][11]), branch-and-bound methods ([2][3][8]), and Lagrangian relaxation methods ([5–7][19]) (plus some combinations of branch-and-bound and dynamic programming methods [18]). Although dynamic programming is conceptually an ideal solution scheme for separable integer programming, the "curse of dimensionality" prevents its direct application to the multiply constrained cases of (P) when m is large. The success of branch-and-bound methods based on continuous relaxation relies on their ability to identify a global optimal solution to continuous relaxation subproblems. Thus, branch-and-bound methods may not be applicable to (P) when a nonconvexity is presented which is often the case in many applications, e.g., concave integer programming and polynomial integer programming. Due to the often existence of a duality gap, Lagrangian relaxation methods in many situations are not used as an exact method to find an optimal solution for (P). Developing an efficient solution scheme for general separable nonlinear integer programming problems in (P) is a challenging task.

Two convergent Lagrangian methods using the objective level cut and domain cut methods have been recently developed in [14–16] for solving separable nonlinear integer programming problems. The purpose of this chapter is to discuss the solution concepts behind these two new methods in order to give new insights and to stimulate further research results.

2.2 Lagrangian Relaxation, Perturbation Function, and Duality

By associating with each constraint in (P) a nonnegative λ_i, the Lagrangian relaxation of (P) is formulated as

$$(L_\lambda) \qquad d(\lambda) = \min_{x \in X} L(x, \lambda) = f(x) + \sum_{i=1}^{m} \lambda_i (g_i(x) - b_i), \qquad (2.2)$$

where $\lambda = (\lambda_1, \dots, \lambda_m)^T \in \mathbb{R}_+^m$ and $L(x, \lambda)$ is called the Lagrangian function of (P). One of the prominent features of the Lagrangian relaxation problem (L_λ) is that it can be decomposed into n one-dimensional problems:

$$\min \ f_j(x_j) + \sum_{i=1}^{m} \lambda_i g_{ij}(x_j) \qquad (2.3)$$

$$\text{s.t. } x_j \in X_j.$$

Notice that (2.3) is a problem of minimizing a univariate function over a finite integer set and its optimal solution can be easily identified. Denote the optimal value of problem (Q) as $v(Q)$. Let the feasible region and the optimal value of (P) be defined as,

$$S = \{x \in X \mid g_i(x) \le b_i, \ i = 1, \ldots, m\},$$
$$v(P) = f^* = \min_{x \in S} f(x).$$

Since $d(\lambda) \le f(x)$ for all $x \in S$ and $\lambda \ge 0$, the dual value $d(\lambda)$ always provides a lower bound for the optimal value of (P) (weak duality):

$$v(P) = f^* \ge d(\lambda), \quad \forall \lambda \ge 0.$$

We assume in the sequel that $\min_{x \in X} f(x) < f^*$, otherwise $\min_{x \in X} = f^*$ must hold and (P) reduces to an unconstrained integer programming problem. The Lagrangian dual problem of (P) is to search for an optimal multiplier vector $\lambda^* \in \mathbb{R}_+^m$ which maximizes $d(\lambda)$ for all $\lambda \ge 0$:

$$(D) \quad d(\lambda^*) = \max_{\lambda \ge 0} d(\lambda). \tag{2.4}$$

By weak duality, $f^* \ge d(\lambda^*)$ holds. The difference $f^* - d(\lambda^*)$ is called the *duality gap* between (P) and (D). Let u be an upper bound of f^*. We denote $u - d(\lambda^*)$ as a *duality bound* between (P) and (D). It is clear that a duality bound is always larger than or equal to the duality gap.

If x^* solves (L_{λ^*}) with $\lambda^* \ge 0$, and, in addition, the following conditions are satisfied:

$$g_i(x^*) \le b_i, \quad i = 1, 2, \ldots, m, \tag{2.5}$$
$$\lambda_i^*(g_i(x^*) - b_i) = 0, \quad i = 1, 2, \ldots, m, \tag{2.6}$$

then x^* solves (P) and $v(P) = v(D)$, i.e., the duality gap is zero. In this situation, the strong Lagrangian duality condition is said to be satisfied. Unfortunately, the strong Lagrangian duality is rarely present in integer programming, and a nonzero duality gap often exists when the Lagrangian relaxation method is adopted.

For any vectors x and $y \in \mathbb{R}^m$, denote $x \le y$ iff $x_i \le y_i$, $i = 1, \ldots, m$. A function h defined on $\mathbb{R}^m$ is said to be nonincreasing if for any x and $y \in \mathbb{R}^m$, $x \le y$ implies $h(x) \ge h(y)$.

Let $g(x) = (g_1(x), \ldots, g_m(x))^T$ and $b = (b_1, \ldots, b_m)^T$. The perturbation function of (P) is defined as follows for $y \in \mathbb{R}^m$,

$$w(y) = \min\{f(x) \mid g(x) \le y, \ x \in X\}, \tag{2.7}$$

where the domain of w is

$$Y = \{y \in \mathbb{R}^m \mid \text{there exists } x \in X \text{ such that } g(x) \le y\}. \tag{2.8}$$

It is easy to see that $w(g(x)) \le f(x)$ for any $x \in X$ and $w(b) = f^*$. By the definition of the perturbation function in (2.7), $w(y)$ is a nonincreasing function.

Moreover, since X is a finite set, w is a piecewise constant function over Y. The domain Y can be decomposed as $Y = \cup_{k=1}^{K} Y_k$, where K is a finite number and Y_k is a subset over which w takes constant value f_k:

$$w(y) = f_k, \quad \forall y \in Y_k, \quad k = 1, \ldots, K. \tag{2.9}$$

For singly constrained cases of (P), i.e., $m = 1$, Y_k has the following form:

$$Y_k = \begin{cases} [c_k, c_{k+1}), & k = 1, \ldots, K-1, \\ [c_k, +\infty), & k = K, \end{cases} \tag{2.10}$$

where $\tilde{c}_1 = \min_{x \in X} g_1(x)$.

For multiply constrained cases of (P), define

$$
\begin{aligned}
c_{ki} &= \min\{y_i \mid y = (y_1, \ldots, y_m)^T \in Y_k\}, \quad k = 1, \ldots, K, \ i = 1, \ldots, m, \\
C &= \{c_k = (c_{k1}, \ldots, c_{km})^T \mid k = 1, \ldots, K\}, \\
\Phi_c &= \{(c_k, f_k) \mid k = 1, \ldots, K\}, \\
\Phi &= \{(y, w(y)) \mid y \in Y\}.
\end{aligned}
$$

It follows from the definition of w that $c_k \in Y_k$ for $k = 1, \ldots, K$ and thus $\Phi_c \subset \Phi$. A point in Φ_c is said to be a *corner point* of the perturbation function w. It is clear that $(y, w(y)) \in \Phi_c$ iff $(y, w(y)) \in \Phi$ and for any $z \in Y$ satisfying $z \leq y$ and $z \neq y$, $w(z) > w(y)$ holds. For all $x \in X$, the points $(g(x), f(x))$ are on or above the image of the perturbation function.

A vector $\lambda \in R^m$ is said to be a subgradient of $w(\cdot)$ at $y = \hat{y}$ if

$$w(y) \geq w(\hat{y}) + \lambda^T(y - \hat{y}), \quad \forall y \in Y.$$

LEMMA 2.1 [17]
Let x^* solve primal problem (P). Then x^* is a solution to Lagrangian relaxation problem $(L_{\hat{\lambda}})$ for some $\hat{\lambda}$ in R_+^m iff $-\hat{\lambda}$ is a subgradient of $w(\cdot)$ at $y = g(x^*)$.

Define the *convex envelope function* of w to be the greatest convex function majorized by w:

$$\psi(y) = \max\{h(y) \mid h(y) \text{ is convex on } Y, \ h(y) \leq w(y), \ \forall y \in Y\}. \tag{2.11}$$

It can be easily seen that ψ is piecewise linear and nonincreasing on Y and $w(y) \geq \psi(y)$ for all $y \in Y$. Moreover, by the convexity of ψ we have

$$\psi(y) = \max\{\lambda^T y + r \mid \lambda \in R^m, \ r \in R, \text{ and } \lambda^T z + r \leq w(z), \ \forall z \in Y\},$$

or equivalently,

$$
\begin{aligned}
\psi(y) = \max \ & \lambda^T y + r \\
\text{s.t. } & \lambda^T c_k + r \leq f_k, \quad k = 1, \ldots, K, \\
& \lambda \in R_-^m, \ r \in \mathbb{R}.
\end{aligned}
\tag{2.12}
$$

For any fixed $y \in Y$, we introduce a dual variable $\mu_k \geq 0$ for each constraint $\lambda^T c_k + r \leq f_k$, $k = 1, \ldots, K$. Dualizing the linear program (2.12) then yields

$$\psi(y) = \min \sum_{k=1}^{K} \mu_k f_k \qquad (2.13)$$

$$\text{s.t. } \sum_{k=1}^{K} \mu_k c_k \leq y,$$

$$\sum_{i=1}^{K} \mu_k = 1, \ \mu_k \geq 0, \ k = 1, \ldots, K.$$

THEOREM 2.1 [15][17]
Let $(-\lambda^*, r^*)$ and μ^* be optimal solutions to (2.12) and (2.13) with $y = b$, respectively. Then

i. *λ^* is an optimal solution to the dual problem (D) and*

$$\psi(b) = \max_{\lambda \geq 0} d(\lambda) = d(\lambda^*). \qquad (2.14)$$

ii. *For each k with $\mu_k^* > 0$, any $\bar{x} \in X$ satisfying $(g(\bar{x}), f(\bar{x})) = (c_k, f_k)$ is an optimal solution to the Lagrangian problem (L_{λ^*}).*

THEOREM 2.2 [17]
Let x^* solve (P). Then, $\{x^*, \lambda^*\}$ is an optimal primal-dual pair iff the hyperplane given by $w = f(x^*) - (\lambda^*)^T[y - g(x^*)]$ is a supporting hyperplane at $[f(x^*), g(x^*)]$ and contains $[\psi(b), b]$.

Now let us investigate the reasons behind the often failures of the traditional Lagrangian method in finding an exact solution of the primal problem. Consider the following example.

Example 2.1

$$\min \ f(x) = -2x_1^2 - x_2 + 3x_3^2$$
$$\text{s.t. } 5x_1 + 3x_2^2 - \sqrt{3}x_3 \leq 7,$$
$$x \in X = \{0 \leq x_i \leq 2, x_i \text{ integer}, i = 1, 2, 3\}.$$

The optimal solution of this example is $x^* = (1, 0, 0)^T$ with $f^* = f(x^*) = -2$. The perturbation function of this problem is illustrated in Figure 2.1. We can see from Figure 2.1 that point C that corresponds to the optimal solution x^* "hides" above the convex envelope of the perturbation function and therefore there does not exist a subgradient of perturbation function w at $g(x^*)$. In other words, it is impossible for x^* to be found by the conventional Lagrangian dual method. The optimal solution to (D) in this example is $\lambda^* = 0.8$ with $d(\lambda^*) = -5.6$. Note the solutions to $(L_{\lambda^*=0.8})$ are $(0, 0, 0)^T$ and $(2, 0, 0)^T$, neither of which is optimal to the primal problem.

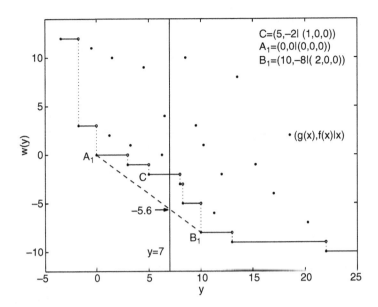

FIGURE 2.1
The perturbation function of Example 2.1.

A vector $\lambda^* \geq 0$ is said to be an *optimal generating multiplier* vector of (P) if an optimal solution x^* to (P) can be generated by solving (L_λ) with $\lambda = \lambda^*$ (see [17]). A pair (x^*, λ^*) is said to be an *optimal primal-dual pair* of (P) if the optimal dual solution λ^* to (D) is an optimal generating multiplier vector for an optimal solution x^* to (P) (see [17]).

The conventional Lagrangian dual method would fail in two critical situations, both of which have been witnessed in the above example. The first situation occurs where no solution of (P) can be generated by problem (L_λ) for any $\lambda \geq 0$. The second situation occurs where no solution to problem (L_{λ^*}), with λ^* being an optimal solution to (D), is a solution to (P).

It is clear that the nonexistence of a linear support at the optimal point leads to a failure of the traditional Lagrangian method. Recognizing the existence of a nonlinear support, nonlinear Lagrangian formulations have been proposed in [12][13][17][20][21] to offer a success guarantee for the dual search in generating an optimal solution of the primal integer programming problem. In contrast to the traditional Lagrangian formulation which is linear with respect to the objective function and constraints, a nonlinear Lagrangian formulation takes nonlinear forms such as pth power or a logarithmic-exponential formulation with respect to the objective function and constraints.

While the nonlinear Lagrangian theory provides a theoretical mechanism to guarantee the success of the dual search, the nonlinear Lagrangian formulation does not lead to a decomposability which is the most prominent feature of the traditional linear Lagrangian. When the original problem is separable, the nonlinear Lagrangian formulation destroys the original separable structure and makes the decomposition impossible. Thus, the computational and implementation issue of nonlinear Lagrangian theory in integer programming remains unsolved since there is no efficient algorithm to solve a general non-separable integer programming problem.

Stimulated by the relationship between the duality gap and the geometry of the perturbation function, two novel convergent Lagrangian methods adopting the objective level cut method and the domain cut method have recently been developed for (P) in [14–16]. They are efficient and convergent Lagrangian solution schemes in a sense that they provide an exact solution to (P) while retaining the decomposability of (P) in the solution process. Both methods are devised to reshape the perturbation function by adding some cuts such that the duality gap can be reduced. A successive reshaping process will eventually expose the optimal solution on the convex envelop of a revised perturbation function and thus the success of convergent dual search can be guaranteed.

2.3 Objective Level Cut Method

We continue to investigate Example 2.1 to motivate the development of the objective level cut method [14] [15].

As we observed from Figure 2.1, the optimal point C hides above the convex envelope of the perturbation and there is no optimal generating vector at C. The duality gap is $f(x^*) - d(\lambda^*) = -2 - (-5.6) = 3.6$ and the current duality bound is $0 - (-5.6) = 5.6$ achieved by the traditional Lagrangian method. A key finding is that point C can be exposed to the convex envelope or the convex hull of the perturbation function by adding an objective cut. As a matter of fact, since A_1 corresponds to a feasible solution $x^0 = (0, 0, 0)^T$, the function value $f(x^0) = 0$ is an upper bound of f^*. Moreover, by the weak duality, the dual value $d(\lambda^*) = -5.6$ is a lower bound of f^*. Therefore, adding an objective cut of $-5.6 \le f(x) \le 0$ to the original problem does not exclude the optimal solution, while the perturbation function will be reshaped due to the modified feasible region. Since the objective function is integer-valued, we can set a stronger objective cut of $-5 \le f(x) \le -1$ after storing the current best feasible solution x^0 as the incumbent. The modified problem then has the following form:

$$\min \ f(x) = -2x_1^2 - x_2 + 3x_3^2 \tag{2.15}$$
$$\text{s.t. } 5x_1 + 3x_2^2 - \sqrt{3}x_3 \le 7,$$
$$x \in X_1 = X \cap \{x \mid -5 \le f(x) \le -1\}.$$

The perturbation function of (2.15) is shown in Figure 2.2. The optimal dual multiplier to (2.15) is 0.7593 with dual value -4.0372. Since $x^1 = (0, 1, 0)^T$ corresponding to A_2 is feasible, the duality bound is now reduced to $f(x^1) - (-4.0372) = -1 + 4.0372 = 3.0372$. Again we can add an objective

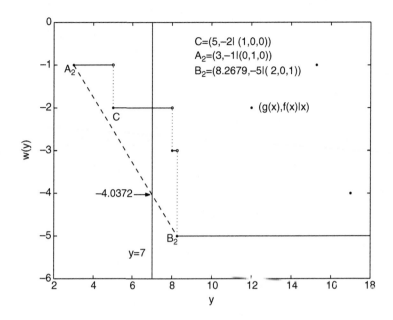

FIGURE 2.2
The perturbation function of (2.15).

cut $-4 \le f(x) \le f(x^1) - 1 = -2$ to (2.15) and obtain the following problem:

$$\min \ f(x) = -2x_1^2 - x_2 + 3x_3^2 \qquad (2.16)$$
$$\text{s.t. } 5x_1 + 3x_2^2 - \sqrt{3}x_3 \le 7,$$
$$x \in X_2 = X \cap \{x \mid -4 \le f(x) \le -2\}.$$

The perturbation function of (2.16) is shown in Figure 2.3. The optimal dual multiplier is 0.3333 with dual value -2.6667. Now point C corresponding to x^* is exposed to the convex hull of the perturbation function and the duality bound is reduced to $f(x^*) - (-2.6667) = -2 + 2.6667 = 0.6667 < 1$. Since the objective function is integer-valued, we claim that $x^* = (1, 0, 0)^T$ is the optimal solution to the original problem.

To reduce the duality gap between the Lagrangian dual problem and the primal problem, we reshape the perturbation function by adding objective cut to the problem. We start with a lower bound derived from the dual value by the conventional dual search and an upper bound by a feasible solution generated in the dual search (if any). The updated lower level cut and upper level cut are imposed to the program successively such that the duality bound (duality gap) is forced to shrink. Note that there is only one lower bound and upper bound constraint during the whole solution process.

How to solve the relaxation problems of the revised problems such as the Lagrangian relaxations of (2.15) and (2.16) is crucial to an efficient implementation of this novel idea. The Lagrangian relaxation of each revised problem is a separable integer programming with a lower bound and upper bound

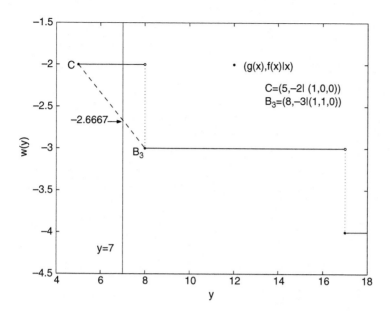

FIGURE 2.3
The perturbation function of (2.16).

constraint for the objective function. Since the objective function is integer-valued, dynamic programming can be used to search for optimal solutions to this kind of problems efficiently. Consider the following modified version of (P) by imposing a lower cut l and an upper cut u:

$$(P(l, u)) \quad \min \ f(x) \tag{2.17}$$
$$\text{s.t. } g_i(x) \le b_i, \ i = 1, \dots, m,$$
$$x \in X(l, u) = \{x \in X \mid l \le f(x) \le u\}.$$

It is obvious that $(P(l, u))$ is equivalent to (P) if $l \le f^* \le u$. The Lagrangian relaxation of $(P(l, u))$ is:

$$(L_\lambda(l, u)) \quad d(\lambda, l, u) = \min_{x \in X(l,u)} \ L(x, \lambda). \tag{2.18}$$

The Lagrangian dual problem of $(P(l, u))$ is then given as

$$(D(l, u)) \quad \max_{\lambda \ge 0} d(\lambda, l, u). \tag{2.19}$$

Notice that $(L_\lambda(l, u))$ can be explicitly written as:

$$d(\lambda, l, u) = \min \left[f(x) + \sum_{i=1}^{m} \lambda_i(g_i(x) - b_i) \right] = \min \left[\sum_{j=1}^{n} \theta_j(x_j, \lambda) - \alpha(\lambda) \right] \tag{2.20}$$

$$\text{s.t. } l \le \sum_{j=1}^{n} f_j(x_j) \le u,$$
$$x \in X,$$

where $\theta_j(x_j, \lambda) = f_j(x_j) + \Sigma_{i=1}^{m}\lambda_i g_{ij}(x_j)$ and $\alpha(\lambda) = \Sigma_{i=1}^{m}\lambda_i b_i$. It is clear that $(L_\lambda(l, u))$ is a separable integer programming problem with a lower bound and upper bound constraint on $f(x)$. By the assumptions in (P), each $f_j(x_j)$ is integer-valued for all $x_j \in X_j$. Therefore, $(L_\lambda(l, u))$ can be efficiently solved by dynamic programming. Let

$$s_k = \sum_{j=1}^{k-1} f_j(x_j), \quad k = 2, \ldots, n+1, \tag{2.21}$$

with an initial condition of $s_1 = 0$. Then $(L_\lambda(l, u))$ can be solved by the following dynamic programming formulation:

$$(DP) \quad \min s_{n+1} + \sum_{j=1}^{n}\left[\sum_{i=1}^{m}\lambda_i g_{ij}(x_j)\right] \tag{2.22}$$

$$\text{s.t. } s_{j+1} = s_j + f_j(x_j), \quad j = 1, 2, \ldots, n,$$
$$s_1 = 0,$$
$$l \le s_{n+1} \le u,$$
$$x_j \in X_j, \quad j = 1, 2, \ldots, n.$$

The state in the above dynamic programming formulation takes finite values at each stage.

THEOREM 2.3
Let λ^* be the optimal solution to (D). Denote by $T(\lambda^*)$ the set of the optimal solutions to the Lagrangian problem (L_{λ^*}). Assume that the duality gap is nonzero, i.e., $d(\lambda^*) < f^*$. Then

    i. There is at least one infeasible solution to (L_{λ^*});
    ii. $\min\{f(x) \mid x \in T(\lambda^*) \setminus S\} \le d(\lambda^*)$.

Based on Theorem 3, the objective cut is always valid when the duality gap is nonzero. More specifically, in each new cut, some infeasible solution that is the solution to the previous Lagrangian relaxation problem will be removed, thus raising the dual value in the current iteration.

THEOREM 2.4

    i. Let $\lambda^*(l, u)$ denote the optimal solution to $(D(l, u))$. The dual optimal value $d(\lambda^*(l, u), l, u)$ is a nondecreasing function of l.
    ii. If $l \le f^* \le u$, then $d(\lambda^*) \le d(\lambda^*(l, u), l, u) \le f^*$. Moreover, let $\sigma = \max\{f(x) \mid f(x) < f^*, x \in X \setminus S\}$. If $\sigma < l \le f^*$, then $\lambda^*(l, u) = 0$ and $d(\lambda^*(l, u), l, u) = f^*$.
    iii. For $l < f^*$, we have $d(\lambda^*(l, u), l, u) \ge l$.

The implication of Theorem 4 is clear: The higher the lower cut, the higher the dual value. The solution process will stop when the duality gap is less than one. From Theorem 2.3 and Theorem 2.4, the following result of finite convergence is evident.

THEOREM 2.5
The objective level cut algorithm either finds an optimal solution of (P) or reports an infeasibility of (P) in at most $u_0 - l_0 + 1$ iterations, where u_0 and l_0 are the upper and lower bounds generated in the first iteration using the conventional Lagrangian method.

Although the objective function is assumed to be integer-valued in the objective level cut algorithm, a rational objective function can be also handled by multiplying the objective function by the least common multiplier of the denominators of all coefficients.

2.4 Domain Cut Method

When the objective function in (P) is a nonincreasing function with respect to all x_is and all the constraints are nondecreasing functions with respect to all x_is, problem (P) becomes a nonlinear knapsack problem. The domain cut method [14][16] is developed for solving nonlinear knapsack problems. Note that the assumption of integer-valued objective function is not needed in the domain cut method.

We illustrate the solution concept of the domain cut method by the following example.

Example 2.2

$$\min \ f(x) = -x_1^2 - 1.5x_2$$
$$\text{s.t. } g(x) = 6x_1 + x_2^2 \leq 23,$$
$$x \in \tilde{X} = \{x \mid 1 \leq x_i \leq 4, \ x_i \text{ integer}, \ i = 1, 2\}.$$

Note that f is nonincreasing and g is nondecreasing. The optimal solution of this example is $x^* = (3, 2)^T$ with $f(x^*) = -12$.

The domain $\tilde{X}$ and the perturbation function $z = w(y)$ of this example are illustrated in Figure 2.4 and Figure 2.5, respectively. It is easy to check that the optimal Lagrangian multiplier is $\lambda^* = 0.8333$ with dual value -15.8333. The duality gap is 3.8333. The Lagrangian problem

$$\min_{x \in \tilde{X}} [f(x) + 0.8333(g(x) - 23)]$$

has a feasible solution $x^0 = (1, 1)^T$ with $f(x^0) = -2.5$ and an infeasible solution $y^0 = (4, 1)^T$. In Figure 2.5, points A, B, C correspond to x^0, y^0, and x^* in $(g(x), f(x))$ plane, respectively. We observe that if points A and B are removed from the plot of the perturbation function, then the duality gap of the revised

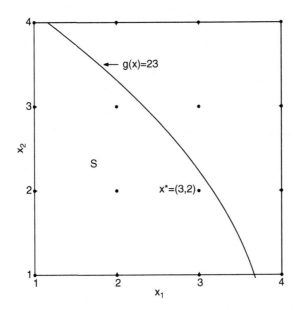

FIGURE 2.4
Domain $\tilde{X}$ and the feasible region of Example 2.2.

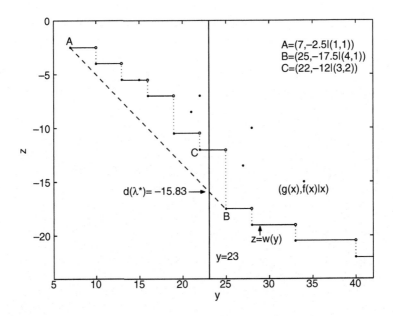

FIGURE 2.5
Perturbation function with domain $\tilde{X}$ of Example 2.2.

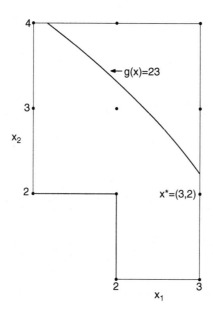

FIGURE 2.6
Domain X^1 in Example 2.2.

problem will be smaller than the original duality gap and thus the "hidden"
point C can be hopefully exposed on the convex envelope of the revised
perturbation function after repeating such a procedure. The monotonicity of
f and g motivates us to cut integer points satisfying $x \leq x^0$ and integer
points satisfying $x \geq y^0$ from box $\tilde{X}$. Denote by X^1 the revised domain of
integer points after such a cut. Figure 2.6 and Figure 2.7 show the integer points
in X^1 and the perturbation function corresponding to the revised problem
by replacing $\tilde{X}$ by X^1. The optimal Lagrangian multiplier for this revised
problem is $\lambda^* = 0.3$ with $d(\lambda^*) = -12.3$. The Lagrangian problem

$$\min_{x \in X^1} [f(x) + 0.3(g(x) - 23)]$$

has a feasible solution $x^1 = (3, 2)^T$ with $f(x^1) = -12$ and an infeasible solution $y^1 = (3, 3)^T$. We continue to do the domain cutting. Denote by X^2 the
integer points in X^1 after removing the integer points satisfying $x \leq x^1$ and
integer points satisfying $x \geq y^1$ from X^1. It is easy to see from Figure 2.6 that
$X^2 = \{(1, 3)^T, (1, 4)^T, (2, 3)^T, (2, 4)^T\}$. Since

$$\min_{x \in X^2} [f(x) + 0.3(g(x) - 23)] = -9.1 > -12 = f(x^1),$$

we conclude by the weak duality that there is no feasible solution in X^2 better
than x^1 and $x^1 = (3, 2)^T$ is therefore the optimal solution of this example.

As we see from the example, to reduce the duality gap between the Lagrangian dual problem and the primal problem, we reshape the perturbation

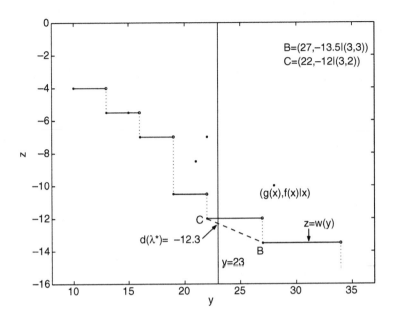

FIGURE 2.7
Perturbation function with domain X^1 of Example 2.2.

function by removing certain points from the domain $\tilde{X}$. The monotonicity of the problem guarantees that an optimal solution either has been found during the search or is still in the remaining feasible region after the domain cut.

Let $\alpha, \beta \in Z^n$, where Z^n denotes the set of integer points in R^n. Let $\langle \alpha, \beta \rangle$ denote the set of integer points in $[\alpha, \beta]$,

$$\langle \alpha, \beta \rangle = \prod_{i=1}^{n} \langle \alpha_i, \beta_i \rangle = \langle \alpha_1, \beta_1 \rangle \times \langle \alpha_2, \beta_2 \rangle \ldots \times \langle \alpha_n, \beta_n \rangle.$$

We call $\langle \alpha, \beta \rangle$ an integer box. Let $l = (l_1, \ldots, l_n)^T$ and $u = (u_1, \ldots, u_n)^T$. Assume that the integer set X in (P) is given by $X = \langle l, u \rangle$. If the objective function is nonincreasing and constraints are nondecreasing, we have the following conclusions:

    i. If $x \in \langle l, u \rangle$ is a feasible solution to (P), then for any $\tilde{x} \in \langle l, x \rangle$, it holds that $f(\tilde{x}) \geq f(x)$.

    ii. If $y \in \langle l, u \rangle$ is an infeasible solution to (P), then any point in $\langle y, u \rangle$ is infeasible.

Therefore, $\langle l, x \rangle$ and $\langle y, u \rangle$ can be cut from the $\langle l, u \rangle$, without missing any optimal solution of (P) after recording the feasible solution x.

A critical issue for an efficient implementation of the domain cut method is how to characterize the nonbox revised domain after performing a cutting process. In [14][16], the following analytical formulas are derived to describe the revised domain as a union of subboxes.

LEMMA 2.2

   i. Let $\tilde{A} = \langle \alpha, \beta \rangle$ and $\tilde{B} = \langle l, \gamma \rangle$, where $\alpha, \beta, l, \gamma \in Z^n$ and $l \leq \alpha \leq \gamma$ $\leq \beta$. Then

$$\tilde{A}\backslash\tilde{B} = \bigcup_{i=1}^{n} \left(\prod_{k=1}^{i-1} \langle \alpha_k, \gamma_k \rangle \times \langle \gamma_i + 1, \beta_i \rangle \times \prod_{k=i+1}^{n} \langle \alpha_k, \beta_k \rangle \right). \quad (2.23)$$

   ii. Let $\tilde{A} = \langle \alpha, \beta \rangle$ and $\tilde{B} = \langle \gamma, u \rangle$, where $\alpha, \beta, \gamma, u \in Z^n$ and $\alpha \leq \gamma \leq \beta$ $\leq u$. Then

$$\tilde{A}\backslash\tilde{B} = \bigcup_{i=1}^{n} \left(\prod_{k=1}^{i-1} \langle \gamma_k, \beta_k \rangle \times \langle \alpha_i, \gamma_i - 1 \rangle \times \prod_{k=i+1}^{n} \langle \alpha_k, \beta_k \rangle \right). \quad (2.24)$$

Thus, the Lagrangian relaxation problem on a nonbox domain can be performed as follows,

$$(L_\lambda) \quad d(\lambda) = \min_{x \in \cup_{j=1}^{k} \tilde{X}^j} L(x, \lambda) = \min_{1 \leq j \leq k} \min_{x \in \tilde{X}^j} L(x, \lambda), \quad (2.25)$$

where $\tilde{X}^j$ is the jth subbox generated in the cutting process. Most importantly, the above problem on a revised domain can be still decomposed into n one-dimensional nonlinear integer programs which can be easily solved.

The domain cut method for nonlinear knapsack problems can be summarized as follows. We start with the conventional Lagrangian method, which generates a solution set with an infeasible solution and a possible feasible solution to the primal problem, a dual value and an incumbent (if there is a feasible solution). We then do the domain cut accordingly. Those subboxes will be discarded if their lower-left corner is infeasible or their upper-right corner is feasible (after comparing the upper-right corner with the incumbent). The dual search will then applied to the revised domain again. In the dual search process, those subboxes whose dual value is larger than the incumbent will be discarded. The next cut will be performed at the subbox where the dual value is achieved. This iterative process repeats until either the duality gap reaches zero or the revised domain is empty. The incumbent is then the optimal solution to the primal problem. The finite convergence is obvious, similar to the objective level cut, as the domain cut is valid each time when the duality gap is not zero.

2.5 Numerical Results

The objective level cut method and the domain cut method discussed in the previous sections have been programmed by Fortran 90 and were run on a Sun workstation (Blade 1000). In this section, we report the numerical results of the proposed algorithms for the following two classes of test problems.

PROBLEM 2.1
Polynomial integer programming (PIP):

$$\min \ f(x) = \sum_{j=1}^{n} \sum_{k=1}^{3} c_{jk} x_j^k$$

$$\text{s.t.} \ g_i(x) = \sum_{j=1}^{n} \sum_{k=1}^{3} a_{ijk} x_j^k \leq b_i, \ i = 1, \ldots, m,$$

$$l_i \leq x_j \leq u_j, \ x_j \text{ integer}, \ j = 1, \ldots, n.$$

The coefficients are generated from uniform distributions: all c_{jk} are integers from $[-10, 10]$, $a_{ijk} \in [-10, 10]$, $i = 1, \ldots, m$, $j = 1, \ldots, n$, $k = 1, 2, 3$. We also take $l = (1, \ldots, 1)^T$ and $u = (5, \ldots, 5)^T$, and $b_i = g_i(l) + 0.5 \times (g_i(u) - g_i(l))$, $i = 1, \ldots, m$. Note that $f(x)$ and $g_i(x)$ in problem (PIP) are not necessarily convex and the objective function $f(x)$ is integer-valued.

PROBLEM 2.2
Polynomial knapsack problems (PKP):

$$\min \ f(x) = \sum_{j=1}^{n} [-c_j x_j + d_j (e_j - x_j)^3]$$

$$\text{s.t.} \ g(x) = Ax \leq b,$$

$$l_i \leq x_j \leq u_j, \ x_j \text{ integer}, \ j = 1, \ldots, n,$$

where $c_j > 0$, $d_j > 0$, $e_j \in (l_j, u_j)$ for $j = 1, \ldots, n$, and $A = (a_{ij})_{m \times n}$ with $a_{ij} \geq 0$ for $i = 1, \ldots, m$, $j = 1, \ldots, n$. The coefficients are generated from uniform distributions: $c_j \in [1, 50]$, $d_j \in [1, 10]$, $e_j \in [1, 5]$ for $j = 1, \ldots, n$, $a_{ij} \in [1, 50]$ for $i = 1, \ldots, m$, $j = 1, \ldots, n$. We also take $l = (1, \ldots, 1)^T$ and $u = (5, \ldots, 5)^T$, and $b = 0.7 \times Au$. We notice that function $f_j(x_j) = -c_j x_j + d_j(e_j - x_j)^3$ is nonincreasing but not necessarily convex or concave on $[l_j, u_j]$ for $j = 1, \ldots, n$. Therefore, problem (PKP) is a nonlinear knapsack problem.

Numerical results of the objective level cut method for problem (PIP) are reported in Table 2.1. We can see from Table 2.1 that the proposed algorithm can solve large-scale multiply constrained separable integer programming problems in reasonable computation time.

Table 2.2 summarizes numerical results of the domain cut algorithm for problem (PKP). It can be seen from Table 2.2 that the domain cut algorithm is efficient for solving large-scale singly-constrained nonlinear knapsack problems in terms of CPU time. Combined with the surrogate constraint technique, the domain cut method can also solve multiply constrained knapsack problems efficiently as shown in Table 2.2.

More computational results of the objective level cut method and the domain cut method for different types of test problems and their comparison with other existing methods in the literature can be found in [14–16]. One evident conclusion from numerical experiments is that while most of the

TABLE 2.1

Numerical Results of Objective Level
Cut Method for (PIP)

n	m	Average Number of Iterations	Average CPU Time (s)
50	5	4	0.5
50	10	12	5.8
50	15	5	1.9
50	20	17	24.2
50	30	15	223.4

TABLE 2.2

Numerical Results of Domain Cut
Method for (PKP)

n	m	Average Number of Integer Boxes	Average CPU Time (s)
500	1	30017	30.7
1000	1	97212	214.5
2000	1	269877	1382.1
20	5	901	2.8
40	5	7042	43.8
60	5	40342	417.0
30	10	5064	34.6
30	20	9251	82.8
30	30	25320	228.6

existing methods suffer from their limitations in nonconvex and multiply-constrained situations, both the objective level cut method and the domain cut method can be applied to general multiply-constrained nonconvex separable nonlinear integer programming problems with promising computational results.

2.6 Summary

Two recently developed convergent Lagrangian methods: objective level cut and domain cut methods for separable nonlinear integer programming have been discussed in this chapter. Due to the discrete nature of the integer programming, the optimal solutions to the Lagrangian relaxation problem corresponding to the optimal multiplier do not necessarily solve the original problem — a duality gap may exist even for linear or convex integer programming. The idea of the new cutting methods is based on a key observation that the duality gap of an integer program can be eliminated by reshaping the perturbation function. Consequently, the optimal solution can be exposed to the convex hull of the revised perturbation function, and thus guaranteeing the convergence of dual search. These two methods have been tested against a variety of large-scale separable nonlinear integer programming problems with up to several thousand variables. Extensive numerical results of these two methods are very promising [14–16].

Acknowledgments

This research was partially supported by the Research Grants Council of Hong Kong, Grant CUHK4214/01E and the National Natural Science Foundation of China under grants 79970107 and 10271073.

References

1. Bell, D.E. and Shapiro, J.F., A convergent duality theory for integer programming. *Operations Research* 25, 419, 1977.
2. Bretthauer, K.M. and Shetty, B., The nonlinear resource allocation problem. *Operations Research* 43, 670, 1995.
3. Bretthauer, K.M. and Shetty, B., A pegging algorithm for the nonlinear resource allocation problem. *Computers & Operations Research* 29, 505, 2002.
4. Cooper, M.W., The use of dynamic programming for the solution of a class of nonlinear programming problems, *Naval Research Logistics Quarterly* 27, 89, 980.
5. Fisher, M.L. and Shapiro, J.F., Constructive duality in integer programming. *SIAM Journal on Applied Mathematics* 27, 31, 1974.
6. Fisher, M.L., The Lagrangian relaxation method for solving integer programming problems. *Management Science* 27, 1, 1981.
7. Geoffrion, A.M., Lagrangian relaxation for integer programming. *Mathematical Programming Study* 2, 82, 1974.
8. Gupta, O.K. and Ravindran, A., Branch and bound experiments in convex nonlinear integer programming. *Management Science* 31, 1533, 1985.
9. Hochbaum, D., A nonlinear knapsack problem. *Operations Research Letters* 17, 103, 1995.
10. Ibaraki, T. and Katoh, N., *Resource Allocation Problems: Algorithmic Approaches*, MIT Press, Cambridge, Massachusetts, 1988.
11. Korner, F., A hybrid method for solving nonlinear knapsack problems. *European Journal of Operational Research* 38, 238, 1989.
12. Li, D., Zero duality gap in integer programming: p-norm surrogate constraint method. *Operations Research Letters* 25, 89, 1999.
13. Li, D. and Sun, X.L., Success guarantee of dual search in integer programming: p-th power Lagrangian method. *Journal of Global Optimization* 18, 235, 2000.
14. Li, D. and Sun, X.L., *Nonlinear Integer Programming*, Springer, to be published in 2005.
15. Li, D., Sun, X.L., Wang, J., and McKinnon, K., A Convergent Lagrangian and domain cut method for nonlinear knapsack problems, Submitted for publication, 2002. Also Technical Report, Serial No: SEEM2002-10, Department of Systems Engineering and Engineering Management, Chinese University of Hong Kong, November 2002.

16. Li, D., Wang, J., and Sun, X.L., Exact solution to separable nonlinear integer programming: Convergent Lagrangian and objective level cut method, Submitted for publication, 2003. Also Technical Report, Serial No: SEEM2003-08, Department of Systems Engineering and Engineering Management, Chinese University of Hong Kong, May 2003.

17. Li, D. and White, D.J., P-th power Lagrangian method for integer programming. *Annals of Operations Research* 98, 151, 2000.

18. Marsten, R.E. and Morin, T.L., A hybrid approach to discrete mathematical programming. *Mathematical Programming* 14, 21, 1978.

19. Shapiro, J.F., A survey of Lagrangian techniques for discrete optimization. *Annals of Discrete Mathematics* 5, 113, 1979.

20. Sun, X.L. and Li, D., Asymptotic strong duality for bounded integer programming: A logarithmic-exponential dual formulation. *Mathematics of Operations Research* 25, 625, 2000.

21. Sun, X.L. and Li, D., New dual formulations in constrained integer programming, X. Q. Yang, Ed., *Progress in Optimization*, Kluwer, 79, 2000.

22. Sun, X.L. and Li, D., Optimality condition and branch and bound algorithm for constrained redundancy optimization in series systems. *Optimization and Engineering* 3, 53, 2002.

3

The Generalized Assignment Problem

Robert M. Nauss

CONTENTS

3.1 Introduction

The *generalized assignment problem* (GAP) has been investigated in numerous research papers over the past 30 years. The problem may be stated as finding a minimum-cost assignment of tasks to agents such that each task is assigned to exactly one agent and such that each agent's resource capacity is not violated. Applications of the GAP range from jobs assigned to computers in computer networks (Balachandran [1]) to loading for flexible manufacturing systems (Mazolla et al. [17]) to facility location (Ross and Soland [24]). A review of applications and algorithms (both exact and heuristic) appears in Cattrysse and Van Wassenhove [5]. Osman [20] describes various heuristic approaches to the GAP.

The chapter is organized as follows. Section 3.2 describes the mathematical formulation of the GAP. Section 3.3 surveys earlier work on the GAP. Solution methodologies employed in our algorithm are described in Section 3.4. Section 3.5 explains the branch-and-bound algorithm and Section 3.6 describes computational results. A concluding section gives ideas for future work and extensions.

3.2   Model Formulation

The GAP may be formulated as an integer linear programming (ILP) model with binary variables. Let n be the number of tasks to be assigned to m agents and define $N = \{1, 2, \ldots, n\}$ and $M = \{1, 2, \ldots, m\}$. Let:

$$c_{ij} = \text{cost of task } j \text{ being assigned to agent } i$$
$$r_{ij} = \text{amount of resource required for task } j \text{ by agent } i$$
$$b_i = \text{resource units available to agent } i.$$

The decision variables are defined as:

$$x_{ij} = \begin{cases} 1, & \text{if task } j \text{ is assigned to agent } i \\ 0, & \text{if not.} \end{cases}$$

The 0-1 ILP model is:

$$(P) \qquad \text{minimize} \sum_{i=1}^{m} \sum_{j=1}^{n} c_{ij}\, x_{ij} \qquad (3.1)$$

subject to:

$$\sum_{j=1}^{n} r_{ij} x_{ij} \leq b_i, \quad \forall i \in M \qquad (3.2)$$

$$\sum_{i=1}^{m} x_{ij} = 1, \quad \forall j \in N \qquad (3.3)$$

$$x_{ij} = 0 \quad \text{or} \quad 1, \quad \forall i \in M, \quad j \in N. \qquad (3.4)$$

The objective function (3.1) totals the costs of the assignments. Constraint (3.2) enforces resource limitations for each agent, and constraint (3.3) assures that each task is assigned to exactly one agent. All data elements are allowed to be real. We note, however, that certain efficiencies follow if the data elements are assumed to be integral.

3.3   Previous Work

Research on the GAP over the past 30 years provides the integer programming student with a smorgasbord of solution approaches devised by numerous researchers. This is indeed propitious for the student since he need only concern himself with mastering the various approaches within the context of a rather simple 0-1 ILP formulation of the GAP.

The first formal computational work addressing the GAP was by Ross and Soland [23]. Its importance revolves around the fact that the authors used an ad hoc approach for solving a tighter (than LP) relaxation which, in effect, was the first published instance of Lagrangian relaxation (LGR) applied to an ILP problem. Although the authors did not refer to Lagrangian relaxation per se, its usage is quite clear. Fisher et al. [8] formalized the use of Lagrangian relaxation and devised a clever dual multiplier adjustment procedure that tightens the initial LGR. This tightening procedure when incorporated in a branch-and-bound algorithm allowed larger, more difficult GAP problems on the order of 100 0-1 variables to be solved to optimality. Jörnsten and Näsberg [13] reformulated the GAP by introducing a set of auxiliary variables and coupling constraints. Using their approach good, feasible solutions were able to be generated in a more straightforward fashion. Guignard and Rosenwein [11] extended the work of tightening the LGR by devising a dual-based approach that also incorporated the use of a surrogate constraint. This relaxation tightening allowed problems of up to 500 variables to be solved to optimality.

Thus through the late 1980s the maximum size GAP problems to be solved to optimality remained at about 500 binary variables. In the early 1990s some researchers started to look at heuristic approaches due to the perceived difficulty in solving large GAP problems to optimality. Cattryese et al. [4] used a column generation/set partitioning approach to generate good feasible solutions, and coupled this with dual ascent and Lagrangian techniques in order to reduce the gap between lower (LB) and upper (UB) bounds. Thus they were able to state that the best feasible solution (UB) was within $(\frac{UB-LB}{LB})\%$ of optimality.

Researchers have continued to devise other heuristic approaches for the GAP. Wilson [26] developed a dual-type algorithm and coupled it with a clever search strategy to find good feasible solutions. Chu and Beasley [6] used a genetic algorithm-based heuristic that generates very good feasible solutions albeit at increased computation time due to the probabilistic search features of genetic algorithms in general. Guignard and Zhu [12] utilized subgradient optimization to tighten the LGR and simultaneously invoked a simple "variables switch" heuristic to obtain good feasible solutions. Yagiura et al. [28] devised a tabu search algorithm utilizing an ejection chain approach to govern the neighborhood search for feasible solutions. They also incorporated a subgradient optimization scheme to generate lower bounds so that solution quality could be measured. Yagiura et al. [27] extended Yagiura et al. [28] by devising a mechanism to combine two or more reference solutions.

Concurrent with the work on heuristics in the 1990s, various researchers continued to devise efficient algorithms to generate provably optimal solutions to the GAP. Savelsburgh [25] employed a set partitioning formulation of the GAP similar to that used by Cattrysee et al. [4] (see above). However, he extended their work by employing both column generation and branch-and-bound in order to generate provably optimal solutions to the GAP. He has been able to extend the size of problems solved to 1000 variables. Cattrysee et al. [3] generated lifted cover cuts to improve the standard LP relaxation, devised two new heuristics for generating good feasible solutions, and incorporated this in a branch-and-bound algorithm. Park et al. [21] utilized subgradient optimization in conjunction with a specialized dual ascent algorithm to tighten the Lagrangian relaxation and incorporated this approach in a branch-and-bound algorithm. Finally Nauss [19] has devised a special-purpose branch-and-bound algorithm that utilizes lifted cover cuts, feasible solution generators, Lagrangian relaxation, and subgradient optimization to solve hard GAP problems with up to 3000 binary variables. A by-product is that good feasible solutions tend to be generated early in the solution process. Thus the algorithm mimics a heuristic approach when the search process is truncated.

3.4 Solution Approach

Over the past 10 years or so, general purpose ILP solvers have become more effective in solving general ILP problems. However certain problem classes, with the GAP being one of them, remain difficult to solve to optimality. Specifically "hard" GAP problems with more than 500 binary variables are very difficult to solve with "off-the-shelf" software such as ILOG's CPLEX solver (Nauss [19]).

Given this reality, the development of a special-purpose branch-and-bound algorithm tailored for the GAP's structure is warranted. In this section we outline a number of techniques and tools used in a special-purpose algorithm. The algorithm itself will be described in Section 3.5. Consider the GAP defined as a minimization problem. A branch-and-bound algorithm may be thought of as a "divide-and-conquer" strategy. In general, strictly decreasing feasible solution values (upper bounds) are found as the algorithm proceeds. We define the best solution known to date as the incumbent. Concurrently the lower bound values of branch-and-bound candidate problems (obtained from valid relaxations) tend to increase as one proceeds deeper in the branch-and-bound tree. Optimality of the incumbent is assured when it can be shown that the lower bounds of all remaining candidate problems are greater than or equal to the upper bound. In the next subsection we proceed to describe a number of methods for improving valid lower bounds (LB) for (P) as well as for candidate problems derived from (P) as part of the branch-and-bound process. In subsection 3.4.2 we describe methods for generating improved feasible solutions (UB).

3.4.1 Methods for Increasing the Lower Bound for (P)

We denote the linear programming relaxation of (P) as $(\overline{P})$ where the binary constraints (3.4) are replaced by $0 \leq x_{ij} \leq 1 \; \forall i \in M, j \in N$. A straightforward way to tighten $(\overline{P})$ is to add additional constraints which "cut" away portions of the feasible region of $(\overline{P})$ while leaving the feasible region of (P) intact. Numerous types of cuts may be utilized. However the special structure of the GAP in terms of the separable knapsack constraints (3.2) suggests minimal cover cuts (Balas and Jeroslow [2]) be applied to the individual knapsack constraints. Let $\bar{x}$ be an optimal solution to $(\bar{P})$ with value $v(\bar{P})$. Suppose a knapsack constraint i in (3.2), namely $\sum_{j=1}^{n} r_{ij} x_{ij} \leq b_i$, has at least one $\bar{x}_{ij}$ fractional. Define S_i to be the index set of $\bar{x}_{ij} > 0$. From the index set S_i, choose variables in decreasing order of r_{ij} and place the indices in set F_i until $\sum_{j \in F_i} r_{ij} > b_i$. Let $C_i = |F_i| - 1$. Then the cut, $\sum_{j \in F_i} x_{ij} \leq C_i$, is a valid cut that the solution $\bar{x}$ violates as long as $\sum_{j \in F_i} \bar{x}_{ij} > C_i$. If $\sum_{j \in F_i} \bar{x}_{ij} \leq C_i$, the cut is not added. If the cut is added, then it is strengthened in the following way. Let $R = \max_{j \in F_i} \{r_{ij}\}$. For all $j \in N - F_i$ and for which $r_{ij} \geq R$, add j to the set F_i (note that the value of C_i remains unchanged). Next, a coefficient-lifting procedure due to Cattrysse et al. [3] is invoked. The lifted cut is written as follows: $\sum_{j \in F_i} a_{ij} x_{ij} \leq C_i$ where the a_{ij}'s are positive and integer-valued. In lieu of reproducing the procedure due to Cattrysee et al. [3] we present a small example where coefficient-lifting occurs. Consider a resource constraint, $11x_{11} + 12x_{12} + 20x_{13} + 22x_{14} \leq 30$ with the corresponding LP solution, $\bar{x}_{11} = \bar{x}_{12} = 1$, $\bar{x}_{13} = 7/20$, $\bar{x}_{14} = 0$. Then $S_1 = \{1, 2, 3\}$, $F_1 = S_1$, and $C_1 = |F_1| - 1 = 2$. Then $x_{11} + x_{12} + x_{13} \leq 2$ is a valid cut that violates the original LP solution. Since $R = 20$, we may strengthen the cut to $x_{11} + x_{12} + x_{13} + x_{14} \leq 2$ and define $F_1 = \{1, 2, 3, 4\}$. We now lift the coefficients for x_{13} and x_{14} to 2 since it is clear that if $x_{13} = 1$ or $x_{14} = 1$, then $x_{11} = x_{12} = 0$ in order to satisfy the original resource constraint.

After these cuts are added to $(\bar{P})$, we resolve the linear program $(\bar{P}|$ cuts appended). A new optimal solution $\bar{x}_c$ is obtained where $v(P) \geq v(\bar{P}|$ cuts appended) $\geq v(\bar{P})$. The procedure for adding cuts is invoked once again with a resolving of the associated LP. It continues until no further cuts can be added that are violated by the "new" LP solution or until some user-specified limit on the maximum number of cuts to be appended is reached. We also note that more than one cut may be added for each knapsack resource constraint due to the continuous resolving of the LP and subsequent addition of cuts.

Lagrangian relaxation is another potentially powerful technique for improving the lower bound for (P). In the description to follow we assume that lifted minimal cover cut constraints have been added to (P). We denote this problem as (P_c). Note of course that $v(P) = v(P_c) \geq v(\bar{P}_c) \geq v(\bar{P})$. We denote $NC(i)$ as the total number of lifted minimal cover cuts added for knapsack resource constraint i and define $NC = \sum_{i=1}^{m} NC(i)$. Then $\sum_{j \in F_{i(k)}} a_{i(k)j} x_{i(k)j} \leq C_{i(k)}$ represents the kth cut for knapsack resource constraint i and $\gamma_{i(k)}$ is the corresponding dual variable for the cut.

By Lagrangianizing the multiple choice constraints (3.3) and the NC cuts we have the Lagrangian relaxation (LGR):

$$LGR(\lambda, \gamma) \equiv \text{minimize} \sum_{i=1}^{m} \sum_{j=1}^{n} c_{ij} x_{ij} + \sum_{j=1}^{n} \lambda_j \left(1 - \sum_{i=1}^{m} x_{ij}\right) \quad (3.5)$$

$$+ \sum_{i=1}^{m} \left[\sum_{k=1}^{NC(i)} \gamma_{i(k)} \left(C_{i(k)} - \sum_{j \in F_{i(k)}} a_{i(k)j} x_{i(k)j} \right) \right]$$

subject to:

$$\sum_{j=1}^{n} r_{ij} x_{ij} \leq b_i, \quad \forall i \in M \tag{3.6}$$

$$x_{ij} = 0 \quad \text{or} \quad 1, \quad \forall i \in M, \quad j \in N. \tag{3.7}$$

Note that $LGR(\lambda, \gamma)$ separates into m independent 0-1 knapsack problems. The individual knapsack problems are solved using an algorithm due to Nauss [18].

If we utilize the optimal dual multipliers λ and γ from the solution of $(\overline{P}_c)$ we are assured that $v(P) \geq v(LGR(\lambda, \gamma)) \geq v(\overline{P}_c) \geq v(\overline{P})$. The LGR may be tightened further by utilizing subgradient optimization to "adjust" λ and γ in the following way:

$$\begin{matrix} \text{maximize} \\ \lambda \text{ unrestricted} \\ \gamma \leq 0 \end{matrix} \left\{ \text{minimize} \sum_{i=1}^{m} \sum_{j=1}^{n} c_{ij} x_{ij} + \sum_{j=1}^{n} \lambda_j \left(1 - \sum_{i=1}^{m} x_{ij} \right) \right.$$

$$\left. + \sum_{i=1}^{m} \left[\sum_{k=1}^{NC(i)} \gamma_{i(k)} \left(C_{i(k)} - \sum_{j \in F_{i(k)}} a_{i(k)j} x_{i(k)j} \right) \right] \right\} \tag{3.8}$$

subject to:

$$\sum_{j=1}^{n} r_{ij} x_{ij} \leq b_i, \quad \forall i \in M \tag{3.9}$$

$$x_{ij} = 0 \quad \text{or} \quad 1, \quad \forall i \in M, j \in N. \tag{3.10}$$

Note that in each subgradient iteration we must solve m 0-1 knapsack problems. Thus we limit the number of iterations to 100 while utilizing a stepsize calculation described in Guignard and Rosenwein [11].

Once the LGR has been solved we introduce the concept of penalties. Penalties are useful in two ways. First in a branch-and-bound algorithm penalties may be used to select a branching variable to separate a candidate problem (CP) under the "divide and conquer" strategy. Variables x_{ij} that can be shown to have relatively large values of $|v(LGR(CP|x_{ij} = 1)) - v(LGR(CP|x_{ij} = 0))|$ are attractive to separate (or branch) on since it is reasonable to assume that the objective value for (P) is quite different if $x_{ij} = 1$ as opposed to if $x_{ij} = 0$. Alternatively if the value is very small, then whether x_{ij} equals 0 or 1 would not seem to have much impact on the optimal objective value of (P). A second use of penalties is to try to fix x_{ij} variables to 0 or 1. This is closely related to branch selection since if it can be shown that x_{ij} cannot be 0, for example, the x_{ij} may be fixed at 1 and a potential branching is eliminated for CP.

For penalty calculation we use the Lagrangian relaxation $LGR(\lambda, \gamma)$ where (λ, γ) are the dual multiplier vectors found via subgradient optimization. Clearly $v(LGR(\lambda, \gamma))$ may be divided into $m+1$ components as follows:

$$\left(\sum_{j=1}^{n} \lambda_j + \sum_{i=1}^{m} \sum_{k=1}^{NC(i)} \gamma_{i(k)} C_{i(k)} \right) + \sum_{i=1}^{m} \left(\sum_{j=1}^{n} \left(c_{ij} - \lambda_j - \sum_{k=1}^{NC(i)} \gamma_{i(k)} \left(\sum_{j \in F_{i(k)}} a_{i(k)j} \right) \right) x_{ij} \right)$$

Letting

$$v(KNAP_i) = \sum_{j=1}^{n} \left(c_{ij} - \lambda_j - \sum_{k=1}^{NC(i)} \gamma_{i(k)} \left(\sum_{j \in F_{i(k)}} a_{i(k)j} \right) \right) x_{ij},$$

we have

$$v(LGR(\lambda, \gamma)) = \sum_{j=1}^{n} \lambda_j + \sum_{i=1}^{m} \sum_{k=1}^{NC(i)} \gamma_{i(k)} C_{i(k)} + \sum_{i=1}^{m} v(KNAP_i).$$

Denoting the optimal LP relaxation solution value for $(KNAP_i)$ as $v(\overline{KNAP_i})$, we have for some i^*,

$$v(LGR(\lambda, \gamma)) \geq \sum_{j=1}^{n} \lambda_j + \sum_{i=1}^{m} \sum_{k=1}^{NC(i)} \gamma_{i(k)} C_{i(k)} + \sum_{\substack{i=1 \\ i \neq i^*}}^{m} v(KNAP_i) + v(\overline{KNAP_{i^*}})$$

The right-hand side of the above inequality is a valid relaxation value for $v(LGR(\lambda, \gamma))$. LP penalties for variables in a knapsack problem are easily calculated using the optimal LP dual multiplier for the resource constraint (Nauss [18]). Note also that the LP penalty for a variable x_{ij} is based on the LP relaxation of a *single* knapsack constraint while the remainder of the knapsack constraints are <u>not</u> relaxed. Thus if Z^* is the incumbent for (P), we may fix x_{i^*j} to 1 if

$$\sum_{j=1}^{n} \lambda_j + \sum_{i=1}^{m} \sum_{k=1}^{NC(i)} \gamma_{i(k)} C_{i(k)} + \left(\sum_{\substack{i=1 \\ i \neq i^*}}^{m} v(KNAP_i) \right)$$

$$+ v(\overline{KNAP_{i^*}}) + PEN(i^*, j | x_{i^*j} = 0) \geq Z^*. \quad (3.11)$$

Note that $PEN(i^*, j | x_{i^*j} = 0) = -c_{i^*j} + \zeta r_{i^*j}$ where ζ is the optimal dual multiplier of $(\overline{KNAP_{i^*}})$. We note also that a stronger penalty may be calculated by finding $v(KNAP_{i^*} | x_{i^*j} = 0)$ and substituting it for $v(\overline{KNAP_{i^*}}) + PEN(i^*, j | x_{i^*j} = 0)$ in (3.11). This of course requires solving a 0-1 knapsack, which may be prohibitive if every variable were so tested. However, the selective use of such enhanced penalties can be effective when a corresponding LP penalty "almost" results in fixing a variable.

Finally we address two logical feasibility-based tests for fixing x_{ij} variables to 0 or 1. The first test looks at the maximum slack allowed in a knapsack resource constraint i for an optimal solution to (P). As such it utilizes an upper bound on $v(P)$ and thus as the upper bound is reduced the maximum slack also tends to be reduced. This, in effect, reduces the feasible region for (P) and potentially allows x_{ij} variables to be fixed to 0 or 1 in candidate problems.

Rewrite the knapsack resource constraints as $\Sigma_{j=1}^n r_{ij} x_{ij} + u_i = b_i, i \in M$ where u_i is defined to be the corresponding nonnegative slack variable. Then for each knapsack resource constraint solve the following LP where Z^* is an upper bound on (P) and $v(LGR(\lambda, \gamma))$ is a lower bound on (P):

$$g_i \equiv \text{maximize } u_i \tag{3.12}$$

subject to:

$$\sum_{j=1}^n r_{ij} x_{ij} + u_i = b_i, \quad \forall i \in M \tag{3.13}$$

$$\sum_{i=1}^m x_{ij} = 1, \quad \forall j \in N \tag{3.14}$$

$$v(LGR(\lambda, \gamma)) \le \sum_{i=1}^m \sum_{j=1}^n c_{ij} x_{ij} \le Z^* \tag{3.15}$$

$$0 \le x_{ij} \le 1 \quad \forall i \in M, \quad j \in N \tag{3.16}$$

$$u_i \ge 0, \quad \forall i \in M. \tag{3.17}$$

We may then add the constraints $0 \le u_i \le g_i \forall i \in M$ to (P) where (3.2) is replaced by (3.13).

Next we calculate the maximum total slack over all knapsack resource constraints by solving:

$$Z_u^* \equiv \text{maximize } \sum_{i=1}^m u_i$$

subject to:

$$(3.13) - (3.16)$$

$$0 \le u_i \le g_i, \quad \forall i \in M.$$

Suppose that for a particular candidate problem a number of x_{ij} variables have been set to 1, and that for some resource constraint ℓ we have: $\Sigma_{j|x_{\ell j}=1} r_{\ell j} x_{\ell j} > b_\ell - g_\ell$ and $\Sigma_{j|x_{\ell j}=1} r_{\ell j} x_{\ell j} \le b_\ell$. For all i define $gcp(i) = \min\{b_i - \Sigma_{j|x_{ij}=1} r_{ij} x_{ij}, g_i\}$. It is easy to see that $gcp(i)$ is a valid bound on u_i for a particular candidate problem. We deduce that $Z_u^* - \Sigma_{\ell=1, \ell \ne i}^m gcp(\ell)$ is also an upper bound on the maximum slack for knapsack i, and if $Z_u^* - \Sigma_{\ell=1*\ell \ne i}^m gcp(\ell) < g_i$, then g_i may be reduced to that value for all descendant candidate problems.

Another method for tightening g_i for a candidate problem is to use LP penalties for each knapsack in the *LGR*. Define $\bar{u}_i$ to be the value of the slack variable in the ith knapsack LP. If $\bar{u}_i = 0$ in an optimal LP solution, it is straightforward to calculate the maximum value that u_i could assume such that the objective value of the *LGR* still exceeds Z^*. These calculations are carried out after all knapsacks in the *LGR* have been solved, since the test depends on $v(LGR(\lambda, \gamma))$, $v(KNAP_i)$, and $v(\overline{KNAP}_i)$. The maximum value that u_i could take on may then replace g_i in the current candidate problem and all its descendants. We note that tightening of the bound on u_i is often more useful in feasibility-based fixing tests than in improving the LB, $v(LGR(\lambda, \gamma))$. This is because the variable to be fixed to 0 is often equal to 0 in the *LGR*. Rather, the benefit of fixing the variable to 0 is important because this may eventually lead to other variables being fixed to 1.

The above procedures for reducing the maximum slack in knapsack resource constraints strengthen the three straightforward logical feasibility test that we now enumerate.

Consider a conditional feasibility test to fix some x_{ik} to 1. Assume x_{ik} is temporarily set to 0. Then if we can show that for some i, $\Sigma_{j|x_{ij}=1} r_{ij} < b_i - g_i$ and $\Sigma_{\ell \neq k \text{ and } \atop \ell \neq j | x_{ij} = 1} r_{i\ell} < b_i - g_i - \Sigma_{j|x_{ij}=1} r_{ij}$, then x_{ik} may be fixed to 1. This follows since we must have $b_i - g_i \leq \Sigma_j r_{ij} x_{ij} \leq b_i$.

Next is a conditional feasibility test where x_{ik} is temporarily set to 1. If it can be shown that no feasible solution exists under this assumption, then x_{ik} may be fixed to 0. For example, if for some i, $\Sigma_{j \neq k \text{ and } \atop j | x_{ij} = 1} r_{ij} + r_{ik} > b_i$, then x_{ik} may be fixed to 0.

Finally, if for some i, $\Sigma_{j|x_{ij}=1} r_{ij} + r_{ik} < b_i - g_i$, and $\Sigma_{j|x_{ij}=1} r_{ij} + r_{ik} + \min_{\ell \neq k \text{ and } \atop \ell \neq j | x_{ij} = 1} r_{i\ell} > b_i$, then x_{ik} may be fixed to 0.

3.4.2 Methods for Decreasing the Upper Bound for (*P*)

In order to decrease the UB, one must find an improved feasible solution to (*P*) with objective value less than that of the current incumbent solution, Z^*. At each node of the branch-and-bound tree a Lagrangian relaxation solution $\hat{x}$ is obtained from (3.8) to (3.10). In this Lagrangian relaxation the multiple choice constraints (3.3) have been relaxed and placed in the objective function. Thus they are the only type of constraints that might be violated. For example, a task ℓ may have $\Sigma_{i=1}^m \hat{x}_{i\ell} \geq 2$ while another task k may have $\Sigma_{i=1}^m \hat{x}_{ik} = 0$. Accordingly, we attempt to modify the *LGR* solution, $\hat{x}$, by attempting to reduce assignments as economically as possible for constraints where $\Sigma_{i=1}^m \hat{x}_{ij} \geq 2$. We find the constraint j^* where $j^* = \arg\max(\Sigma_{i=1}^m \hat{x}_{ij})$. Then all but one $\hat{x}_{ij}$ that are currently equal to 1 are reset to the value 0 in nonincreasing order of the cost coefficients, c_{ij^*}. This creates additional slack in the resource constraints allowing possible assignments to be made for tasks, j, where $\Sigma_{i=1}^m \hat{x}_{ij} = 0$. For each such constraint $\hat{j}$ we attempt to set an $\hat{x}_{ij}$ to 1 where the following two conditions must be met. First the knapsack inequality (3.9) must have sufficient slack to allow r_{ij} to fit *and* the cost coefficient c_{ij} should be as small as possible. If all multiple choice constraints are now satisfied *and* the objective function has value less than Z^*, an improved feasible solution has been found and the UB is reduced.

Whenever an improved feasible solution has been found, we invoke a neighborhood slide/switch heuristic suggested by Ronen [22] to attempt to find an even better solution. Ronen attempts to move each task from its current agent to every other agent. The move is made only if a cost saving is achieved. Next, all pairs of tasks (j_1, j_2) currently assigned to distinct agents (i_1, i_2) where $i_1 \neq i_2$ are tentatively switched so that task j_1 is assigned to agent i_2 and task j_2 is assigned to agent i_1. Once again the switch is made only if a feasible cost saving is achieved. Similar tentative switches of one task for two tasks with a different agent as well as two for three task switches are also attempted. If an improved solution is generated, the entire neighborhood slide/switch and multiple switches are attempted again until a complete cycle is completed without an improved feasible solution being found.

Finally, complete enumeration is used when relatively few tasks (say, five or fewer) remain to be assigned in some candidate problem. Every possible combination of assignments of tasks to agents is evaluated. If an improved feasible solution is found, the incumbent and the upper bound is updated. In any case the candidate problem is discarded from further consideration.

3.5 Branch-and-Bound Algorithm

A special purpose branch-and-bound algorithm for the GAP is described using Geoffrion and Marsten's [9] general framework. We comment on various features of the algorithm at the end of this section.

Step-1. Set FATHOM $= 0$, C $= .014$, and NODE $= 0$.

Step 0. **(Solve root node relaxations)** Solve the LP relaxation $(\overline{P})$. Add violated lifted minimal cover cuts to $(\overline{P})$, namely $(\overline{P}_c)$. Resolve $(\overline{P}_c)$ and add more violated lifted minimal cover cuts. Continue this procedure until no more violated cuts are added. Solve the associated Lagrangian relaxation, (LGR) and apply subgradient optimization. Place (P) in the candidate list. Set $Z^* = Z_F \equiv (1 + C)^* v(LGR)$.

Step 1. **(Tighten resource slack variable bounds)** Solve $m + 1$ linear programs to find the maximum slack for each resource constraint and the maximum total slack.

Step 2. **(Penalty tests)** Attempt to fix variables to 0 or 1 using LP knapsack penalties. Periodically attempt to fix variables with strong IP knapsack penalties. Attempt to fix variables to 0 or 1 using feasibility tests. If any variables have been fixed to 0 or 1, resolve the Lagrangian relaxation and apply subgradient optimization and return to the beginning of step 2. If FATHOM $= 1$, go to step 3. If FATHOM $= 0$, go to step 8.

Step 3. **(Candidate problem selection)** If the candidate list is empty, stop; an optimal solution has been found if $Z^* < Z_F$, or if $Z^* \geq Z_F$ increase C to 2*C and go to step 0. If the candidate list is not empty, select a problem (CP) from the candidate list using a LIFO rule, set NODE = NODE +1, and set FATHOM = 0.

Step 4. **(Complete enumeration)** If (CP) has more than five tasks not assigned, go to step 5. Otherwise completely enumerate (CP). If an improved feasible solution has been found, update Z^*, set FATHOM = 1 and go to step 1. Else go to step 3.

Step 5. **(Feasibility-based tests)** Perform feasibility-based tests to fix variables 0 to 1.

Step 6. **(Solve LGR relaxation)** Solve the LGR of (CP). If $v(LGR) \geq Z^*$, set FATHOM = 1 and go to step 3. If NODE $\leq$ 10 or NODE is a multiple of 5, use subgradient optimization to tighten the LGR of (CP). If $v(LGR) \geq Z^*$ set FATHOM = 1 and go to step 3. If the LGR solution is feasible for (P) update Z^*, set FATHOM = 1 and go to step 1.

Step 7. **(Feasible solution generator)** Attempt to modify the Lagrangian relaxation to find an improved feasible solution to (P). If one is found, update Z^*. If $Z^* = v(LGR)$ set FATHOM = 1 and go to step 1.

Step 8. **(Separation)** Choose a free variable x_{ij} in (CP) and add the candidate problems $(CP \mid x_{ij} = 0)$ and $(CP \mid x_{ij} = 1)$ to the bottom of the candidate list. Go to step 3.

The indicator FATHOM is set to 1 whenever a node is fathomed. When a new node is chosen in step 3, FATHOM is reset to 0 and the node counter, NODE is incremented. The value of C is initially set to .014. We use this parameter in order to artificially set an upper bound Z_F at 1.4% larger than $v(LGR)$. If it is proven that no feasible solution to (P) exists with value less than Z_F, we double the value of C and restart the algorithm with Z_F at a value 2.8% larger than $v(LGR)$. This technique is used since generally the gaps between $v(LGR)$ and optimal value Z^* are less than 1.4% The initialization of Z_F allows us to benefit in two ways. First, Guignard and Rosenwein [11] suggest that a good upper bound target be used for subgradient optimization in order to reduce "overshooting." Second, variables may be fixed at 0 or 1 using the penalty tests described in earlier sections thus reducing unnecessary branching or separation. In the next section we comment further on the use of Z_F.

In Step 8 the variable to be separated on is determined in the following way. Recall that a LIFO rule is used to select candidate problems and thus the "1" branch is chosen before "0" branch. Let J_{FIX1} be the index set of all tasks j that have an x_{ij} fixed to 1 in the current candidate problem. Over all $j \notin J_{FIX1}$, place j in the index set JB if $\Sigma_{i=1}^{m} \hat{x}_{ij} = 1$ (where $\hat{x}$ is the current LGR solution). The branch variable x_{ij} is selected by finding

$$\max_{\substack{j \in JB \\ \text{and} \hat{x}_{ij}=1}} PEN\ (i,\ j \mid x_{i*j} = 0)$$

where *PEN* (*) is the knapsack LP penalty. In effect we select a variable x_{ij} with the largest penalty for setting x_{ij} to 0 such that the corresponding multiple choice constraint $(\Sigma_{i=1}^{m} x_{ij} = 1)$ is not violated in the Lagrangian relaxation.

3.6 Computational Results

The algorithm was coded in FORTRAN and compiled using Compaq Visual FORTRAN 6.6 (Compaq [7]). XMP (Marsten [16]) was used to solve the linear programs. A Dell 340 Precision Workstation (2.53 GHz Pentium 4 with 512 MB of PC800 ECC RDRAM) was used for the computational work.

A dataset consisting of 540 random problems was generated according to guidelines outlined in Chu and Beasley [6] and in Laguna et al. [15]. The dataset consists entirely of "hard" GAP problems of types D and E ranging from 750 to 4000 binary variables. Both problem types D and E have objective function coefficients, c_{ij}, that are inversely correlated with resource coefficients, r_{ij}.

We present the computational results in two tables. Table 3.1 gives results for 270 type D problems and Table 3.2 gives results for 270 Type E problems. Each line in the tables gives the average results for ten problems of a particular size. Six averages for each problem set are reported.

They are:

lpgap(%) –   the percentage gap between the LP and best solution found
lgrgap(%) – the percentage gap between the original LGR and the best
             solution found
bstlgr(%) – the percentage gap between the LGR that reflects variables
             that were fixed to 0 or 1 based on penalty tests calculated
             after every feasible solution found and the best solution found
             node
node –   the number of branch-and-bound nodes evaluated
tmbst (sec) – the CPU time in seconds until the best feasible solution was
             found
tottime(sec) – the CPU time in seconds to either prove optimality or until
             the time limit was reached

A time limit of 1200 CPU seconds was enforced and a suboptimality tolerance of 0% was used. In both tables the problems are given in order of increasing size. Within each set of problems with the same number of variables (but differing number of agents and tasks), the ratio of tasks/agents is decreasing.

Results for types D and E are dissimilar for problems of size 1250 or greater. In fact only 8 of 240 D type problems with variable sizes of 1250 to 4000 are solved to optimality. On the other hand 220 of 240 E type problems with variable sizes of 1250 to 4000 are solved to optimality. A key variable that seems to explain problem difficulty is bstlgr(%). As can be seen in Table 3.1 and

TABLE 3.1
Type D Problems

Agents	Tasks	Variables	Tasks/Agents	lpgap(%)	lgrgap(%)	bstlgr(%)	Nodes	Tmbst (sec)	Tottime (sec)					Problems Requiring Larger C
5	150	750	30.0	0.06	0.06	0.03	71,286	175.1	185.4	10	of	10	proven optimal	0
10	75	750	7.5	0.49	0.41	0.15	92,681	300.9	414.6	10	of	10	proven optimal	0
15	50	750	3.3	1.74	1.07	0.29	12,900	61.5	71.7	10	of	10	proven optimal	0
10	125	1250	12.5	0.24	0.21	0.12	214,629	695.9	1200.4	0	of	10	proven optimal	0
10	150	1500	15.0	0.21	0.20	0.13	192,837	574.7	1200.6	0	of	10	proven optimal	0
20	75	1500	3.8	1.17	0.84	0.24	110,330	533.8	1049.9	2	of	10	proven optimal	0
10	175	1750	17.5	0.21	0.20	0.15	174,244	738.8	1200.2	0	of	10	proven optimal	0
25	70	1750	2.8	1.78	1.10	0.25	64,667	538.3	786.0	6	of	10	proven optimal	0
10	200	2000	20.0	0.18	0.17	0.14	165,770	703.8	1200.2	0	of	10	proven optimal	0
20	100	2000	5.0	0.78	0.64	0.30	98,685	739.3	1200.2	0	of	10	proven optimal	0
25	80	2000	3.2	1.39	0.97	0.29	87,615	802.6	1200.4	0	of	10	proven optimal	0
10	225	2250	22.5	0.17	0.17	0.14	148,146	915.7	1200.1	0	of	10	proven optimal	0
15	150	2250	10.0	0.34	0.31	0.21	110,815	683.0	1200.2	0	of	10	proven optimal	0
25	90	2250	3.6	1.16	0.85	0.33	79,220	753.8	1200.1	0	of	10	proven optimal	0
10	250	2500	25.0	0.17	0.17	0.15	135,247	700.8	1200.2	0	of	10	proven optimal	0
20	125	2500	6.3	0.57	0.49	0.28	87,157	521.9	1200.2	0	of	10	proven optimal	0
25	100	2500	4.0	0.97	0.76	0.33	75,179	783.9	1200.1	0	of	10	proven optimal	0
10	275	2750	27.5	0.15	0.14	0.13	133,488	760.9	1200.1	0	of	10	proven optimal	0
25	110	2750	4.4	0.83	0.68	0.34	68,447	689.5	1200.2	0	of	10	proven optimal	0
10	300	3000	30.0	0.16	0.16	0.15	115,690	818.3	1200.1	0	of	10	proven optimal	0
15	200	3000	13.3	0.26	0.25	0.20	89,744	906.6	1200.1	0	of	10	proven optimal	0
20	150	3000	7.5	0.46	0.42	0.28	68,476	642.4	1200.1	0	of	10	proven optimal	0
25	120	3000	4.8	0.71	0.59	0.32	62,377	710.4	1200.1	0	of	10	proven optimal	0
30	100	3000	3.3	1.26	0.94	0.42	56,638	845.6	1200.1	0	of	10	proven optimal	0
10	400	4000	40.0	0.18	0.18	0.17	89,290	691.4	1200.2	0	of	10	proven optimal	0
20	200	4000	10.0	0.37	0.35	0.28	57,581	717.9	1200.2	0	of	10	proven optimal	0
25	160	4000	6.4	0.54	0.49	0.33	48,692	774.4	1200.2	0	of	10	proven optimal	0
Overall Averages				0.61	0.47	0.23	100,438	658.6	1070.8	1.4	of	10	proven optimal	0
										38	of	270	proven optimal	

TABLE 3.2
Type E Problems

Agents	Tasks	Variables	Tasks/Agents	lpgap(%)	lgrgap(%)	bstlgr(%)	Nodes	Tmbst (secs)	Tottime (secs)				Problems Requiring Larger C
5	150	750	30.0	0.07	0.05	0.03	1,285	3.0	3.7	10	of	10 proven optimal	0
10	75	750	7.5	0.50	0.10	0.09	1,434	4.8	5.8	10	of	10 proven optimal	0
15	50	750	3.3	2.85	0.34	0.33	1,113	5.2	7.2	10	of	10 proven optimal	8
10	125	1250	12.5	0.14	0.05	0.03	4,263	17.7	19.6	10	of	10 proven optimal	0
10	150	1500	15.0	0.11	0.04	0.02	6,028	27.3	30.0	10	of	10 proven optimal	0
20	75	1500	3.8	1.79	0.19	0.18	4,789	29.5	42.0	10	of	10 proven optimal	0
10	175	1750	17.5	0.08	0.04	0.02	8,933	30.5	42.1	10	of	10 proven optimal	0
25	70	1750	2.8	3.65	0.24	0.22	3,575	37.2	48.0	10	of	10 proven optimal	10
10	200	2000	20.0	0.05	0.03	0.01	4,898	27.0	29.9	10	of	10 proven optimal	0
20	100	2000	5.0	0.85	0.09	0.08	19,653	181.7	196.8	10	of	10 proven optimal	0
25	80	2000	3.2	2.50	0.21	0.20	6,425	45.3	78.6	10	of	10 proven optimal	5
10	225	2250	22.5	0.05	0.03	0.01	6,054	30.1	36.1	10	of	10 proven optimal	0
15	150	2250	10.0	0.17	0.05	0.03	14,343	102.8	117.6	10	of	10 proven optimal	0
25	90	2250	3.6	1.91	0.21	0.20	17,646	171.7	229.5	10	of	10 proven optimal	0
10	250	2500	25.0	0.03	0.02	0.01	8,246	49.0	52.6	10	of	10 proven optimal	0
20	125	2500	6.3	0.42	0.06	0.05	17,173	146.6	197.0	10	of	10 proven optimal	0
25	100	2500	4.0	1.36	0.17	0.16	31,460	250.8	424.9	8	of	10 proven optimal	0
10	275	2750	27.5	0.03	0.02	0.01	11,861	79.1	87.1	10	of	10 proven optimal	0
25	110	2750	4.4	1.04	0.12	0.11	23,077	198.5	341.2	8	of	10 proven optimal	0
10	300	3000	30.0	0.03	0.02	0.01	27,383	125.4	182.8	10	of	10 proven optimal	0
15	200	3000	13.3	0.09	0.03	0.02	19,581	146.3	205.2	10	of	10 proven optimal	0
20	150	3000	7.5	0.29	0.05	0.03	37,947	424.9	537.0	9	of	10 proven optimal	0
25	120	3000	4.8	0.81	0.08	0.07	22,587	269.6	392.8	8	of	10 proven optimal	0
30	100	3000	3.3	2.21	0.20	0.19	27,238	310.5	496.5	8	of	10 proven optimal	3
10	400	4000	40.0	0.03	0.02	0.01	87,607	650.6	724.3	7	of	10 proven optimal	0
20	200	4000	10.0	0.15	0.04	0.03	54,610	594.5	813.3	6	of	10 proven optimal	0
25	160	4000	6.4	0.46	0.13	0.12	37,090	509.1	714.1	6	of	10 proven optimal	0
Overall Averages			6.4	**0.80**	**0.10**	**0.08**	**18,752**	**165.5**	**224.3**	**9.3**	**of**	**10 proven optimal**	**26**
										250	of	270 proven optimal	

Table 3.2 the bstlgr(%) in Table 3.1 is generally larger than the corresponding (same number of agents and tasks) value of bstlgr(%) in Table 3.2. In fact over all 270 type E problems the average bstlgr(%) is .08 while the corresponding value is .23 over all 270 type D problems. Also note that in numerous instances in Table 3.1 and Table 3.2 there are average lpgap(%) values in excess of 1. Somewhat perplexing is the fact that the overall average lpgap(%) for type D is .61 while it is .80 for type E. However the overall average lgrgap(%) for type D is .47 while only .10 for type E. Thus we have the interesting behavior of type D problems having tighter LP relaxations, but looser LGR relaxations when compared to type E problems. We also note that 26 of 270 Type E problems required that the value of C be increased from .014 to .028. Put another way the value of Z_F was set too low in these cases. The CPU time and node averages include the time and nodes evaluated when C was .014 for those problems.

We also note that as the tasks/agents ratio becomes smaller (for a fixed number of variables), it is almost uniformly true that the lpgap(%), lgrgap(%), and bstlgr(%) grow larger. The only exception is in Table 3.1 when comparing the bstlgr(%) for 2000 variable problems. This behavior appears to make sense since as the expected number of tasks per agent decreases it is more likely that slack occurs in the resource constraints. This implies that available resources are not being efficiently utilized and hence the lpgap(%), lgrgap(%), and bstlgr(%) will tend to grow larger.

Finally we note that Yagiura et al. [27] analyzed the "distance" between good feasible solutions for individual problems. Interestingly they found that average distances are larger for Type D problems. This suggests two things. First, Type D problems appear to have good feasible solutions more highly dispersed over the feasible region. Second, this dispersal suggests that a "depth first" search (which we use) may not be as effective as a "breadth first" search for Type D problems. This appears to be confirmed by the excellent results that were generated for Type E problems where good feasible solutions tend to be "closer together."

3.7 Conclusions

While our results are favorable for Type E problems up to 4000 variables, Type D problems of 1250 or more variables still remain difficult for our algorithm. However, it should be pointed out that the largest average bestlgr(%) gap for any Type D problem set is .42% with an overall average for all 270 problems of .23%. Thus excellent feasible solutions are generated with the shortcomings being the difficulty in further tightening of the lower bounds and in the use of a LIFO approach for selection of candidate problems (*cf.* Section 3.6). Further work on Type D problems might include a "breadth first" approach for the search process as well as additional work on tightening the relaxed feasible region of candidate problems. For example the incorporation of additional cuts as developed by Gottlieb and Rao [10] might result in a Lagrangian relaxation that is easier (and faster) to solve in the subgradient optimization phase. In line with this it should be noted that since 30 to 60% of CPU time is devoted to solving the 0-1 knapsack problems, an alternate knapsack code (Kellerer et al. [14]) might be incorporated. Such an implementation might require that all data (c_{ij} and r_{ij}) be integer valued. Secondly, the use of a state-of-the-art LP solver such as CPLEX would result in reduced computational times. Finally with the advent of parallel processing it would be instructive to "marry" the branch-and-bound code with a heuristic such as Yagiura et al. [27]. In such a marriage multiple processors might be allocated for separate copies of the heuristic (where different feasible region neighborhoods are examined). Then when an improved feasible solution is found it is stored in shared memory to be used by multiple processors that are running the branch-and-bound algorithm. (Of course improved feasible solutions generated by the branch-and-bound algorithm would also be stored in this shared memory.) Alternatively the neighborhood search heuristics might be "called" to explore various feasible region neighborhoods on different processors as the branch-and-bound algorithm generates sufficiently different candidate problems.

References

1. Balachandran, V., An integer generalized transportation model for optimal job assignment in computer networks, Working Paper 34-72-3, Graduate School of Industrial Administration, Carnegie Mellon University, Pittsburgh, 1972.
2. Balas, E. and Jeroslow, R., Canonical cuts in the unit hypercube, *SIAM J. Appl. Math.*, 23, 61, 1972.
3. Cattrysse, D., Degraeve Z., and Tistaert, J., Solving the generalized assignment problem using polyhedral results, *European J. Operational Research*, 108, 618, 1998.
4. Cattrysse, D., Salomon, M., and Van Wassenhove, L., A set partitioning heuristic for the generalized assignment problem, *European J. Operational Research*, 72, 167, 1994.
5. Cattrysse, D. and Van Wassenhove, L., A survey of algorithms for the generalized assignment problem, *European J. Operational Research*, 60, 260, 1992.
6. Chu, P.C. and Beasley, J.B., A genetic algorithm for the generalized assignment problem, *Computers and Operations Research*, 24, 17, 1997.
7. *COMPAQ Fortran Language Reference Manual*, Compaq Computer Corporation, Houston, 1999.
8. Fisher, M. and Jaikumar, R., A generalized assignment heuristic for vehicle routing, *Networks*, 11, 109, 1981.
9. Geoffrion, A.M. and Marsten, R.E., Integer programming algorithms: a framework and state-of-the-art survey, *Management Science*, 18, 465, 1972.
10. Gottlieb, E. and Rao, M., The generalized assignment problem: valid inequalities and facets, *Mathematical Programming*, 46, 31, 1990.
11. Guignard, M. and Rosenwein, M., An improved dual based algorithm for the generalized assignment problem, *Operations Research*, 37, 658, 1989.

12. Guignard, M. and Zhu, S., *A Two Phase Dual Algorithm for Solving Lagrangian Duals in Mixed Integer Programming*, working paper, The Wharton School, University of Pennsylvania, Philadelphia, 1996.
13. Jörnsten, K. and Näsberg, M., A new Lagrangian relaxation approach to the generalized assignment problem, *European Journal of Operational Research*, 27, 313, 1986.
14. Kellerer, H., Pferschy, U., and Pisinger, D., *Knapsack Problems*, Springer, 2004.
15. Laguna, M., Kelly, J., Gonzalez-Velarde, J., and Glover, F., Tabu search for the multilevel generalized assignment problem, *European J. Operational Research*, 82, 176, 1995.
16. Marsten, R., The design of the XMP linear programming library, technical report, Department of Management Information Systems, University of Arizona, Tucson, 80-2, 1980.
17. Mazzola, J., Neebe, A., and Dunn, C., Production planning of a flexible manufacturing system in a material requirements planning environment, *International Journal of Flexible Manufacturing Systems*, Vol. 1 of 2, 115, 1989.
18. Nauss, R.M., An efficient algorithm for the 0-1 knapsack problem, *Management Science*, 23, 27, 1976.
19. Nauss, R.M., Solving the generalized assignment problem: an optimizing and heuristic approach, *INFORMS Journal on Computing*, 15, 249, 2003.
20. Osman, I.H., Heuristics for the generalized assignment problem, *OR Spektrum*, 17, 211, 1995.
21. Park, J.S., Lim, B.H., and Lee, Y., A Lagrangian dual-based branch-and-bound algorithm for the generalized multi-assignment problem, *Management Science*, 44, S271, 1998.
22. Ronen, D., Allocation of trips to trucks operating from a single terminal, *Computers and Operations Research*, 19, 129, 1992.
23. Ross, G.T. and Soland, R.M., A branch and bound algorithm for the generalized assignment problem, *Mathematical Programming*, 8, 91, 1975.
24. Ross, G.T. and Soland, R.M., Modeling facility location problems as generalized assignment problems, *Management Science*, 24, 345, 1977.
25. Savelsbergh, M., A branch-and-price algorithm for the generalized assignment problem, *Operations Research*, 45, 831, 1997.
26. Wilson, J.M., A simple dual algorithm for the generalized assignment problem, *Journal of Heuristics*, 2, 303, 1997.
27. Yagiura, M., Ibaraki, T. and Glover, F., A path relinking approach with ejection chains for the generalized assignment problem, *European Journal of Operational Research*, in press, 2005.
28. Yagiura, M., Ibaraki, T., and Glover, F., An ejection chain approach for the generalized assignment problem, *INFORMS Journal on Computing*, 16, No. 2, 2004.

4

Decomposition in Integer Linear Programming

Ted K. Ralphs and Matthew V. Galati

CONTENTS

4.1 Introduction

In this chapter, we discuss the principle of decomposition as it applies to the computation of bounds on the value of an optimal solution to an integer linear program (ILP). Most bounding procedures for ILP are based on the generation of a polyhedron that approximates $\mathcal{P}$, the convex hull of feasible solutions.

Solving an optimization problem over such a polyhedral approximation, provided it fully contains $\mathcal{P}$, produces a bound that can be used to drive a branch-and-bound algorithm. The effectiveness of the bounding procedure depends largely on how well $\mathcal{P}$ can be approximated. The most straightforward approximation is the *continuous approximation*, consisting simply of the linear constraints present in the original ILP formulation. The bound resulting from this approximation is frequently too weak to be effective, however. In such cases, it can be improved by dynamically generating additional polyhedral information that can be used to augment the approximation.

Traditional dynamic procedures for augmenting the continuous approximation can be grouped roughly into two categories. *Cutting plane methods* improve the approximation by dynamically generating half-spaces containing $\mathcal{P}$, i.e., valid inequalities, to form a second polyhedron, and then intersect this second polyhedron with the continuous approximation to yield a final approximating polyhedron. With this approach, the valid inequalities are generated by solution of an associated *separation problem*. Generally, the addition of each valid inequality reduces the hypervolume of the approximating polyhedron, resulting in a potentially improved bound. Because they dynamically generate part of the description of the final approximating polyhedron as the intersection of half-spaces (an *outer representation*), we refer to cutting plane methods as *outer approximation methods*.

Traditional decomposition methods, on the other hand, improve the approximation by dynamically generating the extreme points of a polyhedron containing $\mathcal{P}$, which is again intersected with the continuous approximation, as in the cutting plane method, to yield a final approximating polyhedron. In this case, each successive extreme point is generated by solution of an associated *optimization problem* and at each step, the hypervolume of the approximating polyhedron is increased. Because decomposition methods dynamically generate part of the description of the approximating polyhedron as the convex hull of a finite set (an *inner representation*), we refer to these methods as *inner approximation methods*.

Both inner and outer methods work roughly by alternating between a procedure for computing solution and bound information (the *master problem*) and a procedure for augmenting the current approximation (the *subproblem*). The two approaches, however, differ in important ways. Outer methods require that the master problem produce "primal" solution information, which then becomes the input to the subproblem, a *separation problem*. Inner methods require "dual" solution information, which is then used as the input to the subproblem, an *optimization problem*. In this sense, the two approaches can be seen as "dual" to one another. A more important difference, however, is that the valid inequalities generated by an inner method can be valid with respect to *any* polyhedron containing $\mathcal{P}$ (see Section 4.5), whereas the extreme points generated by an inner method must ostensibly be from a single polyhedron. Procedures for generating new valid inequalities can also take advantage of knowledge of previously generated valid inequalities to further improve the approximation, whereas with inner methods, such "backward-looking" procedures do not appear to be possible. Finally, the separation procedures used in the cutting plane method can be heuristic in nature as long as it can be proven that the resulting half-spaces do actually contain $\mathcal{P}$. Although heuristic methods can be employed in solving the optimization problem required of an inner method, valid bounds are only obtained when using exact optimization. On the whole, outer methods have proven to be more flexible and powerful and this is reflected in their position as the approach of choice for solving most ILPs.

As we will show, however, inner methods do still have an important role to play. Although inner and outer methods have traditionally been considered separate and distinct, it is possible, in principle, to integrate them in a straightforward way. By doing so, we obtain bounds at least as good as those yielded by either approach alone. In such an integrated method, one alternates between a master problem that produces both primal and dual information, and either one of two subproblems, one an optimization problem and the other a separation problem. This may result in significant synergy between the subproblems, as information generated by solving the optimization subproblem can be used to generate cutting planes and vice versa.

The remainder of the chapter is organized as follows. In Section 4.2, we introduce definitions and notation. In Section 4.3, we describe the principle of decomposition and its application to integer linear programming in a traditional setting. In Section 4.4, we extend the traditional framework to show how the cutting plane method can be integrated with either the Dantzig-Wolfe method or the Lagrangian method to yield improved bounds. In Section 4.5, we discuss solution of the separation subproblem and introduce an extension of the well-known template paradigm, called *structured separation*, inspired by the fact that separation of structured solutions is frequently easier than separation of arbitrary real vectors. We also introduce a decomposition-based separation algorithm called *decompose and cut* that exploits structured separation. In Section 4.6, we discuss some of the algorithms that can be used to solve the master problem. In Section 4.7, we describe a software framework for implementing the algorithms presented in the paper. Finally, in Section 4.8, we present applications that illustrate the principles discussed herein.

4.2 Definitions and Notation

For ease of exposition, we consider only pure integer linear programs with bounded, nonempty feasible regions, although the methods presented herein can be extended to more general settings. For the remainder of the chapter, we consider an ILP whose feasible set is the integer vectors contained in the polyhedron $\mathcal{Q} = \{x \in \mathbb{R}^n \mid Ax \geq b\}$, where $A \in \mathbb{Q}^{m \times n}$ is the constraint matrix and $b \in \mathbb{Q}^m$ is the vector of requirements. Let $\mathcal{F} = \mathcal{Q} \cap \mathbb{Z}^n$ be the feasible set and let $\mathcal{P}$ be the convex hull of $\mathcal{F}$. The canonical *optimization problem* for $\mathcal{P}$ is that of determining

$$z_{IP} = \min_{x \in \mathbb{Z}^n}\{c^\top x \mid Ax \geq b\} = \min_{x \in \mathcal{F}}\{c^\top x\} = \min_{x \in \mathcal{P}}\{c^\top x\} \tag{4.1}$$

for a given cost vector $c \in \mathbb{Q}^n$, where $z_{IP} = \infty$ if $\mathcal{F}$ is empty. We refer to such an ILP by the notation $ILP(\mathcal{P}, c)$. In what follows, we also consider the equivalent decision version of this problem, which is to determine, for a given upper bound U, whether there is a member of $\mathcal{P}$ with objective function value strictly better than U. We denote by $OPT(\mathcal{P}, c, U)$ a subroutine for solving this decision problem. The subroutine is assumed to return either the empty set, or a set of one *or more* (depending on the situation) members of $\mathcal{P}$ with objective value better than U.

A related problem is the *separation problem* for $\mathcal{P}$, which is typically already stated as a decision problem. Given $x \in \mathbb{R}^n$, the problem of separating x from $\mathcal{P}$ is that of deciding whether $x \in \mathcal{P}$ and if not, determining $a \in \mathbb{R}^n$ and $\beta \in \mathbb{R}$ such that $a^\top y \geq \beta \ \forall y \in \mathcal{P}$ but $a^\top x < \beta$. A pair $(a, \beta) \in \mathbb{R}^{n+1}$ such that $a^\top y \geq \beta \ \forall y \in \mathcal{P}$ is a *valid inequality* for $\mathcal{P}$ and is said to be *violated* by $x \in \mathbb{R}^n$ if $a^\top x < \beta$. We denote by $SEP(\mathcal{P}, x)$ a subroutine that separates an arbitrary vector $x \in \mathbb{R}^n$ from polyhedron $\mathcal{P}$, returning either the empty set or a set of one or more violated valid inequalities. Note that the optimization form of the separation problem is that of finding the *most violated* inequality and is equivalent to the decision form stated here.

A closely related problem is the *facet identification problem*, which restricts the generated inequalities to only those that are *facet-defining* for $\mathcal{P}$. In [32], it was shown that the facet identification problem for $\mathcal{P}$ is polynomially equivalent to the optimization problem for $\mathcal{P}$ (in the worst case sense). However, a theme that arises in what follows is that the complexity of optimization and separation can vary significantly if either the input or the output must have known structure. If the solution to an optimization problem is required to be integer, the problem generally becomes much harder to solve. On the other hand, if the input vector to a separation problem is an integral vector, then the separation problem frequently becomes much *easier* to solve in the worst case. From the dual point of view, if the input cost vector of an optimization problem has known structure, such as being integral, this may make the problem easier. Requiring the output of the separation problem to have known structure is known as the template paradigm and may also make the separation problem easier, but such a requirement is essentially equivalent to enlarging $\mathcal{P}$. These concepts are discussed in more detail in Section 4.5.

4.3 The Principle of Decomposition

We now formalize some of the notions described in the introduction. Implementing a branch-and-bound algorithm for solving an ILP requires a procedure that will generate a lower bound as close as possible to the optimal value z_{IP}. The most commonly used method of bounding is to solve the linear programming (LP) relaxation obtained by removing the integrality requirement from the ILP formulation. The *LP Bound* is given by

$$z_{LP} = \min_{x \in \mathbb{R}^n}\{c^\top x \mid Ax \geq b\} = \min_{x \in \mathcal{Q}}\{c^\top x\}, \tag{4.2}$$

and is obtained by solving a linear program with the original objective function c over the polyhedron $\mathcal{Q}$. It is clear that $z_{LP} \leq z_{IP}$ since $\mathcal{P} \subseteq \mathcal{Q}$. This LP relaxation is usually much easier to solve than the original ILP, but z_{LP} may be arbitrarily far away from z_{IP} in general, so we need to consider more effective procedures.

In most cases, the description of $\mathcal{Q}$ is small enough that it can be represented explicitly and the bound computed using a standard linear programming algorithm. To improve the LP bound, decomposition methods construct a second approximating polyhedron that can be intersected with $\mathcal{Q}$ to form a better approximation. Unlike $\mathcal{Q}$, this second polyhedron usually has a description of exponential size, and we must generate portions of its description dynamically. Such a dynamic procedure is the basis for both cutting plane methods, which generate an outer approximation, and for traditional decomposition methods, such as the Dantzig-Wolfe method [19] and the Lagrangian method [22, 14], which generate inner approximations.

For the remainder of this section, we consider the relaxation of (4.1) defined by

$$\min_{x \in \mathbb{Z}^n}\{c^\top x \mid A'x \geq b'\} = \min_{x \in \mathcal{F}'}\{c^\top x\} = \min_{x \in \mathcal{P}'}\{c^\top x\}, \tag{4.3}$$

where $\mathcal{F} \subset \mathcal{F}' = \{x \in \mathbb{Z}^n \mid A'x \geq b'\}$ for some $A' \in \mathbb{Q}^{m' \times n}, b' \in \mathbb{Q}^{m'}$ and $\mathcal{P}'$ is the convex hull of $\mathcal{F}'$. Along with $\mathcal{P}'$ is associated a set of *side constraints* $[A'', b''] \in \mathbb{Q}^{m'' \times (n+1)}$ such that $\mathcal{Q} = \{x \in \mathbb{R}^n \mid A'x \geq b', A''x \geq b''\}$. We denote by $\mathcal{Q}'$ the polyhedron described by the inequalities $[A', b']$ and by $\mathcal{Q}''$ the polyhedron described by the inequalities $[A'', b'']$. Thus, $\mathcal{Q} = \mathcal{Q}' \cap \mathcal{Q}''$ and $\mathcal{F} = \{x \in \mathbb{Z}^n \mid x \in \mathcal{P}' \cap \mathcal{Q}''\}$. For the decomposition to be effective, we must have that $\mathcal{P}' \cap \mathcal{Q}'' \subset \mathcal{Q}$, so that the bound obtained by optimizing over $\mathcal{P}' \cap \mathcal{Q}''$ is at least as good as the LP bound and strictly better for some objective functions. The description of $\mathcal{Q}''$ must also be "small" so that we can construct it explicitly. Finally, we assume that there exists an *effective* algorithm for optimizing over $\mathcal{P}'$ and thereby, for separating arbitrary real vectors from $\mathcal{P}'$. We are deliberately using the term *effective* here to denote an algorithm that has an acceptable average-case running time, since this is more relevant than worst-case behavior in our computational framework.

Traditional decomposition methods can all be viewed as techniques for iteratively computing the bound

$$z_D = \min_{x \in \mathcal{P}'}\{c^\top x \mid A''x \geq b''\} = \min_{x \in \mathcal{F}' \cap \mathcal{Q}''}\{c^\top x\} = \min_{x \in \mathcal{P}' \cap \mathcal{Q}''}\{c^\top x\}. \tag{4.4}$$

In Section 4.3.1 to Section 4.3.3 below, we review the cutting plane method, the Dantzig-Wolfe method, and the Lagrangian method, all classical approaches that can be used to compute this bound. The common perspective motivates Section 4.4, where we consider a new class of decomposition methods called *integrated decomposition methods*, in which both inner *and* outer approximation techniques are used in tandem. In both this section and the next, we describe the methods at a high level and leave until later sections the discussion of how the master problem and subproblems are solved. To illustrate the effect of applying the decomposition principle, we now introduce two examples that we build on throughout the chapter. The first is a simple generic ILP.

Example 4.1

Let the following be the formulation of a given ILP:

$$\min x_1,$$
$$7x_1 - x_2 \geq 13, \tag{4.5}$$
$$x_2 \geq 1, \tag{4.6}$$
$$-x_1 + x_2 \geq -3, \tag{4.7}$$
$$-4x_1 - x_2 \geq -27, \tag{4.8}$$
$$-x_2 \geq -5, \tag{4.9}$$
$$0.2x_1 - x_2 \geq -4, \tag{4.10}$$
$$-x_1 - x_2 \geq -8, \tag{4.11}$$
$$-0.4x_1 + x_2 \geq 0.3, \tag{4.12}$$
$$x_1 + x_2 \geq 4.5, \tag{4.13}$$
$$3x_1 + x_2 \geq 9.5, \tag{4.14}$$
$$0.25x_1 - x_2 \geq -3, \tag{4.15}$$
$$x \in \mathbb{Z}^2. \tag{4.16}$$

In this example, we let

$$\mathcal{P} = \mathrm{conv}\{x \in \mathbb{R}^2 \mid x \text{ satisfies } (4.5) - (4.16)\},$$
$$\mathcal{Q}' = \{x \in \mathbb{R}^2 \mid x \text{ satisfies } (4.5) - (4.10)\},$$
$$\mathcal{Q}'' = \{x \in \mathbb{R}^2 \mid x \text{ satisfies } (4.11) - (4.15)\}, \text{ and}$$
$$\mathcal{P}' = \mathrm{conv}(\mathcal{Q}' \cap \mathbb{Z}^2).$$

In Figure 4.1(a), we show the associated polyhedra, where the set of feasible solutions $\mathcal{F} = \mathcal{Q}' \cap \mathcal{Q}'' \cap \mathbb{Z}^2 = \mathcal{P}' \cap \mathcal{Q}'' \cap \mathbb{Z}^2$ and $\mathcal{P} = \mathrm{conv}(\mathcal{F})$. Figure 4.1(b) depicts the continuous approximation $\mathcal{Q}' \cap \mathcal{Q}''$, while Figure 4.1(c) shows the improved approximation $\mathcal{P}' \cap \mathcal{Q}''$. For the objective function in this example, optimization over $\mathcal{P}' \cap \mathcal{Q}''$ leads to an improvement over the LP bound obtained by optimization over $\mathcal{Q}$.

In our second example, we consider the classical *Traveling Salesman Problem* (TSP), a well-known combinatorial optimization problem. The TSP is in

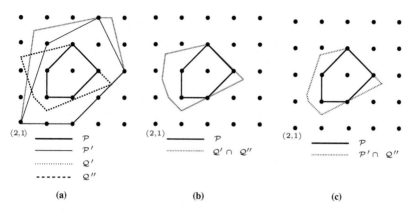

FIGURE 4.1
Polyhedra (Example 4.1).

the complexity class $\mathcal{N}P$-hard, but lends itself well to the application of the principle of decomposition, as the standard formulation contains an exponential number of constraints and has a number of well-solved combinatorial relaxations.

Example 4.2

The Traveling Salesman Problem is that of finding a minimum cost tour in an undirected graph G with vertex set $V = \{0, 1, \ldots, |V|-1\}$ and edge set E. We assume without loss of generality that G is complete. A *tour* is a connected subgraph for which each node has degree 2. The TSP is then to find such a subgraph of minimum cost, where the cost is the sum of the costs of the edges comprising the subgraph. With each edge $e \in E$, we therefore associate a binary variable x_e, indicating whether edge e is part of the subgraph, and a cost $c_e \in \mathbb{R}$. Let $\delta(S) = \{\{i, j\} \in E \mid i \in S, j \notin S\}$, $E(S : T) = \{\{i, j\} \mid i \in S, j \in T\}$, $E(S) = E(S : S)$ and $x(F) = \sum_{e \in F} x_e$. Then an ILP formulation of the TSP is as follows:

$$\min \sum_{e \in E} c_e x_e,$$

$$x(\delta(\{i\})) = 2 \quad \forall \, i \in V, \tag{4.17}$$

$$x(E(S)) \leq |S| - 1 \quad \forall \, S \subset V, \; 3 \leq |S| \leq |V| - 1, \tag{4.18}$$

$$0 \leq x_e \leq 1 \quad \forall \, e \in E, \tag{4.19}$$

$$x_e \in \mathbb{Z} \quad \forall \, e \in E. \tag{4.20}$$

The continuous approximation, referred to as the *TSP polyhedron*, is then

$$\mathcal{P} = \text{conv}\{x \in \mathbb{R}^E \mid x \text{ satisfies } (4.17) - (4.20)\}.$$

The equations (4.17) are the *degree constraints*, which ensure that each vertex has degree two in the subgraph, while the inequalities (4.18) are known as the *subtour elimination constraints* (SECs) and enforce connectivity. Since there are an exponential number of SECs, it is impossible to explicitly construct the LP relaxation of TSP for large graphs. Following the pioneering work of Held and Karp [35], however, we can apply the principle of decomposition by employing the well-known *Minimum 1-Tree Problem*, a combinatorial relaxation of TSP.

A 1-tree is a tree spanning $V \setminus \{0\}$ plus two edges incident to vertex 0. A 1-tree is hence a subgraph containing exactly one cycle through vertex 0. The Minimum 1-Tree Problem is to find a 1-tree of minimum cost and can thus be formulated as follows:

$$\min \sum_{e \in E} c_e x_e,$$

$$x(\delta(\{0\})) = 2, \tag{4.21}$$

$$x(E(V \setminus \{0\})) = |V| - 2, \tag{4.22}$$

$$x(E(S)) \leq |S| - 1 \quad \forall S \subset V \setminus \{0\}, 3 \leq |S| \leq |V| - 1, \tag{4.23}$$

$$x_e \in \{0, 1\} \quad \forall e \in E. \tag{4.24}$$

A minimum cost 1-tree can be obtained easily as the union of a minimum cost spanning tree of $V \setminus \{0\}$ plus two cheapest edges incident to vertex 0. For this example, we thus let $\mathcal{P}' = \text{conv}(\{x \in \mathbb{R}^E \mid x \text{ satisfies } (4.21) - (4.24)\})$ be the 1-Tree Polyhedron, while the degree and bound constraints comprise the polyhedron $\mathcal{Q}'' = \{x \in \mathbb{R}^E \mid x \text{ satisfies } (4.17) \text{ and } (4.19)\}$ and $\mathcal{Q}' = \{x \in \mathbb{R}^E \mid x \text{ satisfies } (4.18)\}$. Note that the bound constraints appear in the descriptions of both polyhedra for computational convenience. The set of feasible solutions to TSP is then $\mathcal{F} = \mathcal{P}' \cap \mathcal{Q}'' \cap \mathbb{Z}^E$. ∎

4.3.1 Cutting Plane Method

Using the cutting plane method, the bound z_D can be obtained by dynamically generating portions of an outer description of $\mathcal{P}'$. Let $[D, d]$ denote the set of facet-defining inequalities of $\mathcal{P}'$, so that

$$\mathcal{P}' = \{x \in \mathbb{R}^n \mid Dx \geq d\}. \tag{4.25}$$

Then the cutting plane formulation for the problem of calculating z_D can be written as

$$z_{CP} = \min_{x \in \mathcal{Q}''} \{c^\top x \mid Dx \geq d\}. \tag{4.26}$$

This is a linear program, but since the set of valid inequalities $[D, d]$ is potentially of exponential size, we dynamically generate them by solving a separation problem. An outline of the method is presented in Figure 4.2.

In Step 2, the master problem is a linear program whose feasible region is the current outer approximation $\mathcal{P}_O^t$, defined by a set of initial valid inequalities plus those generated dynamically in Step 3. Solving the master problem in iteration t, we generate the relaxed (primal) solution x_{CP}^t and a valid lower bound. In the figure, the initial set of inequalities is taken to be those of $\mathcal{Q}''$, since it is

---

Cutting Plane Method

Input: An instance $ILP(\mathcal{P}, c)$.
Output: A lower bound z_{CP} on the optimal solution value for the instance, and $\hat{x}_{CP} \in \mathbb{R}^n$ such that $z_{CP} = c^\top \hat{x}_{CP}$.

1. **Initialize:** Construct an initial outer approximation

$$\mathcal{P}_O^0 = \{x \in \mathbb{R}^n \mid D^0 x \geq d^0\} \supseteq \mathcal{P}, \qquad (4.27)$$

where $D^0 = A''$ and $d^0 = b''$, and set $t \leftarrow 0$.

2. **Master Problem:** Solve the linear program

$$z_{CP}^t = \min_{x \in \mathbb{R}^n} \{c^\top x \mid D^t x \geq d^t\} \qquad (4.28)$$

to obtain the optimal value $z_{CP}^t = \min_{x \in \mathcal{P}_O^t} \{c^\top x\} \leq z_{IP}$ and optimal primal solution x_{CP}^t.

3. **Subproblem:** Call the subroutine $SEP(\mathcal{P}, x_{CP}^t)$ to generate a set of potentially *improving* valid inequalities $[\check{D}, \check{d}]$ for $\mathcal{P}$, violated by x_{CP}^t.

4. **Update:** If violated inequalities were found in Step 3, set $[D^{t+1}, d^{t+1}] \leftarrow \begin{bmatrix} D^t & d^t \\ \check{D} & \check{d} \end{bmatrix}$ to form a new outer approximation

$$\mathcal{P}_O^{t+1} = \{x \in \mathbb{R}^n \mid D^{t+1} x \leq d^{t+1}\} \supseteq \mathcal{P}, \qquad (4.29)$$

and set $t \leftarrow t + 1$. Go to Step 2.

5. If no violated inequalities were found, output $z_{CP} = z_{CP}^t \leq z_{IP}$ and $\hat{x}_{CP} = x_{CP}^t$.

FIGURE 4.2
Outline of the cutting plane method.

assumed that the facet-defining inequalities for $\mathcal{P}'$, which dominate those of $\mathcal{Q}'$, can be generated dynamically. In practice, however, this initial set may be chosen to include those of $\mathcal{Q}'$ or some other polyhedron, on an empirical basis.

In Step 3, we solve the subproblem, which is to generate a set of *improving* valid inequalities, i.e., valid inequalities that improve the bound when added to the current approximation. This step is usually accomplished by applying one of the many known techniques for separating x_{CP}^t from $\mathcal{P}$. The algorithmic details of the generation of valid inequalities are covered more thoroughly in Section 4.5, so the unfamiliar reader may wish to refer to this section for background or to [1] for a complete survey of techniques. It is well known that violation of x_{CP}^t is a necessary condition for an inequality to be improving, and hence, we generally use this condition to judge the potential effectiveness of generated valid inequalities. However, this condition is not sufficient and unless the inequality separates the entire optimal face of $\mathcal{P}_O^t$, it will not actually be improving. Because we want to refer to these results later in the chapter, we state them formally as theorem and corollary without proof. See [59] for a thorough treatment of the theory of linear programming that leads to this result.

THEOREM 4.1
Let F be the face of optimal solutions to an LP over a nonempty, bounded polyhedron X with objective function vector f. Then (a, β) is an improving inequality for X with respect to f, i.e.,

$$\min\{f^\top x \mid x \in X, a^\top x \geq \beta\} > \min\{f^\top x \mid x \in X\}, \qquad (4.30)$$

if and only if $a^\top y < \beta$ for all $y \in F$.

COROLLARY 4.1
If (a, β) is an improving inequality for X with respect to f, then $a^\top \hat{x} < \beta$, where $\hat{x}$ is any optimal solution to the linear program over X with objective function vector f.

Even in the case when the optimal face cannot be separated in its entirety, the augmented cutting plane LP must have a different optimal solution, which in turn may be used to generate more potential improving inequalities. Since the condition of Theorem 4.1 is difficult to verify, one typically terminates the bounding procedure when increases resulting from additional inequalities become "too small."

If we start with the continuous approximation $\mathcal{P}_O^0 = \mathcal{Q}''$ and generate only facet-defining inequalities of $\mathcal{P}'$ in Step 3, then the procedure described here terminates in a finite number of steps with the bound $z_{CP} = z_D$ (see [52]). Since $\mathcal{P}_O^t \supseteq \mathcal{P}' \cap \mathcal{Q}'' \supseteq \mathcal{P}$, each step yields an approximation for $\mathcal{P}$, along with a valid bound. In Step 3, we are permitted to generate any valid inequality for $\mathcal{P}$, however, not just those that are facet-defining for $\mathcal{P}'$. In theory, this means that the cutting plane method can be used to compute the bound z_{IP} exactly. However, this is rarely practical.

To illustrate the cutting plane method, we show how it could be applied to generate the bound z_D for the ILPs of Example 4.1 and Example 4.2. Since we are discussing the computation of the bound z_D, we only generate facet-defining inequalities for $\mathcal{P}'$ in these examples. We discuss more general scenarios later in the chapter.

Example 4.1 (Continued)
We define the initial outer approximation to be $\mathcal{P}_O^0 = \mathcal{Q}' \cap \mathcal{Q}'' = \{x \in \mathbb{R}^2 \mid x$ satisfies (4.5) − (4.15)}, the continuous approximation.

Iteration 0: Solving the master problem over $\mathcal{P}_O^0$, we find an optimal primal solution $x_{CP}^0 = (2.25, 2.75)$ with bound $z_{CP}^0 = 2.25$, as shown in Figure 4.3(a). We then call the subroutine $SEP(\mathcal{P}, x_{CP}^0)$, generating facet-defining inequalities of $\mathcal{P}'$ that are violated by x_{CP}^0. One such facet-defining inequality, $3x_1 - x_2 \geq 5$, is pictured in Figure 4.3(a). We add this inequality to form a new outer approximation $\mathcal{P}_O^1$.

Iteration 1: We again solve the master problem, this time over $\mathcal{P}_O^1$, to find an optimal primal solution $x_{CP}^1 = (2.42, 2.25)$ and bound $z_{CP}^1 = 2.42$, as shown in

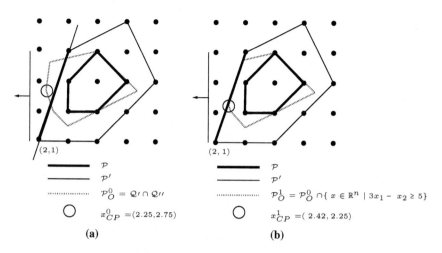

FIGURE 4.3
Cutting plane method (Example 4.1).

Figure 4.3(b). We then call the subroutine $SEP(\mathcal{P}, x^1_{CP})$. However, as illustrated in Figure 4.3(b), there are no more facet-defining inequalities violated by x^1_{CP}. In fact, further improvement in the bound would necessitate the addition of valid inequalities violated by points in $\mathcal{P}'$. Since we are only generating facets of $\mathcal{P}'$ in this example, the method terminates with bound $z_{CP} = 2.42 = z_D$. ∎

We now consider the use of the cutting plane method for generating the bound z_D for the TSP of Example 4.2. Once again, we only generate facet-defining inequalities for $\mathcal{P}'$, the 1-tree polyhedron.

Example 4.2   (Continued)
We define the initial outer approximation to be comprised of the degree con-straints and the bound constraints, so that

$$\mathcal{P}^0_O = \mathcal{Q}'' = \{x \in \mathbb{R}^E \mid x \text{ satisfies (4.17) and (4.19)}\}.$$

The bound z_D is then obtained by optimizing over the intersection of the 1-tree polyhedron with the polyhedron $\mathcal{Q}''$ defined by constraints (4.17) and (4.19). Note that because the 1-tree polyhedron has integer extreme points, we have that $z_D = z_{LP}$ in this case. To calculate z_D, however, we must dynamically generate violated facet-defining inequalities (the SECs (4.23)) of the 1-tree polyhedron $\mathcal{P}'$ defined earlier. Given a vector $\hat{x} \in \mathbb{R}^E$ satisfying (4.17) and (4.19), the problem of finding an inequality of the form (4.23) violated by $\hat{x}$ is equivalent to the well-known minimum cut problem, which can be nominally solved in $O(|V|^4)$ [53]. We can use this approach to implement Step 3 of the cutting plane method and hence compute the bound z_D effectively. As an example, consider the vector $\hat{x}$ pictured graphically in Figure 4.4, obtained in Step 2 of the cutting plane method. In the figure, only edges e for which $\hat{x}_e > 0$ are shown. Each edge e is labeled with the value $\hat{x}_e$, except for edges e

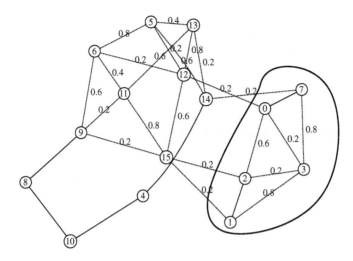

FIGURE 4.4
Finding violated inequalities in the cutting plane method (Example 4.2).

with $\hat{x}_e = 1$. The circled set of vertices $S = \{0, 1, 2, 3, 7\}$ define a SEC violated by $\hat{x}$, since $\hat{x}(E(S)) = 4.6 > 4.0 = |S| - 1$. ∎

4.3.2 Dantzig-Wolfe Method

In the Dantzig-Wolfe method, the bound z_D can be obtained by dynamically generating portions of an inner description of $\mathcal{P}'$ and intersecting it with $\mathcal{Q}''$. Consider Minkowski's Theorem, which states that every bounded polyhedron is finitely generated by its extreme points [52]. Let $\mathcal{E} \subseteq \mathcal{F}'$ be the set of extreme points of $\mathcal{P}'$, so that

$$\mathcal{P}' = \left\{ x \in \mathbb{R}^n \;\middle|\; x = \sum_{s \in \mathcal{E}} s\lambda_s, \sum_{s \in \mathcal{E}} \lambda_s = 1, \lambda_s \geq 0 \; \forall s \in \mathcal{E} \right\}. \qquad (4.31)$$

Then the Dantzig-Wolfe formulation for computing the bound z_D is

$$z_{DW} = \min_{x \in \mathbb{R}^n} \left\{ c^\top x \;\middle|\; A''x \geq b'', x = \sum_{s \in \mathcal{E}} s\lambda_s, \sum_{s \in \mathcal{E}} \lambda_s = 1, \lambda_s \geq 0 \; \forall s \in \mathcal{E} \right\}. \qquad (4.32)$$

By substituting out the original variables, this formulation can be rewritten in the more familiar form

$$z_{DW} = \min_{\lambda \in \mathbb{R}_+^{\mathcal{E}}} \left\{ c^\top \left(\sum_{s \in \mathcal{E}} s\lambda_s \right) \;\middle|\; A'' \left(\sum_{s \in \mathcal{E}} s\lambda_s \right) \geq b'', \sum_{s \in \mathcal{E}} \lambda_s = 1 \right\}. \qquad (4.33)$$

This is a linear program, but since the set of extreme points $\mathcal{E}$ is potentially of exponential size, we dynamically generate those that are relevant by solving an optimization problem over $\mathcal{P}'$. An outline of the method is presented in Figure 4.5.

Dantzig-Wolfe Method

<u>Input</u>: An instance $ILP(\mathcal{P}, c)$.
<u>Output</u>: A lower bound z_{DW} on the optimal solution value for the instance, a primal solution $\hat{\lambda}_{DW} \in \mathbb{R}^{\mathcal{E}}$, and a dual solution $(\hat{u}_{DW}, \hat{\alpha}_{DW}) \in \mathbb{R}^{m''+1}$.

1. **Initialize**: Construct an initial inner approximation

$$\mathcal{P}_I^0 = \left\{ \sum_{s \in \mathcal{E}^0} s\lambda_s \ \middle| \ \sum_{s \in \mathcal{E}^0} \lambda_s = 1, \lambda_s \geq 0 \ \forall s \in \mathcal{E}^0, \lambda_s = 0 \ \forall s \in \mathcal{E} \setminus \mathcal{E}^0 \right\} \subseteq \mathcal{P}'$$

(4.34)

from an initial set $\mathcal{E}^0$ of extreme points of $\mathcal{P}'$ and set $t \leftarrow 0$.

2. **Master Problem**: Solve the Dantzig-Wolfe reformulation

$$\bar{z}_{DW}^t = \min_{\lambda \in \mathbb{R}_+^{\mathcal{E}}} \left\{ c^\top \left(\sum_{s \in \mathcal{E}} s\lambda_s \right) \ \middle| \ A'' \left(\sum_{s \in \mathcal{E}} s\lambda_s \right) \geq b'', \sum_{s \in \mathcal{E}} \lambda_s = 1, \lambda_s = 0 \ \forall s \in \mathcal{E} \setminus \mathcal{E}^t \right\}$$

(4.35)

to obtain the optimal value $\bar{z}_{DW}^t = \min_{\mathcal{P}_I^t \cap \mathcal{Q}''} c^\top x \geq z_{DW}$, an optimal primal solution $\lambda_{DW}^t \in \mathbb{R}_+^{\mathcal{E}}$, and an optimal dual solution $(u_{DW}^t, \alpha_{DW}^t) \in \mathbb{R}^{m''+1}$.

3. **Subproblem**: Call the subroutine $OPT(c^\top - (u_{DW}^t)^\top A'', \mathcal{P}', \alpha_{DW}^t)$, generating a set of $\tilde{\mathcal{E}}$ of *improving* members of $\mathcal{E}$ with negative reduced cost, where the reduced cost of $s \in \mathcal{E}$ is

$$rc(s) = \left(c^\top - \left(u_{DW}^t \right)^\top A'' \right)s - \alpha_{DW}^t.$$

(4.36)

If $\tilde{s} \in \tilde{\mathcal{E}}$ is the member of $\mathcal{E}$ with smallest reduced cost, then $\underline{z}_{DW}^t = rc(\tilde{s}) + \alpha_{DW}^t + (u_{DW}^t)^\top b'' \leq z_{DW}$ provides a valid lower bound.

4. **Update**: If $\tilde{\mathcal{E}} \neq \emptyset$, set $\mathcal{E}^{t+1} \leftarrow \mathcal{E}^t \cup \tilde{\mathcal{E}}$ to form the new inner approximation

$$\mathcal{P}_I^{t+1} = \left\{ \sum_{s \in \mathcal{E}^{t+1}} s\lambda_s \ \middle| \ \sum_{s \in \mathcal{E}^{t+1}} \lambda_s = 1, \lambda_s \geq 0 \ \forall s \in \mathcal{E}^{t+1}, \lambda_s = 0 \ \forall s \in \mathcal{E} \setminus \mathcal{E}^{t+1} \right\} \subseteq \mathcal{P}'$$

(4.37)

and set $t \leftarrow t + 1$. Go to Step 2.

5. If $\tilde{\mathcal{E}} = \emptyset$, output the bound $z_{DW} = \bar{z}_{DW}^t = \underline{z}_{DW}^t$, $\hat{\lambda}_{DW} = \lambda_{DW}^t$, and $(\hat{u}_{DW}, \hat{\alpha}_{DW}) = (u_{DW}^t, \alpha_{DW}^t)$.

FIGURE 4.5
Outline of the Dantzig-Wolfe method.

In Step 2, we solve the master problem, which is a restricted linear program obtained by substituting $\mathcal{E}^t$ for $\mathcal{E}$ in (4.33). In Section 4.6, we discuss several alternatives for solving this LP. In any case, solving it results in a primal solution λ_{DW}^t, and a dual solution consisting of the dual multipliers u_{DW}^t on the constraints corresponding to $[A'', b'']$ and the multiplier α_{DW}^t on the convexity constraint. The dual solution is needed to generate the improving columns in Step 3. In each iteration, we are generating an inner approximation, $\mathcal{P}_I^t \subseteq \mathcal{P}'$, the convex hull of $\mathcal{E}^t$. Thus $\mathcal{P}_I^t \cap \mathcal{Q}''$ may or may not contain $\mathcal{P}$ and the bound returned from the master problem in Step 2, $\bar{z}_{DW}^t$, provides an *upper* bound on z_{DW}. Nonetheless, it is easy to show (see Section 4.3.3) that an optimal solution to the subproblem solved in Step 3 yields a valid lower bound. In particular, if $\tilde{s}$ is a member of $\mathcal{E}$ with the smallest reduced cost in Step 3, then

$$\underline{z}_{DW}^t = c^\top \tilde{s} + \left(u_{DW}^t\right)^\top (b'' - A''\tilde{s}) \tag{4.38}$$

is a valid lower bound. This means that, in contrast to the cutting plane method, where a valid lower bound is always available, the Dantzig-Wolfe method only yields a valid lower bound when the subproblem is solved to optimality, i.e., the optimization version is solved, as opposed to the decision version. This need not be done in every iteration, as described below.

In Step 3, we search for *improving* members of $\mathcal{E}$, where, as in the previous section, this means members that when added to $\mathcal{E}^t$ yield an improved bound. It is less clear here, however, which bound we would like to improve, $\bar{z}_{DW}^t$ or $\underline{z}_{DW}^t$. A necessary condition for improving $\bar{z}_{DW}^t$ is the generation of a column with negative reduced cost. In fact, if one considers (4.38), it is clear that this condition is also necessary for improvement of $\underline{z}_{DW}^t$. However, we point out again that the subproblem must be solved to optimality in order to update the bound $\underline{z}_{DW}^t$. In either case, however, we are looking for members of $\mathcal{E}$ with negative reduced cost. If one or more such members exist, we add them to $\mathcal{E}^t$ and iterate.

An area that deserves some deeper investigation is the relationship between the solution obtained by solving the reformulation (4.35) and the solution that would be obtained by solving an LP directly over $\mathcal{P}_I^t \cap \mathcal{Q}''$ with the objective function c. Consider the primal optimal solution λ_{DW}^t, which we refer to as an *optimal decomposition*. If we combine the members of $\mathcal{E}^t$ using λ_{DW}^t to obtain an *optimal fractional solution*

$$x_{DW}^t = \sum_{s \in \mathcal{E}^t} s \left(\lambda_{DW}^t\right)_s, \tag{4.39}$$

then we see that $\bar{z}_{DW}^t = c^\top x_{DW}^t$. In fact, $x_{DW}^t \in \mathcal{P}_I^t \cap \mathcal{Q}''$ is an optimal solution to the linear program solved directly over $\mathcal{P}_I^t \cap \mathcal{Q}''$ with objective function c.

The optimal fractional solution plays an important role in the integrated methods to be introduced later. To illustrate the Dantzig-Wolfe method and the role of the optimal fractional solution in the method, we show how to apply it to generate the bound z_D for the ILP of Example 4.1.

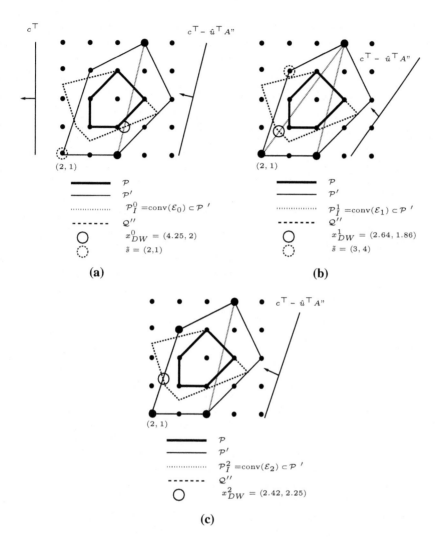

FIGURE 4.6
Dantzig-Wolfe method (Example 4.1).

Example 4.1 (Continued)

For the purposes of illustration, we begin with a randomly generated initial set of points $\mathcal{E}_0 = \{(4, 1), (5, 5)\}$. Taking their convex hull, we form the initial inner approximation $\mathcal{P}_I^0 = \text{conv}(\mathcal{E}^0)$, as illustrated in Figure 4.6(a).

Iteration 0. Solving the master problem with inner polyhedron $\mathcal{P}_I^0$, we obtain an optimal primal solution $(\lambda_{DW}^0)_{(4,1)} = 0.75$, $(\lambda_{DW}^0)_{(5,5)} = 0.25$, $x_{DW}^0 = (4.25, 2)$, and bound $\bar{z}_{DW}^0 = 4.25$. Since constraint (4.12) is binding at x_{DW}^0, the only nonzero component of u_{DW}^0 is $(u_{DW}^0)_{(4.12)} = 0.28$, while the dual variable associated with the convexity constraint has value $\alpha_{DW}^0 = 4.17$. All other dual variables have value zero. Next, we search for an extreme point of $\mathcal{P}'$ with negative reduced cost, by solving the subproblem $OPT(c^\top - \mathcal{P}', (u_{DW}^t)^\top A''$, $\alpha_{DW}^0)$. From Figure 4.6(a), we see that $\tilde{s} = (2, 1)$. This gives a valid lower bound $\bar{z}_{DW}^0 = 2.03$. We add the corresponding column to the restricted master and set $\mathcal{E}^1 = \mathcal{E}^0 \cup \{(2, 1)\}$.

Iteration 1. The next iteration is depicted in Figure 4.6(b). First, we solve the master problem with inner polyhedron $\mathcal{P}_I^1 = \text{conv}(\mathcal{E}^1)$ to obtain $(\lambda_{DW}^1)_{(5,5)} = 0.21$, $(\lambda_{DW}^1)_{(2,1)} = 0.79$, $x_{DW}^1 = (2.64, 1.86)$, and bound and $\bar{z}_{DW}^1 = 2.64$. This also provides the dual solution $(u_{DW}^1)_{(4.13)} = 0.43$ and $\alpha_{DW}^1 = 0.71$ (all other dual values are zero). Solving $OPT(c^\top - \mathcal{P}', u_{DW}^1 A'', \alpha_{DW}^1)$, we obtain $\tilde{s} = (3, 4)$, and $\bar{z}_{DW}^1 = 1.93$. We add the corresponding column to the restricted master and set $\mathcal{E}^2 = \mathcal{E}^1 \cup \{(3, 4)\}$.

Iteration 2. The final iteration is depicted in Figure 4.6(c). Solving the master problem once more with inner polyhedron $\mathcal{P}_I^2 = \text{conv}(\mathcal{E}^2)$, we obtain $(\lambda_{DW}^2)_{(2,1)} = 0.58$ and $(\lambda_{DW}^2)_{(3,4)} = 0.42$, $x_{DW}^2 = (2.42, 2.25)$, and bound $\bar{z}_{DW}^2 = 2.42$. This also provides the dual solution $(u_{DW}^2)_{(4.14)} = 0.17$ and $\alpha_{DW}^2 = 0.83$. Solving $OPT(c^\top - \mathcal{P}', u_{DW}^2 A'', \alpha_{DW}^2)$, we conclude that $\tilde{\mathcal{E}} = \emptyset$. We therefore terminate with the bound $z_{DW} = 2.42 = z_D$. ∎

As a further brief illustration, we return to the TSP example introduced earlier.

Example 4.2   (Continued)
As we noted earlier, the Minimum 1-Tree Problem can be solved by computing a minimum cost spanning tree on vertices $V \setminus \{0\}$, and then adding two cheapest edges incident to vertex 0. This can be done in $O(|E| \log |V|)$ using standard algorithms. In applying the Dantzig-Wolfe method to compute z_D using the decomposition described earlier, the subproblem to be solved in Step 3 is a Minimum 1-Tree Problem. Because we can solve this problem effectively, we can apply the Dantzig-Wolfe method in this case. As an example of the result of solving the Dantzig-Wolfe master problem (4.35), Figure 4.7 depicts an optimal fractional solution (a) to a Dantzig-Wolfe master LP and the six extreme points in Figure 4.7(b to g) of the 1-tree polyhedron $\mathcal{P}'$, with nonzero weight comprising an optimal decomposition. We return to this figure later in Section 4.4.

Now consider the set $\mathcal{S}(u, \alpha)$, defined as

$$\mathcal{S}(u, \alpha) = \{s \in \mathcal{E} \mid (c^\top - u^\top A'')s = \alpha\}, \tag{4.40}$$

where $u \in \mathbb{R}^{m''}$ and $\alpha \in \mathbb{R}$. The set $\mathcal{S}(u_{DW}^t, \alpha_{DW}^t)$ is the set of members of $\mathcal{E}$ with reduced cost zero at optimality for (4.35) in iteration t. It follows that $\text{conv}(\mathcal{S}(u_{DW}^t, \alpha_{DW}^t))$ is in fact the face of optimal solutions to the linear program solved over $\mathcal{P}_I^t$ with objective function $c^\top - u^\top A''$. This line of reasoning culminates in the following theorem tying together the set $\mathcal{S}(u_{DW}^t, \alpha_{DW}^t)$ defined above, the vector x_{DW}^t, and the optimal face of solutions to the LP over the polyhedron $\mathcal{P}_I^t \cap \mathcal{Q}''$.

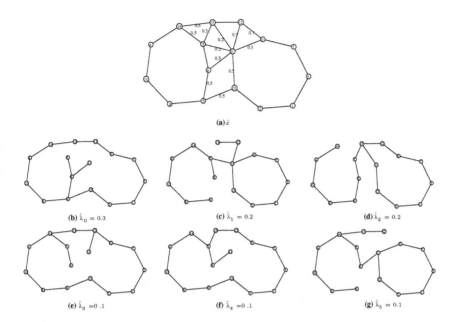

(a) $\hat{x}$

(b) $\hat{\lambda}_0 = 0.3$ **(c)** $\hat{\lambda}_1 = 0.2$ **(d)** $\hat{\lambda}_2 = 0.2$

(e) $\hat{\lambda}_3 = 0.1$ **(f)** $\hat{\lambda}_4 = 0.1$ **(g)** $\hat{\lambda}_5 = 0.1$

FIGURE 4.7
Dantzig-Wolfe method (Example 4.2).

THEOREM 4.2
$conv(S(u^t_{DW}, \alpha^t_{DW}))$ is a face of $\mathcal{P}^t_I$ and contains x^t_{DW}.

PROOF   We first show that $conv(S(u^t_{DW}, \alpha^t_{DW}))$ is a face of $\mathcal{P}^t_I$. Observe that

$$\left(c^\top - \left(u^t_{DW}\right)^\top A'', \alpha^t_{DW}\right)$$

defines a valid inequality for $\mathcal{P}^t_I$ since α^t_{DW} is the optimal value for the problem of minimizing over $\mathcal{P}^t_I$ with objective function $c^\top - (u^t_{DW})^\top A''$. Thus, the set

$$G = \left\{x \in \mathcal{P}^t_I \,\middle|\, (c^\top - \left(u^t_{DW}\right)^\top A'')x = \alpha^t_{DW}\right\}, \qquad (4.41)$$

is a face of $\mathcal{P}^t_I$ that contains $S(u^t_{DW}, \alpha^t_{DW})$. We will show that $conv(S(u^t_{DW}, \alpha^t_{DW})) = G$. Since G is convex and contains $S(u^t_{DW}, \alpha^t_{DW})$, it also contains $conv(S(u^t_{DW}, \alpha^t_{DW}))$, so we just need to show that $conv(S(u^t_{DW}, \alpha^t_{DW}))$ contains G. We do so by observing that the extreme points of G are elements of $S(u^t_{DW}, \alpha^t_{DW})$. By construction, all extreme points of $\mathcal{P}^t_I$ are members of $\mathcal{E}$ and the extreme points of G are also extreme points of $\mathcal{P}^t_I$. Therefore, the extreme points of G must be members of $\mathcal{E}$ and contained in $S(u^t_{DW}, \alpha^t_{DW})$. The claim follows and $conv(S(u^t_{DW}, \alpha^t_{DW}))$ is a face of $\mathcal{P}^t_I$.

The fact that $x^t_{DW} \in conv(S(u^t_{DW}, \alpha^t_{DW}))$ follows from the fact that x^t_{DW} is a convex combination of members of $S(u^t_{DW}, \alpha^t_{DW})$. ∎

An important consequence of Theorem 4.2 is that the face of optimal solutions to the LP over the polyhedron $\mathcal{P}^t_I \cap \mathcal{Q}''$ is actually contained in $conv(S(u^t_{DW}, \alpha^t_{DW})) \cap \mathcal{Q}''$, as stated in the following corollary.

COROLLARY 4.2
If F is the face of optimal solutions to the linear program solved directly over $\mathcal{P}_I^t \cap \mathcal{Q}''$
with objective function vector c, then $F \subseteq conv(\mathcal{S}(u_{DW}^t, \alpha_{DW}^t)) \cap \mathcal{Q}''$.

PROOF Let $\hat{x} \in F$ be given. Then we have that $\hat{x} \in \mathcal{P}_I^t \cap \mathcal{Q}''$ by definition,
and

$$c^\top \hat{x} = \alpha_{DW}^t + \left(u_{DW}^t\right)^\top b'' = \alpha_{DW}^t + \left(u_{DW}^t\right)^\top A'' \hat{x}, \qquad (4.42)$$

where the first equality in this chain is a consequence of strong duality and
the last is a consequence of complementary slackness. Hence, it follows that
$(c^\top - (u_{DW}^t)^\top A'')\hat{x} = \alpha_{DW}^t$ and the result is proven. ■

Hence, each iteration of the method not only produces the primal solu-
tion $x_{DW}^t \in \mathcal{P}_I^t \cap \mathcal{Q}''$, but also a dual solution $(u_{DW}^t, \alpha_{DW}^t)$ that defines a face
$conv(\mathcal{S}(u_{DW}^t, \alpha_{DW}^t))$ of $\mathcal{P}_I^t$ that contains the entire optimal face of solutions to
the LP solved directly over $\mathcal{P}_I^t \cap \mathcal{Q}''$ with the original objective function vector c.

When no column with negative reduced cost exists, the two bounds must be
equal to z_D and we stop, outputting both the primal solution $\hat{\lambda}_{DW}$, and the dual
solution $(\hat{u}_{DW}, \hat{\alpha}_{DW})$. It follows from the results proven above that in the final
iteration, any column of (4.35) with reduced cost zero must in fact have a cost
of $\hat{\alpha}_{DW} = z_D - \hat{u}_{DW}^\top b''$ when evaluated with respect to the modified objective
function $c^\top - \hat{u}_{DW}^\top A''$. In the final iteration, we can therefore strengthen the
statement of Theorem 4.2, as follows.

THEOREM 4.3
$conv(\mathcal{S}(\hat{u}_{DW}, \hat{\alpha}_{DW}))$ is a face of $\mathcal{P}'$ and contains $\hat{x}_{DW}$.

The proof follows along the same lines as Theorem 4.2. As before, we can also
state the following important corollary.

COROLLARY 4.3
If F is the face of optimal solutions to the linear program solved directly over $\mathcal{P}' \cap \mathcal{Q}''$
with objective function vector c, then $F \subseteq conv(\mathcal{S}(\hat{u}_{DW}, \hat{\alpha}_{DW})) \cap \mathcal{Q}''$.

Thus, $conv(\mathcal{S}(\hat{u}_{DW}, \hat{\alpha}_{DW}))$ is actually a face of $\mathcal{P}'$ that contains $\hat{x}_{DW}$ and the
entire face of optimal solutions to the LP solved over $\mathcal{P}' \cap \mathcal{Q}''$ with objective
function c. This fact provides strong intuition regarding the connection be-
tween the Dantzig-Wolfe method and the cutting plane method and allows
us to regard Dantzig-Wolfe decomposition as either a procedure for producing the bound $z_D = c^\top \hat{x}_{DW}$ from primal solution information or the bound
$z_D = c^\top \hat{s} + \hat{u}_{DW}^\top (b'' - A'' \hat{s})$, where $\hat{s}$ is any member of $\mathcal{S}(\hat{u}_{DW}, \hat{\alpha}_{DW})$, from dual
solution information. This fact is important in the next section, as well as later
when we discuss integrated methods.

The exact relationship between $\mathcal{S}(\hat{u}_{DW}, \hat{\alpha}_{DW})$, the polyhedron $\mathcal{P}' \cap \mathcal{Q}''$, and
the face F of optimal solutions to an LP solved over $\mathcal{P}' \cap \mathcal{Q}''$ can vary for
different polyhedra and even for different objective functions. Figure 4.8

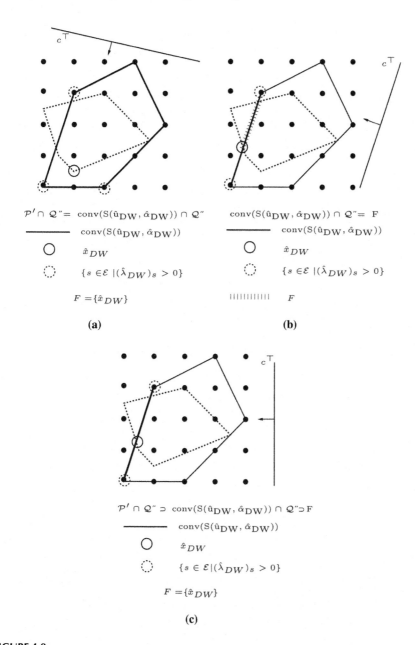

FIGURE 4.8
The relationship of $\mathcal{P}' \cap \mathcal{Q}''$, $\mathrm{conv}(\mathcal{S}(\hat{u}_{DW}, \hat{\alpha}_{DW})) \cap \mathcal{Q}''$, and the face F.

shows the polyhedra of Example 4.1 with three different objective functions indicated. The convex hull of $\mathcal{S}(\hat{u}_{DW}, \hat{\alpha}_{DW})$ is typically a proper face of $\mathcal{P}'$, but it is possible for $\hat{x}_{DW}$ to be an inner point of $\mathcal{P}'$, in which case we have the following result.

THEOREM 4.4
If $\hat{x}_{DW}$ is an inner point of $\mathcal{P}'$, then $\mathrm{conv}(\mathcal{S}(\hat{u}_{DW}, \hat{\alpha}_{DW})) = \mathcal{P}'$.

PROOF We prove the contrapositive. Suppose $\mathrm{conv}(\mathcal{S}(\hat{u}_{DW}, \hat{\alpha}_{DW}))$ is a proper face of $\mathcal{P}'$. Then there exists a facet-defining valid inequality $(a, \beta) \in \mathbb{R}^{n+1}$ such that $\mathrm{conv}(\mathcal{S}(\hat{u}_{DW}, \hat{\alpha}_{DW})) \subseteq \{x \in \mathbb{R}^n \mid ax = \beta\}$. By Theorem 4.3, $\hat{x}_{DW} \in \mathrm{conv}(\mathcal{S}(\hat{u}_{DW}, \hat{\alpha}_{DW}))$ and $\hat{x}_{DW}$ therefore cannot satisfy the definition of an inner point. ∎

In this case, illustrated graphically in Figure 4.8(a) with the polyhedra from Example 4.1, $z_{DW} = z_{LP}$ and Dantzig-Wolfe decomposition does not improve the bound. All columns of the Dantzig-Wolfe LP have reduced cost zero and any member of $\mathcal{E}$ can be given positive weight in an optimal decomposition. A necessary condition for an optimal fractional solution to be an inner point of $\mathcal{P}'$ is that the dual value of the convexity constraint in an optimal solution to the Dantzig-Wolfe LP be zero. This condition indicates that the chosen relaxation may be too weak.

A second case of potential interest is when $F = \mathrm{conv}(\mathcal{S}(\hat{u}_{DW}, \hat{\alpha}_{DW})) \cap \mathcal{Q}''$, illustrated graphically in Figure 4.8(b). In this case, all constraints of the Dantzig-Wolfe LP *other than* the convexity constraint must have dual value zero, since removing them does not change the optimal solution value. This condition can be detected by examining the objective function values of the members of $\mathcal{E}$ with positive weight in the optimal decomposition. If they are all identical, any such member that is contained in $\mathcal{Q}''$ (if one exists) must be optimal for the original ILP, since it is feasible and has objective function value equal to z_{IP}. The more typical case, in which F is a proper subset of $\mathrm{conv}(\mathcal{S}(\hat{u}_{DW}, \hat{\alpha}_{DW})) \cap \mathcal{Q}''$, is shown in Figure 4.8(c).

4.3.3 Lagrangian Method

The Lagrangian method [14, 22] is a general approach for computing z_D that is closely related to the Dantzig-Wolfe method, but is focused primarily on producing dual solution information. The Lagrangian method can be viewed as a method for producing a particular face of $\mathcal{P}'$, as in the Dantzig-Wolfe method, but no explicit approximation of $\mathcal{P}'$ is maintained. Although there are implementations of the Lagrangian method that *do* produce approximate primal solution information similar to the solution information that the Dantzig-Wolfe method produces (see Section 4.3.2), our viewpoint is that the main difference between the Dantzig-Wolfe method and the Lagrangian method is the type of solution information they produce. This distinction is important when we discuss integrated methods in Section 4.4. When exact primal solution information is not required, faster algorithms for determining the dual solution are possible. By employing a Lagrangian framework instead of a Dantzig-Wolfe framework, we can take advantage of this fact.

For a given vector $u \in \mathbb{R}_+^{m''}$, the *Lagrangian relaxation* of (4.1) is given by

$$z_{LR}(u) = \min_{s \in \mathcal{F'}} \{c^\top s + u^\top (b'' - A''s)\}. \qquad (4.43)$$

It is easily shown that $z_{LR}(u)$ is a lower bound on z_{IP} for any $u \geq 0$. The elements of the vector u are called *Lagrange multipliers* or *dual multipliers* with respect to the rows of $[A'', b'']$. Note that (4.43) is the same subproblem solved in the Dantzig-Wolfe method to generate the most negative reduced cost column. The problem

$$z_{LD} = \max_{u \in \mathbb{R}_+^{m''}} \{z_{LR}(u)\} \qquad (4.44)$$

of maximizing this bound over all choices of dual multipliers is a dual to (4.1) called the *Lagrangian dual* and also provides a lower bound z_{LD}, which we call the *LD bound*. A vector of multipliers $\hat{u}$ that yield the largest bound are called *optimal (dual) multipliers*.

It is easy to see that $z_{LR}(u)$ is a piecewise linear concave function and can be maximized by any number of methods for nondifferentiable optimization. In Section 4.6, we discuss some alternative solution methods (for a complete treatment, see [34]). In Figure 4.9 we give an outline of the steps involved in the

---

Lagrangian Method

Input: An instance $ILP(\mathcal{P}, c)$.
Output: A lower bound z_{LD} on the optimal solution value for the instance and a dual solution $\hat{u}_{LD} \in \mathbb{R}^{m''}$.

1. Let $s_{LD}^0 \in \mathcal{E}$ define some initial extreme point of $\mathcal{P'}$, u_{LD}^0 some initial setting for the dual multipliers and set $t \leftarrow 0$.

2. **Master Problem:** Using the solution information gained from solving the pricing subproblem, and the previous dual setting u_{LD}^t, update the dual multipliers u_{LD}^{t+1}.

3. **Subproblem:** Call the subroutine $OPT(c^\top - \mathcal{P'}, (u_{LD}^t)^\top A'', (c - (u_{LD}^t)^\top A'')s_{LD}^t)$, to solve

$$z_{LD}^t = \min_{s \in \mathcal{F'}} \{(c^\top - (u_{LD}^t)^\top A'')s + b''^\top u_{LD}^t\}. \qquad (4.45)$$

Let $s_{LD}^{t+1} \in \mathcal{E}$ be the optimal solution to this subproblem, if one is found.

4. If a prespecified stopping criterion is met, then output $z_{LD} = z_{LD}^t$ and $\hat{u}_{LD} = u_{LD}^t$, otherwise, go to Step 2.

FIGURE 4.9
Outline of the Lagrangian method.

Lagrangian method. As in Dantzig-Wolfe, the main loop involves updating the dual solution and then generating an *improving* member of $\mathcal{E}$ by solving a subproblem. Unlike the Dantzig-Wolfe method, there is no approximation and hence no update step, but the method can nonetheless be viewed in the same frame of reference.

To more clearly see the connection to the Dantzig-Wolfe method, consider the dual of the Dantzig-Wolfe LP (4.33),

$$z_{DW} = \max_{\alpha \in \mathbb{R}, u \in \mathbb{R}_+^{m''}} \{\alpha + b''^\top u \mid \alpha \le (c^\top - u^\top A'')s \;\forall s \in \mathcal{E}\}. \tag{4.46}$$

Letting $\eta = \alpha + b''^\top u$ and rewriting, we see that

$$z_{DW} = \max_{\eta \in \mathbb{R}, u \in \mathbb{R}_+^{m''}} \{\eta \mid \eta \le (c^\top - u^\top A'')s + b''^\top u \;\forall s \in \mathcal{E}\} \tag{4.47}$$

$$= \max_{\eta \in \mathbb{R}, u \in \mathbb{R}_+^{m''}} \{\min_{s \in \mathcal{E}}\{(c^\top - u^\top A'')s + b''^\top u\}\} = z_{LD}. \tag{4.48}$$

Thus, we have that $z_{LD} = z_{DW}$ and that (4.44) is another formulation for the problem of calculating z_D. It is also interesting to observe that the set $\mathcal{S}(u_{LD}^t, z_{LD}^t - b''^\top u_{LD}^t)$ is the set of alternative optimal solutions to the subproblem solved at iteration t in Step 3. The following theorem is a counterpart to Theorem 4.3 that follows from this observation.

THEOREM 4.5
$conv(\mathcal{S}(\hat{u}_{LD}, z_{LD} - b''^\top \hat{u}_{LD}))$ is a face of $\mathcal{P}'$. Also, if F is the face of optimal solutions to the linear program solved directly over $\mathcal{P}' \cap \mathcal{Q}''$ with objective function vector c, then $F \subseteq conv(\mathcal{S}(\hat{u}_{LD}, z_{LD} - b''^\top \hat{u}_{LD})) \cap \mathcal{Q}''$.

Again, the proof is similar to that of Theorem 4.3. This shows that while the Lagrangian method does not maintain an explicit approximation, it does produce a face of $\mathcal{P}'$ containing the optimal face of solutions to the linear program solved over the approximation $\mathcal{P}' \cap \mathcal{Q}''$.

4.4 Integrated Decomposition Methods

In Section 4.3, we demonstrated that traditional decomposition approaches can be viewed as utilizing dynamically generated polyhedral information to improve the LP bound by either building an inner or an outer approximation of an implicitly defined polyhedron that approximates $\mathcal{P}$. The choice between inner and outer methods is largely an empirical one, but recent computational research has favored outer methods. In what follows, we discuss three methods for integrating inner and outer methods. In principle, this is not difficult to do and can result in bounds that are improved over those achieved by either approach alone.

While traditional decomposition approaches build either an inner *or* an outer approximation, *integrated decomposition methods* build both an inner *and* an outer approximation. These methods follow the same basic loop as traditional decomposition methods, except that the master problem is required to generate both primal *and* dual solution information and the subproblem can be either a separation problem *or* an optimization problem. The first two techniques we describe integrate the cutting plane method with either the Dantzig-Wolfe method or the Lagrangian method. The third technique, described in Section 4.5, is a cutting plane method that uses an inner approximation to perform separation.

4.4.1 Price and Cut

The integration of the cutting plane method with the Dantzig-Wolfe method results in a procedure that alternates between a subproblem that generates improving columns (the *pricing* subproblem) and a subproblem that generates improving valid inequalities (the *cutting* subproblem). Hence, we call the resulting method *price and cut*. When employed in a branch-and-bound framework, the overall technique is called *branch, price, and cut*. This method has already been studied previously by a number of authors [11, 12, 39, 60, 61] and more recently by Arãgao and Uchoa [21].

As in the Dantzig-Wolfe method, the bound produced by price and cut can be thought of as resulting from the intersection of two approximating polyhedra. However, the Dantzig-Wolfe method required one of these, Q'', to have a short description. With integrated methods, both polyhedra can have descriptions of exponential size. Hence, price and cut allows partial descriptions of both an inner polyhedron $\mathcal{P}_I$ *and* an outer polyhedron $\mathcal{P}_O$ to be generated dynamically. To optimize over the intersection of $\mathcal{P}_I$ and $\mathcal{P}_O$, we use a Dantzig-Wolfe reformulation as in (4.33), except that the $[A'', b'']$ is replaced by a matrix that changes dynamically. The outline of this method is shown in Figure 4.10.

In examining the steps of this generalized method, the most interesting question that arises is how methods for generating improving columns and valid inequalities translate to this new dynamic setting. Potentially troublesome is the fact that column generation results in a reduction of the bound $\bar{z}_{PC}^t$ produced by (4.51), while generation of valid inequalities is aimed at increasing it. Recall again, however, that while it is the bound $\bar{z}_{PC}^t$ that is directly produced by solving (4.51), it is the bound $\underline{z}_{PC}^t$ obtained by solving the pricing subproblem that one might claim is more relevant to our goal and this bound can be potentially improved by generation of either valid inequalities or columns.

Improving columns can be generated in much the same way as they were in the Dantzig-Wolfe method. To search for new columns, we simply look for those that have negative reduced cost, where reduced cost is defined to be the usual LP reduced cost with respect to the current reformulation. Having a negative reduced cost is still a necessary condition for a column to be improving. However, it is less clear how to generate improving valid inequalities. Consider an optimal fractional solution x_{PC}^t obtained by combining the members of $\mathcal{E}$ according to weights yielded by the optimal decomposition λ_{PC}^t

Price and Cut Method

<u>Input</u>: An instance $ILP(\mathcal{P}, c)$.

<u>Output</u>: A lower bound z_{PC} on the optimal solution value for the instance, a primal solution $\hat{x}_{PC} \in \mathbb{R}^n$, an optimal decomposition $\hat{\lambda}_{PC} \in \mathbb{R}^{\mathcal{E}}$, a dual solution $(\hat{u}_{PC}, \hat{\alpha}_{PC}) \in \mathbb{R}^{m'+1}$, and the inequalities $[D_{PC}, d_{PC}] \in \mathbb{R}^{m'' \times (n+1)}$.

1. **Initialize:** Construct an initial inner approximation

$$\mathcal{P}_I^0 = \left\{ \sum_{s \in \mathcal{E}^0} s\lambda_s \;\middle|\; \sum_{s \in \mathcal{E}^0} \lambda_s = 1, \lambda_s \geq 0 \; \forall s \in \mathcal{E}^0, \lambda_s = 0 \; \forall s \in \mathcal{E} \setminus \mathcal{E}^0 \right\} \subseteq \mathcal{P}'$$

(4.49)

from an initial set $\mathcal{E}^0$ of extreme points of $\mathcal{P}'$ *and* an initial outer approximation

$$\mathcal{P}_O^0 = \{ x \in \mathbb{R}^n \mid D^0 x \geq d^0 \} \supseteq \mathcal{P},$$

(4.50)

where $D^0 = A''$ and $d^0 = b''$, and set $t \leftarrow 0, m^0 = m''$.

2. **Master Problem:** Solve the Dantzig-Wolfe reformulation

$$\bar{z}_{PC}^t = \min_{\lambda \in \mathbb{R}_+^{\mathcal{E}}} \left\{ c^\top \left(\sum_{s \in \mathcal{E}} s\lambda_s \right) \;\middle|\; D^t \left(\sum_{s \in \mathcal{E}} s\lambda_s \right) \geq d^t, \sum_{s \in \mathcal{E}} \lambda_s = 1, \lambda_s = 0 \; \forall s \in \mathcal{E} \setminus \mathcal{E}^t \right\}$$

(4.51)

of the LP over the polyhedron $\mathcal{P}_I^t \cap \mathcal{P}_O^t$ to obtain the optimal value $\bar{z}_{PC}^t$, an optimal primal solution $\lambda_{PC}^t \in \mathbb{R}^{\mathcal{E}}$, an optimal fractional solution $x_{PC}^t = \sum_{s \in \mathcal{E}} s(\lambda_{PC}^t)_s$, and an optimal dual solution $(u_{PC}^t, \alpha_{PC}^t) \in \mathbb{R}^{m^t+1}$.

3. Do either (a) or (b).

   (a) **Pricing Subproblem and Update:** Call the subroutine *OPT* $(c^\top - \mathcal{P}', (u_{PC}^t)^\top D^t, \alpha_{PC}^t)$, generating a set $\tilde{\mathcal{E}}$ of *improving* members of $\mathcal{E}$ with negative reduced cost (defined in Figure 4.5). If $\tilde{\mathcal{E}} \neq \emptyset$, set $\mathcal{E}^{t+1} \leftarrow \mathcal{E}^t \cup \tilde{\mathcal{E}}$ to form a new inner approximation $\mathcal{P}_I^{t+1}$. If $\tilde{s} \in \mathcal{E}$ is the member of $\mathcal{E}$ with smallest reduced cost, then $\underline{z}_{PC}^t = rc(\tilde{s}) + \alpha_{PC}^t + (d^t)^\top u_{PC}^t$ provides a valid lower bound. Set $[D^{t+1}, d^{t+1}] \leftarrow [D^t, d^t], \mathcal{P}_O^{t+1} \leftarrow \mathcal{P}_O^t, m^{t+1} \leftarrow m^t, t \leftarrow t+1$, and go to Step 2.

   (b) **Cutting Subproblem and Update:** Call the subroutine *SEP* $(\mathcal{P}, x_{PC}^t)$ to generate a set of *improving* valid inequalities $[\tilde{D}, \tilde{d}] \in \mathbb{R}^{\tilde{m} \times n+1}$ for $\mathcal{P}$, violated by x_{PC}^t. If violated inequalities were found, set $[D^{t+1}, d^{t+1}] \leftarrow \left[\begin{smallmatrix} D^t & d^t \\ \tilde{D} & \tilde{d} \end{smallmatrix} \right]$ to form a new outer approximation $\mathcal{P}_O^{t+1}$. Set $m^{t+1} \leftarrow m^t + \tilde{m}, \mathcal{E}^{t+1} \leftarrow \mathcal{E}^t, \mathcal{P}_I^{t+1} \leftarrow \mathcal{P}_I^t, t \leftarrow t+1$, and go to Step 2.

4. If $\tilde{\mathcal{E}} = \emptyset$ and no valid inequalities were found, output the bound $z_{PC} = \bar{z}_{PC}^t = \underline{z}_{PC}^t = c^\top x_{PC}^t, \hat{x}_{PC} = x_{PC}^t, \hat{\lambda}_{PC} = \lambda_{PC}^t, (\hat{u}_{PC}, \hat{\alpha}_{PC}) = (u_{PC}^t, \alpha_{PC}^t)$, and $[D_{PC}, d_{PC}] = [D^t, d^t]$.

FIGURE 4.10
Outline of the price and cut method.

in iteration t. Following a line of reasoning similar to that followed in analyzing the results of the Dantzig-Wolfe method, we can conclude that x_{PC}^t is in fact an optimal solution to an LP solved directly over $\mathcal{P}_I^t \cap \mathcal{P}_O^t$ with objective function vector c and that therefore, it follows from Theorem 4.1 that any improving inequality must be violated by x_{PC}^t. It thus seems sensible to consider separating x_{PC}^t from $\mathcal{P}$. This is the approach taken in the method of Figure 4.10.

To demonstrate how the price and cut method works, we return to Example 4.1.

Example 4.1 (Continued)
We pick up the example at the last iteration of the Dantzig-Wolfe method and show how the bound can be further improved by dynamically generating valid inequalities.

Iteration 0. Solving the master problem with $\mathcal{E}^0 = \{(4, 1), (5, 5), (2, 1), (3, 4)\}$ and the initial inner approximation $\mathcal{P}_I^0 = \text{conv}(\mathcal{E}^0)$ yields $(\lambda_{PC}^0)_{(2,1)} = 0.58$ and $(\lambda_{PC}^0)_{(3,4)} = 0.42$, $x_{PC}^0 = (2.42, 2.25)$, bound $\underline{z}_{PC}^0 = \bar{z}_{PC}^0 = 2.42$. Next, we solve the cutting subproblem $SEP(\mathcal{P}, x_{PC}^0)$, generating facet-defining inequalities of $\mathcal{P}$ that are violated by x_{PC}^0. One such facet-defining inequality, $x_1 \geq 3$, is illustrated in Figure 4.11(a). We add this inequality to the current set $D^0 = [A'', b'']$ to form a new outer approximation $\mathcal{P}_O^1$, defined by the set D^1.

Iteration 1. Solving the new master problem, we obtain an optimal primal solution $(\lambda_{PC}^1)_{(4,1)} = 0.42$, $(\lambda_{PC}^1)_{(2,1)} = 0.42$, $(\lambda_{PC}^1)_{(3,4)} = 0.17$, $x_{PC}^1 = (3, 1.5)$, bound $\bar{z}_{PC}^1 = 3$, as well as an optimal dual solution $(u_{PC}^1, \alpha_{PC}^1)$. Next, we consider the pricing subproblem. Since x_{PC}^1 is in the interior of $\mathcal{P}'$, every extreme point of $\mathcal{P}'$ has reduced cost 0 by Theorem 4.4. Therefore, there are no negative reduced cost columns and we switch again to the cutting subproblem $SEP(x_{PC}^1, \mathcal{P})$. As illustrated in Figure 4.11(b), we find another facet-defining inequality of $\mathcal{P}$ violated by x_{PC}^1, $x_2 \geq 2$. We then add this inequality to form D^2 and further tighten the outer approximation, now $\mathcal{P}_O^2$.

Iteration 2. In the final iteration, we solve the master problem again to obtain $(\lambda_{PC}^2)_{(4,1)} = 0.33$, $(\lambda_{PC}^2)_{(2,1)} = 0.33$, $(\lambda_{PC}^2)_{(3,4)} = 0.33$, $x_{PC}^2 = (3, 2)$, bound $\bar{z}_{PC}^2 = 3$. Now, since the primal solution is integral, and is contained in $\mathcal{P}' \cap \mathcal{Q}''$, we know that $\bar{z}_{PC}^2 = z_{IP}$ and we terminate.

Let us now return to the TSP example to further explore the use of the price and cut method.

Example 4.2 (Continued)
As described earlier, application of the Dantzig-Wolfe method along with the 1-tree relaxation for the TSP allows us to compute the bound z_D obtained by optimizing over the intersection of the 1-tree polyhedron (the inner polyhedron) with the polyhedron $\mathcal{Q}''$ (the outer polyhedron) defined by constraints (4.17) and (4.19). With price and cut, we can further improve the bound by allowing both the inner and outer polyhedra to have large descriptions. For this

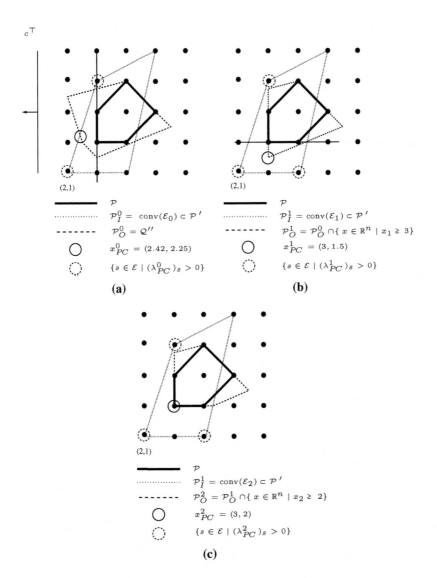

FIGURE 4.11
Price-and-cut method (Example 4.1).

purpose, let us now introduce the well-known *comb inequalities* [30, 31], which we will generate to improve our outer approximation. A comb is defined by a set $H \subset V$, called the *handle* and sets $T_1, T_2, \ldots, T_k \subset V$, called the *teeth*, which satisfy

$$H \cap T_i \neq \emptyset \text{ for } i = 1, \ldots, k,$$
$$T_i \setminus H \neq \emptyset \text{ for } i = 1, \ldots, k,$$
$$T_i \cap T_j = \emptyset \text{ for } 1 \leq i < j \leq k,$$

for some odd $k \geq 3$. Then, for $|V| \geq 6$ the comb inequality,

$$x(E(H)) + \sum_{i=1}^{k} x(E(T_i)) \leq |H| + \sum_{i=1}^{k}(|T_i| - 1) - \lceil k/2 \rceil \qquad (4.52)$$

is valid and facet-defining for the TSP. Let the comb polyhedron be defined by constraints (4.17), (4.19), and (4.52).

There are no known efficient algorithms for solving the general facet identification problem for the comb polyhedron. To overcome this difficulty, one approach is to focus on comb inequalities with special forms. One subset of the comb inequalities, known as the *blossom inequalities*, is obtained by restricting the teeth to have exactly two members. The facet identification for the polyhedron comprised of the blossom inequalities and constraints (4.17) and (4.19) can be solved in polynomial time, a fact we return to shortly. Another approach is to use heuristic algorithms not guaranteed to find a violated comb inequality when one exists (see [4] for a survey). These heuristic algorithms could be applied in price and cut as part of the cutting subproblem in Step 3b to improve the outer approximation.

In Figure 4.7 of Section 4.3.2, we showed an optimal fractional solution $\hat{x}$ that resulted from the solution of a Dantzig-Wolfe master problem and the corresponding optimal decomposition, consisting of six 1-trees. In Figure 4.12, we show the sets $H = \{0, 1, 2, 3, 6, 7, 9, 11, 12, 15\}$, $T_1 = \{5, 6\}$, $T_2 = \{8, 9\}$, and $T_3 = \{12, 13\}$ forming a comb that is violated by this fractional solution, since

$$\hat{x}(E(H)) + \sum_{i=1}^{k} \hat{x}(E(T_i)) = 11.3 > 11 = |H| + \sum_{i=1}^{k}(|T_i| - 1) - \lceil k/2 \rceil.$$

Such a violated comb inequality, if found, could be added to the description of the outer polyhedron to improve on the bound z_D. This shows the additional

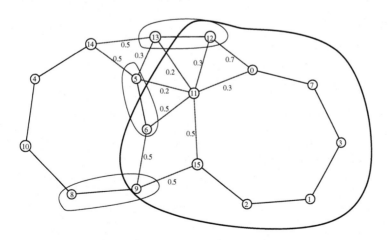

FIGURE 4.12
Price-and-cut method (Example 4.2).

power of price and cut over the Dantzig-Wolfe method. Of course, it should be noted that it is also possible to generate such inequalities in the standard cutting plane method and to achieve the same bound improvement.

The choice of relaxation has a great deal of effect on the empirical behavior of decomposition algorithms. In Example 4.2, we employed an inner polyhedron with integer extreme points. With such a polyhedron, the integrality constraints of the inner polyhedron have no effect and $z_D = z_{LP}$. In Example 4.3, we consider a relaxation for which the bound z_D may be strictly improved over z_{LP} by employing an inner polyhedron that is not integral.

Example 4.3

Let G be a graph as defined in Example 4.2 for the TSP. A *2-matching* is a subgraph in which every vertex has degree two. Every TSP tour is hence a 2-matching. The *Minimum 2-Matching Problem* is a relaxation of TSP whose feasible region is described by the degree (4.17), bound (4.19), and integrality constraints (4.20) of the TSP. Interestingly, the 2-matching polyhedron, which is implicitly defined to be the convex hull of the feasible region just described, can also be described by replacing the integrality constraints (4.20) with the blossom inequalities. Just as the SEC constraints provide a complete description of the 1-tree polyhedron, the blossom inequalities (plus degree and bound) constraints provide a complete description of the 2-matching polyhedron. Therefore, we could use this polyhedron as an outer approximation to the TSP polyhedron. In [50], Müller-Hannemann and Schwartz present several polynomial algorithms for optimizing over the 2-matching polyhedron. We can therefore also use the 2-matching relaxation in the context of price and cut to generate an inner approximation of the TSP polyhedron. Using integrated methods, it would then be possible to simultaneously build up an outer approximation of the TSP polyhedron consisting of the SECs (4.18). Note that this simply reverses the roles of the two polyhedra from Example 4.2 and thus would yield the same bound.

Figure 4.13 shows an optimal fractional solution arising from the solution of the master problem and the 2-matchings with positive weight in a corresponding optimal decomposition. Given this fractional subgraph, we could employ the separation algorithm discussed in Example 4.2 of Section 4.3.1 to generate the violated subtour $S = \{0, 1, 2, 3, 7\}$. ∎

Another approach to generating improving inequalities in price and cut is to try to take advantage of the information contained in the optimal decomposition to aid in the separation procedure. This information, though computed by solving (4.51) is typically ignored. Consider the fractional solution x_{PC}^t generated in iteration t of the method in Figure 4.10. The optimal decomposition for the master problem in iteration t, λ_{PC}^t, provides a decomposition of x_{PC}^t into a convex combination of members of $\mathcal{E}$. We refer to elements of $\mathcal{E}$ that have a positive weight in this combination as *members of the decomposition*. The following theorem shows how such a decomposition can be used to derive

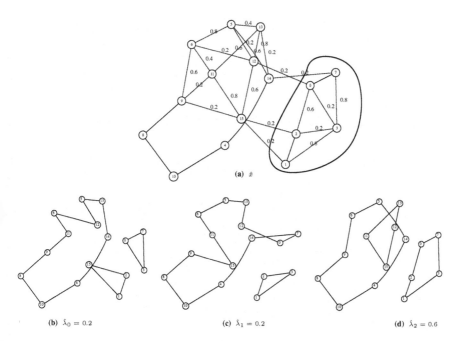

(a) $\hat{x}$

(b) $\hat{\lambda}_0 = 0.2$ (c) $\hat{\lambda}_1 = 0.2$ (d) $\hat{\lambda}_2 = 0.6$

FIGURE 4.13
Finding violated inequalities in price-and-cut (Example 4.3).

an alternate necessary condition for an inequality to be improving. Because we apply this theorem in a more general context later in the paper, we state it in a general form.

THEOREM 4.6
If $\hat{x} \in \mathbb{R}^n$ violates the inequality $(a, \beta) \in \mathbb{R}^{(n+1)}$ and $\hat{\lambda} \in \mathbb{R}_+^{\mathcal{E}}$ is such that $\Sigma_{s \in \mathcal{E}} \hat{\lambda}_s = 1$ and $\hat{x} = \Sigma_{s \in \mathcal{E}} s \hat{\lambda}_s$, then there must exist an $s \in \mathcal{E}$ with $\hat{\lambda}_s > 0$ such that s also violates the inequality (a, β).

PROOF Let $\hat{x} \in \mathbb{R}^n$ and $(a, \beta) \in \mathbb{R}^{(n+1)}$ be given such that $a^\top \hat{x} < \beta$. Also, let $\hat{\lambda} \in \mathbb{R}_+^{\mathcal{E}}$ be given such that $\Sigma_{s \in \mathcal{E}} \hat{\lambda}_s = 1$ and $\hat{x} = \Sigma_{s \in \mathcal{E}} s \hat{\lambda}_s$. Suppose that $a^\top s \geq \beta$ for all $s \in \mathcal{E}$ with $\hat{\lambda}_s > 0$. Since $\Sigma_{s \in \mathcal{E}} \hat{\lambda}_s = 1$, we have $a^\top (\Sigma_{s \in \mathcal{E}} s \hat{\lambda}_s) \geq \beta$. Hence, $a^\top \hat{x} = a^\top (\Sigma_{s \in \mathcal{E}} s \hat{\lambda}_s) \geq \beta$, which is a contradiction. ∎

In other words, an inequality can be improving only if it is violated by at least one member of the decomposition. If $\mathcal{I}$ is the set of all improving inequalities in iteration t, then the following corollary is a direct consequence of Theorem 4.6.

COROLLARY 4.4
$\mathcal{I} \subseteq \mathcal{V} = \{(a, \beta) \in \mathbb{R}^{(n+1)} : a^\top s < \beta \text{ for some } s \in \mathcal{E} \text{ such that } (\lambda_{PC}^t)_s > 0\}.$

The importance of these results is that, in many cases, separation of members of $\mathcal{F}'$ from $\mathcal{P}$ is easier than separation of arbitrary real vectors. There are a number of well-known polyhedra for which the problem of separating an arbitrary real vector is difficult, but the problem of separating a solution to a given relaxation is easy. This concept is formalized in Section 4.5 and some examples are discussed in Section 4.8. In Figure 4.14, we propose a new separation procedure that can be embedded in price and cut that takes advantage of this fact. The procedure takes as input an arbitrary real vector $\hat{x}$ that has been previously decomposed into a convex combination of vectors with known structure. In price and cut, the arbitrary real vector is x_{PC}^t and it is decomposed into a convex combination of members of $\mathcal{E}$ by solving the master problem (4.51). Rather than separating x_{PC}^t directly, the procedure consists of separating each one of the members of the decomposition in turn, then checking each inequality found for violation against x_{PC}^t.

The running time of this procedure depends in part on the cardinality of the decomposition. Carathéodory's Theorem assures us that there exists a decomposition with less than or equal to $\dim(\mathcal{P}_I^t)+1$ members. Unfortunately, even if we limit our search to a particular known class of valid inequalities, the number of such inequalities violated by each member of $\mathcal{D}$ in Step 2 may be extremely large and these inequalities may not be violated by x_{PC}^t (such an inequality cannot be improving). Unless we enumerate *every* inequality in the set $\mathcal{V}$ from Corollary 4.4, either implicitly or explicitly, the procedure does not guarantee that an improving inequality will be found, even if one exists. In cases where examination of the set $\mathcal{V}$ in polynomial time is possible, the worst-case complexity of the entire procedure is polynomially equivalent to that of optimizing over $\mathcal{P}'$. Obviously, it is unlikely that the set $\mathcal{V}$ can be examined in polynomial time in situations when separating x_{PC}^t is itself an $\mathcal{NP}$-complete problem. In such cases, the procedure to select inequalities that are likely to

---

Separation Using a Decomposition

<u>Input</u>: A decomposition $\lambda \in \mathbb{R}^{\mathcal{E}}$ of $\hat{x} \in \mathbb{R}^n$.
<u>Output</u>: A set $[D, d]$ of potentially improving inequalities.

1. Form the set $\mathcal{D} = \{s \in \mathcal{E} \mid \lambda_s > 0\}$.
2. For each $s \in \mathcal{D}$, call the subroutine $SEP(\mathcal{P}, s)$ to obtain a set $[\breve{D}, \breve{d}]$ of violated inequalities.
3. Let $[D, d]$ be composed of the inequalities found in Step 2 that are also violated by $\hat{x}$, so that $D\hat{x} < d$.
4. Return $[D, d]$ as the set of potentially improving inequalities.

FIGURE 4.14
Solving the cutting subproblem with the aid of a decomposition.

be violated by x_{PC}^t in Step 2 is necessarily a problem-dependent heuristic. The effectiveness of such heuristics can be improved in a number of ways, some of which are discussed in [57].

Note that members of the decomposition in iteration t must belong to the set $S(u_{PC}^t, \alpha_{PC}^t)$, as defined by (4.40). It follows that the convex hull of the decomposition is a subset of $conv(S(u_{PC}^t, \alpha_{PC}^t))$ that contains x_{PC}^t and can be thought of as a surrogate for the face of optimal solutions to an LP solved directly over $\mathcal{P}_I^t \cap \mathcal{P}_O^t$ with objective function vector c. Combining this corollary with Theorem 4.1, we conclude that separation of $S(u_{PC}^t, \alpha_{PC}^t)$ from $\mathcal{P}$ is a sufficient condition for an inequality to be improving. Although this sufficient condition is difficult to verify in practice, it does provide additional motivation for the method described in Figure 4.14.

Example 4.1 (Continued)
Returning to the cutting subproblem in iteration 0 of the price and cut method, we have a decomposition $x_{PC}^0 = (2.42, 2.25) = 0.58(2, 1) + 0.42(3, 4)$, as depicted in Figure 4.11(a). Now, instead of trying to solve the subproblem $SEP(\mathcal{P}, x_{PC}^0)$, we instead solve $SEP(\mathcal{P}, s)$, for each $s \in \mathcal{D} = \{(2, 1), (3, 4)\}$. In this case, when solving the separation problem for $s = (2, 1)$, we find the same facet-defining inequality of $\mathcal{P}$ as we did by separating x_{PC}^0 directly.

Similarly, in iteration 1, we have a decomposition of $x_{PC}^2 = (3, 1.5)$ into a convex combination of $\mathcal{D} = \{(4, 1), (2, 1), (3, 4)\}$. Clearly, solving the separation problem for either $(2, 1)$ or $(4, 1)$ produces the same facet-defining inequality as with the original method. ∎

Example 4.2 (Continued)
Returning again to Example 4.2, recall the optimal fractional solution and the corresponding optimal decomposition arising during solution of the TSP by the Dantzig-Wolfe method in Figure 4.7. Figure 4.12 shows a comb inequality violated by this fractional solution. By Theorem 4.6, at least one of the members of the optimal decomposition shown in Figure 4.7 must also violate this inequality. In fact, the member with index 0, also shown in Figure 4.15, is the only such member. Note that the violation is easy to discern from the structure of this integral solution. Let $\hat{x} \in \{0, 1\}^E$ be the incidence vector of a 1-tree. Consider a subset H of V whose induced subgraph in the 1-tree is a path with edge set P. Consider also an odd set O of edges of the 1-tree of cardinality at least 3 and disjoint from P, such that each edge has one endpoint in H and one endpoint in $V \setminus H$. Taking the set H to be the handle and the endpoints of each member of O to be the teeth, it is easy to verify that the corresponding comb inequality will be violated by the 1-tree, since

$$\hat{x}(E(H)) + \sum_{i=1}^{k} \hat{x}(E(T_i)) = |H| - 1 + \sum_{i=1}^{k}(|T_i| - 1) > |H| + \sum_{i=1}^{k}(|T_i| - 1) - \lceil k/2 \rceil.$$

Hence, searching for such configurations in the members of the decomposition, as suggested in the procedure of Figure 4.14, may lead to the discovery

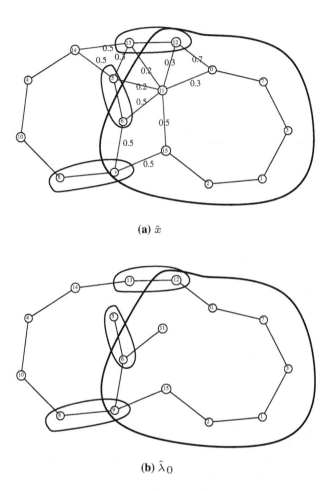

(a) $\hat{x}$

(b) $\hat{\lambda}_0$

FIGURE 4.15
Using the optimal decomposition to find violated inequalities in price and cut (Example 4.2).

of comb inequalities violated by the optimal fractional solution. In this case, such a configuration does in fact lead to discovery of the previously indicated comb inequality. Note that we have restricted ourselves in the above discussion to the generation of blossom inequalities. The teeth, as well as the handles can have more general forms that may lead to the discovery of more general forms of violated combs. ■

Example 4.3 (Continued)
Returning now to Example 4.3, recall the optimal fractional solution and the corresponding optimal decomposition, consisting of the 2-matchings shown in Figure 4.13. Previously, we produced a set of vertices defining a SEC violated by the fractional point by using a minimum cut algorithm with the optimal fractional solution as input. Now, let us consider applying the procedure

of Figure 4.14 by examining the members of the decomposition in order to discover inequalities violated by the optimal fractional solution. Let $\hat{x} \in \{0, 1\}^E$ be the incidence vector of a 2-matching. If the corresponding subgraph does not form a tour, then it must be disconnected. The vertices corresponding to any connected component thus define a violated SEC. By determining the connected components of each member of the decomposition, it is easy to find violated SECs. In fact, for any 2-matching, every component of the 2-matching forms a SEC that is violated by exactly 1. For the 2-matching corresponding to $\hat{s}$, we have $\hat{x}(E(S)) = |S| > |S| - 1$. Figure 4.16(b) shows the third member of the decomposition along with a violated SEC defined by one

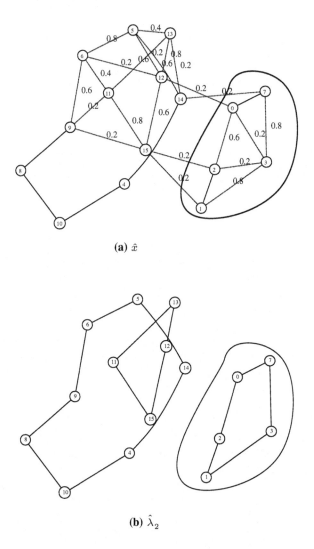

(a) $\hat{x}$

(b) $\hat{\lambda}_2$

FIGURE 4.16
Using the optimal decomposition to find violated inequalities in price and cut (Example 4.3).

of its components. This same SEC is also violated by the optimal fractional solution.

There are many variants of the price-and-cut method shown in Figure 4.10. Most significant is the choice of which subproblem to execute during Step 3. It is easy to envision a number of heuristic rules for deciding this. For example, one obvious rule is to continue generating columns until no more are available and then switch to valid inequalities for one iteration, then generate columns again until none are available. This can be seen as performing a "complete" dual solution update before generating valid inequalities. Further variants can be obtained by not insisting on a "complete" dual update before solving the pricing problem [17, 29]. This rule could easily be inverted to generate valid inequalities until no more are available and then generate columns. A hybrid rule in which some sort of alternation occurs is a third option. The choice between these options is primarily empirical.

4.4.2   Relax and Cut

Just as with the Dantzig-Wolfe method, the Lagrangian method of Figure 4.9 can be integrated with the cutting plane method to yield a procedure several authors have termed *relax and cut*. This is done in much the same fashion as in price and cut, with a choice in each iteration between solving a pricing subproblem and a cutting subproblem. In each iteration that the cutting subproblem is solved, the generated valid inequalities are added to the description of the outer polyhedron, which is explicitly maintained as the algorithm proceeds. As with the traditional Lagrangian method, no explicit inner polyhedron is maintained, but the algorithm can again be seen as one that computes a face of the implicitly defined inner polyhedron that contains the optimal face of solutions to a linear program solved over the intersection of the two polyhedra. When employed within a branch and bound framework, we call the overall method *branch, relax, and cut*.

An outline of the relax-and-cut method is shown in Figure 4.17. The question again arises as to how to ensure that the inequalities being generated in the cutting subproblem are improving. In the case of the Lagrangian method, this is a much more difficult issue because we cannot assume the availability of the same primal solution information available within price and cut. Furthermore, we cannot verify the condition of Corollary 4.1, which is the best available necessary condition for an inequality to be improving. Nevertheless, *some* primal solution information is always available in the form of the solution s_{RC}^t to the last pricing subproblem that was solved. Intuitively, separating s_{RC}^t makes sense since the infeasibilities present in s_{RC}^t may possibly be removed through the addition of valid inequalities violated by s_{RC}^t.

As with both the cutting plane and price-and-cut methods, the difficulty is that the valid inequalities generated by separating s_{RC}^t from $\mathcal{P}$ may not be improving, as Guignard first observed in [33]. To deepen understanding of the potential effectiveness of the valid inequalities generated, we further

---

Relax and Cut Method

<u>Input</u>: An instance $ILP(\mathcal{P}, c)$.
<u>Output</u>: A lower bound z_{RC} on the optimal solution value for the instance and a dual solution $\hat{u}_{RC} \in \mathbb{R}^{m^t}$.

1. Let $s^0_{RC} \in \mathcal{E}$ define some initial extreme point of $\mathcal{P}'$ and construct an initial outer approximation

$$\mathcal{P}^0_O = \{x \in \mathbb{R}^n \mid D^0 x \geq d^0\} \supseteq \mathcal{P}, \tag{4.53}$$

where $D^0 = A''$ and $d^0 = b''$. Let $u^0_{RC} \in \mathbb{R}^{m''}$ be some initial set of dual multipliers associated with the constraints $[D^0, d^0]$. Set $t \leftarrow 0$ and $m^t = m''$.

2. **Master Problem:** Using the solution information gained from solving the pricing subproblem, and the previous dual solution u^t_{RC}, update the dual solution (if the pricing problem was just solved) or initialize the new dual multipliers (if the cutting subproblem was just solved) to obtain $u^{t+1}_{RC} \in \mathbb{R}^{m^t}$.

3. Do either (a) or (b).

   (a) **Pricing Subproblem:** Call the subroutine $OPT(c - \mathcal{P}'(u^t_{RC})^\top D^t, (c - (u^t_{RC})^\top D^t)s^t_{RC})$ to obtain

   $$z^t_{RC} = \min_{s \in \mathcal{F}'} \{(c^\top - (u^t_{RC})D^t)s + d^t(u^t_{RC})\}. \tag{4.54}$$

   Let $s^{t+1}_{RC} \in \mathcal{E}$ be the optimal solution to this subproblem. Set $[D^{t+1}, d^{t+1}] \leftarrow [D^t, d^t]$, $\mathcal{P}^{t+1}_O \leftarrow \mathcal{P}^t_O$, $m^{t+1} \leftarrow m^t$, $t \leftarrow t + 1$, and go to Step 2.

   (b) **Cutting Subproblem:** Call the subroutine $SEP(\mathcal{P}, s^t_{RC})$ to generate a set of *improving* valid inequalities $[\tilde{D}, \tilde{d}] \in \mathbb{R}^{\tilde{m} \times n + 1}$ for $\mathcal{P}$, violated by s^t_{RC}. If violated inequalities were found, set $[D^{t+1}, d^{t+1}] \leftarrow \begin{bmatrix} D^t & d^t \\ \tilde{D} & \tilde{d} \end{bmatrix}$ to form a new outer approximation $\mathcal{P}^{t+1}_O$. Set $m^{t+1} \leftarrow m^t + \tilde{m}$, $s^{t+1}_{RC} \leftarrow s^t_{RC}$, $t \leftarrow t + 1$, and go to Step 2.

4. If a prespecified stopping criterion is met, then output $z_{RC} = z^t_{RC}$ and $\hat{u}_{RC} = u^t_{RC}$.

5. Otherwise, go to Step 2.

FIGURE 4.17
Outline of the relax and cut method.

examine the relationship between s_{RC}^t and x_{PC}^t by recalling again the results from Section 4.3.2. Consider the set $\mathcal{S}(u_{RC}^t, z_{RC}^t)$, where z_{RC}^t is obtained by solving the pricing subproblem (4.54) from Figure 4.17 and the set $\mathcal{S}(\cdot, \cdot)$ is as defined in (4.40). In each iteration where the pricing subproblem is solved, s_{RC}^{t+1} is a member of $\mathcal{S}(u_{RC}^t, z_{RC}^t)$. In fact, $\mathcal{S}(u_{RC}^t, z_{RC}^t)$ is exactly the set of alternative solutions to this pricing subproblem. In price and cut, a number of members of this set are available, one of which must be violated in order for a given inequality to be improving. This yields a verifiable necessary condition for a generated inequality to be improving. Relax and cut, in its most straightforward incarnation, produces one member of this set. Even if improving inequalities exist, it is possible that none of them are violated by the member of $\mathcal{S}(u_{RC}^t, z_{RC}^t)$ so produced, especially if it would have had a small weight in the optimal decomposition produced by the corresponding iteration of price and cut.

It is important to note that by keeping track of the solutions to the Lagrangian subproblem that are produced while solving the Lagrangian dual, one can approximate the optimal decomposition and the optimal fractional solution produced by solving (4.51). This is the approach taken by the volume algorithm [9] and a number of other subgradient-based methods. As in price and cut, when this fractional solution is an inner point of $\mathcal{P}'$, all members of $\mathcal{F}'$ are alternative optimal solutions to the pricing subproblem and the bound is not improved over what the cutting plane method alone would produce. In this case, solving the cutting subproblem to obtain additional inequalities is unlikely to yield further improvement.

As with price and cut, there are again many variants of the algorithm shown in Figure 4.17, depending on the choice of subproblem to execute at each step. One such variant is to alternate between each of the subproblems, first solving one and then the other [46]. In this case, the Lagrangian dual is not solved to optimality before solving the cutting subproblem. Alternatively, another approach is to solve the Lagrangian dual all the way to optimality before generating valid inequalities. Again, the choice is primarily empirical.

4.5 Solving the Cutting Subproblem

In this section, we formalize some notions that have been introduced in our examples and provide more details regarding how the cutting subproblem is solved in practice in the context of the various methods we have outlined. We review the well-known *template paradigm* for separation and introduce a new concept called *structured separation*. Finally, we describe a separation algorithm called *decompose and cut* that is closely related to the integrated decomposition methods we have already described and utilizes several of the concepts introduced earlier.

4.5.1 The Template Paradigm

The ability to generate valid inequalities for $\mathcal{P}$ violated by a given real vector is a crucial step in many of the methods discussed in this paper. Ideally, we would be able to solve the general facet identification problem for $\mathcal{P}$, allowing us to generate a violated valid inequality whenever one exists. This is clearly not practical in most cases, since the complexity of this problem is the same as that of solving the original ILP. In practice, the subproblem $SEP(x_{CP}^t, \mathcal{P})$ in Step 3 of the cutting plane method pictured in Figure 4.2 is usually solved by dividing the valid inequalities for $\mathcal{P}$ into *template classes* with known structure. Procedures are then designed and executed for identifying violated members of each class individually.

A template class (or simply *class*) of valid inequalities for $\mathcal{P}$ is a set of related valid inequalities that describes a polyhedron containing $\mathcal{P}$, so we can identify each class with its associated polyhedron. In Example 4.2, we described two well-known classes of valid inequalities for the TSP, the *subtour elimination constraints* and the *comb inequalities*. Both classes have an identifiable coefficient structure and describe polyhedra containing $\mathcal{P}$. Consider a polyhedron $\mathcal{C}$ described by a class of valid inequalities for $\mathcal{P}$. The separation problem for the class $\mathcal{C}$ of valid inequalities for $\mathcal{P}$ is defined to be the facet identification problem over the polyhedron $\mathcal{C}$. In other words, the separation problem for a class of valid inequalities depends on the form of the inequality and is independent of the polyhedron $\mathcal{P}$. It follows that the worst case running time for solving the separation problem is also independent of $\mathcal{P}$. In particular, the separation problem for a particular class of inequalities may be much easier to solve than the general facet identification problem for $\mathcal{P}$. Therefore, in practice, the separation problem is usually attempted over "easy" classes first, and more difficult classes are only attempted when needed. In the case of the TSP, the separation problem for the SECs is solvable in polynomial time, whereas there is no known efficient algorithm for solving the separation problem for comb inequalities. In general, the intersection of the polyhedra associated with the classes of inequalities for which the separation problem can be reasonably solved is not equal to $\mathcal{P}$.

4.5.2 Separating Solutions with Known Structure

In many cases, the complexity of the separation problem is also affected by the structure of the real vector being separated. In Section 4.4, we informally introduced the notion that a solution vector with known structure may be easier to separate from a given polyhedron than an arbitrary one and illustrated this phenomenon in Example 4.2 and Example 4.3. This is a concept called *structured separation* that arises quite frequently in the solution of combinatorial optimization problems where the original formulation is of exponential size. When using the cutting plane method to solve the LP relaxation of the TSP, for example, as described in Example 4.2, we must generate the SECs dynamically. It is thus possible that the intermediate solutions are integer-valued, but nonetheless not feasible because they violate some SEC that is not present in the current approximation. When the current solution is optimal, however, it is easy to determine whether it violates a SEC — we simply examine the connected components of the underlying support graph, as described earlier. This process can be done in $O(|V| + |E|)$ time. For an arbitrary real vector, the separation problem for SECs is more difficult, taking $O(|V|^4)$ time.

It is also frequently the case that when applying a sequence of separation routines for progressively more difficult classes of inequalities, routines for the more difficult classes assume implicitly that the solution to be separated satisfies all inequalities of the the easier classes. In the case of the TSP, for instance, any solution passed to the subroutine for separating the comb inequalities is generally assumed to satisfy the degree and subtour elimination constraints. This assumption can allow the separation algorithms for subsequent classes to be implemented more efficiently.

For the purposes of the present work, our main concern is with separating solutions that are known to be integral, in particular, members of $\mathcal{F}'$. In our framework, the concept of structured separation is combined with the template paradigm in specifying template classes of inequalities for which separation of integral solutions is much easier, in a complexity sense, than separation of arbitrary real vectors over that same class. A number of examples of problems and classes of valid inequalities for which this situation occurs are examined in Section 4.8. We now examine a separation paradigm called *decompose and cut* that can take advantage of our ability to easily separate solutions with structure.

4.5.3   Decompose and Cut

The use of a decomposition to aid in separation, as is described in the procedure of Figure 4.14, is easy to extend to a traditional branch-and-cut framework using a technique we call *decompose and cut*, originally proposed in [56] and further developed in [40] and [57]. Suppose now that we are given an optimal fractional solution x_{CP}^t obtained during iteration t of the cutting plane method and suppose that for a given $s \in \mathcal{F}'$, we can determine effectively whether $s \in \mathcal{F}$ and if not, generate a valid inequality (a, β) violated by s. By first *decomposing* x_{CP}^t (i.e., expressing x_{CP}^t as a convex combination of members of $\mathcal{E} \subseteq \mathcal{F}'$) and then separating each member of this decomposition from $\mathcal{P}$ in the fashion described in Figure 4.14, we may be able to find valid inequalities for $\mathcal{P}$ that are violated by x_{CP}^t.

The difficult step is finding the decomposition of x_{CP}^t. This can be accomplished by solving a linear program whose columns are the members of $\mathcal{E}$, as described in Figure 4.18. This linear program is reminiscent of (4.33) and in fact can be solved using an analogous column-generation scheme, as described in Figure 4.19. This scheme can be seen as the "inverse" of the method described in Section 4.4.1, since it begins with the fractional solution x_{CP}^t and tries to compute a decomposition, instead of the other way around. By the equivalence of optimization and facet identification, we can conclude that the

---

Separation in Decompose and Cut

<u>Input:</u> $\hat{x} \in \mathbb{R}^n$
<u>Output:</u> A valid inequality for $\mathcal{P}$ violated by $\hat{x}$, if one is found.

1. Apply standard separation techniques to separate $\hat{x}$. If one of these returns a violated inequality, then STOP and output the violated inequality.

2. Otherwise, solve the linear program

$$\max_{\lambda \in \mathbb{R}_+^{\mathcal{E}}} \left\{ 0^\top \lambda \,\middle|\, \sum_{s \in \mathcal{E}} s\lambda_s = \hat{x}, \ \sum_{s \in \mathcal{E}} \lambda_s = 1 \right\}, \tag{4.55}$$

as in Figure 4.19.

3. The result of Step 2 is either (1) a subset $\mathcal{D}$ of members of $\mathcal{E}$ participating in a convex combination of $\hat{x}$, or (2) a valid inequality (a, β) for $\mathcal{P}$ that is violated by $\hat{x}$. In the first case, go to Step 4. In the second case, STOP and output the violated inequality.

4. Attempt to separate each member of $\mathcal{D}$ from $\mathcal{P}$. For each inequality violated by a member of $\mathcal{D}$, check whether it is also violated by $\hat{x}$. If an inequality violated by $\hat{x}$ is encountered, STOP and output it.

FIGURE 4.18
Separation in the decompose-and-cut method.

Column Generation in Decompose and Cut

<u>Input:</u> $\hat{x} \in \mathbb{R}^n$
<u>Output:</u> Either (1) a valid inequality for $\mathcal{P}$ violated by $\hat{x}$; or (2) a subset $\mathcal{D}$ of $\mathcal{E}$ and a vector $\hat{\lambda} \in \mathbb{R}_+^{\mathcal{E}}$ such that $\Sigma_{s \in \mathcal{D}} \lambda_s s = \hat{x}$ and $\Sigma_{s \in \mathcal{D}} \lambda_s = 1$.

2.0 Generate an initial subset $\mathcal{E}^0$ of $\mathcal{E}$ and set $t \leftarrow 0$.

2.1 Solve (4.55), replacing $\mathcal{E}$ by $\mathcal{E}^t$. If this linear program is feasible, then the elements of $\mathcal{E}^t$ corresponding to the nonzero components of $\hat{\lambda}$, the current solution, comprise the set $\mathcal{D}$, so STOP.

2.2 Otherwise, let (a, β) be a valid inequality for $\text{conv}(\mathcal{E}^t)$ violated by $\hat{x}$ (i.e., the proof of infeasibility). Solve $OPT(\mathcal{P}', a, \beta)$ and let $\tilde{\mathcal{E}}$ be the resulting set of solutions. If $\tilde{\mathcal{E}} \neq \emptyset$, then set $\mathcal{E}^{t+1} \leftarrow \mathcal{E}^t \cup \tilde{\mathcal{E}}, t \rightarrow t+1$, and go to 2.1. Otherwise, (a, β) is an inequality valid for $\mathcal{P}' \supseteq \mathcal{P}$ and violated by $\hat{x}$, so STOP.

FIGURE 4.19
Column generation for the decompose-and-cut method.

problem of finding a decomposition of x_{CP}^t is polynomially equivalent to that of optimizing over $\mathcal{P}'$.

Once the decomposition is found, it can be used as before to locate a violated valid inequality. In contrast to price and cut, however, it is possible that $x_{CP}^t \notin \mathcal{P}'$. This could occur, for instance, if exact separation methods for $\mathcal{P}'$ are too expensive to apply consistently. In this case, it is obviously not possible to find a decomposition in Step 2 of Figure 4.18. The proof of infeasibility for the linear program (4.55), however, provides an inequality separating x_{CP}^t from $\mathcal{P}'$ at no additional expense. Hence, even if we fail to find a decomposition, we still find an inequality valid for $\mathcal{P}$ and violated by x_{CP}^t. This idea was originally suggested in [56] and was further developed in [40]. A similar concept was also discovered and developed independently by Applegate, et al. [3].

Applying decompose and cut in every iteration as the sole means of separation is theoretically equivalent to price and cut. In practice, however, the decomposition is only computed when needed, i.e., when less expensive separation heuristics fail to separate the optimal fractional solution. This could give decompose and cut an advantage in terms of computational efficiency. In other respects, the computations performed in each method are similar.

4.6   Solving the Master Problem

The choice of a proper algorithm for solving the master problem is important for these methods, both because a significant portion of the computational effort is spent solving the master problem and because the solver must be capable of returning the solution information required by the method. In this section, we would like to briefly give the reader a taste for the issues involved and summarize the existing methodology. The master problems we have discussed are linear programs, or can be reformulated as linear programs. Hence, one option for solving them is to use either simplex or interior point methods. In the case of solving a Lagrangian dual, subgradient methods may also be employed.

Simplex methods have the advantage of providing accurate primal solution information. They are therefore well-suited for algorithms that utilize primal solution information, such as price and cut. The drawback of these methods is that updates to the dual solution at each iteration are relatively expensive. In their most straightforward implementations, they also tend to converge slowly when used with column generation. This is primarily due to the fact that they produce basic (extremal) dual solutions, which tend to change substantially from one iteration to the next, causing wide oscillations in the input to the column-generation subproblem. This problem can be addressed by implementing one of a number of stabilization methods that prevent the dual solution from changing "too much" from one iteration to the next (for a survey, see [43]).

Subgradient methods, on the other hand, do not produce primal solution information in their most straightforward form, so they are generally most appropriate for Lagrangian methods such as relax and cut. It is possible, however, to obtain approximate primal solutions from variants of subgradient such as the volume algorithm [9]. Subgradient methods also have convergence issues without some form of stabilization. A recent class of algorithms that has proven effective in this regard is bundle methods [18].

Interior point methods may provide a middle ground by providing accurate primal solution information and more stable dual solutions [28, 58]. In addition, hybrid methods that alternate between simplex and subgradient methods for updating the dual solution have also shown promise [10, 36].

4.7 Software

The theoretical and algorithmic framework proposed in Section 4.3 to Section 4.5 lends itself nicely to a wide-ranging and flexible generic software framework. All of the techniques discussed can be implemented by combining a set of basic algorithmic building blocks. DECOMP is a C++ framework designed with the goal of providing a user with the ability to easily utilize various traditional and integrated decomposition methods while requiring only the provision of minimal problem-specific algorithmic components [25]. With DECOMP, the majority of the algorithmic structure is provided as part of the framework, making it easy to compare various algorithms directly and determine which option is the best for a given problem setting. In addition, DECOMP is extensible — each algorithmic component *can* be overridden by the user, if they so wish, in order to develop sophisticated variants of the aforementioned methods.

The framework is divided into two separate user interfaces, an applications interface DecompApp, in which the user must provide implementations of problem-specific methods (e.g., solvers for the subproblems), and an algorithms interface DecompAlgo, in which the user can modify DECOMP's internal algorithms, if desired. A DecompAlgo object provides implementations of all of the methods described in Section 4.3 and Section 4.4, as well as options for solving the master problem, as discussed in Section 4.6. One important feature of DECOMP is that the problem is always represented in the original space, rather than in the space of a particular reformulation. The user has only to provide subroutines for separation and column generation in the original space without considering the underlying method. The framework performs all of the necessary bookkeeping tasks, including automatic reformulation in the Dantzig-Wolfe master, constraint dualization for relax and cut, cut and variable pool management, as well as, row and column expansion.

In order to develop an application, the user must provide implementations of the following two methods:

- `DecompApp::createCore()`. The user must define the initial set of constraints $[A'', b'']$.
- `DecompApp::solveRelaxedProblem()`. The user must provide a solver for the relaxed problem $OPT(\mathcal{P}', c, U)$ that takes a cost vector $c \in \mathbb{R}^n$ as its input and returns a set of solutions as `DecompVar` objects. Alternatively, the user has the option to provide the inequality set $[A', b']$ and solve the relaxed problem using the built-in ILP solver.

If the user wishes to invoke the traditional cutting plane method using problem-specific methods, then the following method must also be implemented:

- `DecompApp::generateCuts(x)`. A method for solving the separation problem $SEP(\mathcal{P}, x)$, given an arbitrary real vector, $x \in \mathbb{R}^n$, which returns a set of `DecompCut` objects.

Alternatively, various generic separation algorithms are also provided. The user might also wish to implement separation routines specifically for members of $\mathcal{F}'$ that can take advantage of the structure of such solutions, as was described in Section 4.5:

- `DecompApp::generateCuts(s)`. A method for solving the separation problem $SEP(\mathcal{P}, s)$, given members of $\mathcal{F}'$, which returns a set of `DecompCut` objects.

At a high level, the main loop of the base algorithm provided in `DecompAlgo` follows the paradigm described earlier, alternating between solving a master problem to obtain solution information, followed by a subproblem to generate new polyhedral information. Each of the methods described in this paper have its own separate interface derived from `DecompAlgo`. For example, the base class for the price-and-cut method is `DecompAlgo::DecompAlgoPC`. In this manner, the user can override a specific subroutine common to all methods (in `DecompAlgo`) or restrict it to a particular method.

4.8 Applications

In this section, we further illustrate the concepts presented with three more examples. We focus here on the application of integrated methods, a key component of which is the paradigm of structured separation introduced in Section 4.5. For each example, we discuss three key polyhedra: (1) an original ILP defined by a polyhedron $\mathcal{P}$ and associated feasible set $\mathcal{F} = \mathcal{P} \cap \mathbb{Z}^n$; (2) an effective relaxation of the original ILP with feasible set $\mathcal{F}' \supseteq \mathcal{F}$ such that optimization over the polyhedron $\mathcal{P}_I = \text{conv}(\mathcal{F}')$ is possible; and (3) a polyhedron $\mathcal{P}_O$, such that $\mathcal{F} = \mathcal{P}_I \cap \mathcal{P}_O \cap \mathbb{Z}^n$. In each case, the polyhedron $\mathcal{P}_O$

is comprised of a known class or classes of valid inequalities that could be generated during execution of the cutting subproblem of one of the integrated methods discussed in Section 4.4. As before, $\mathcal{P}_I$ is a polyhedron with an inner description generated dynamically through the solution of an optimization problem, while $\mathcal{P}_O$ is a polyhedron with an outer description generated dynamically through the solution of a separation problem. We do not discuss standard methods of solving the separation problem for $\mathcal{P}_O$, i.e., unstructured separation, as these are well-covered in the literature. Instead, we focus here on problems and classes of valid inequalities for which structured separation, i.e., separation of a member of $\mathcal{F}'$, is much easier than unstructured separation. A number of ILPs that have appeared in the literature have relaxations and associated classes of valid inequalities that fit into this framework, such as the Generalized Assignment Problem [54], the Edge-Weighted Clique Problem [37], the Knapsack Constrained Circuit Problem [42], the Rectangular Partition Problem [16], the Linear Ordering Problem [15], and the Capacitated Minimum Spanning Tree Problem [24].

4.8.1 Vehicle Routing Problem

We first consider the *Vehicle Routing Problem* (VRP) introduced by Dantzig and Ramser [20]. In this $\mathcal{NP}$-hard optimization problem, a fleet of k vehicles with uniform capacity C must service known customer demands for a single commodity from a common depot at minimum cost. Let $V = \{1, \ldots, |V|\}$ index the set of customers and let the depot have index 0. Associated with each customer $i \in V$ is a demand d_i. The cost of travel from customer i to j is denoted c_{ij} and we assume that $c_{ij} = c_{ji} > 0$ if $i \neq j$ and $c_{ii} = 0$.

By constructing an associated complete undirected graph G with vertex set $N = V \cup \{0\}$ and edge set E, we can formulate the VRP as an integer program. A *route* is a set of vertices $R = \{i_1, i_2, \ldots, i_m\}$ such that the members of R are distinct. The edge set of R is $E_R = \{\{i_j, i_{j+1}\} \mid j \in 0, \ldots, m\}$, where $i_0 = i_{m+1} = 0$. A feasible solution is then any subset of E that is the union of the edge sets of k disjoint routes $R_i, i \in [1, \ldots, k]$, each of which satisfies the capacity restriction, i.e., $\Sigma_{j \in R_i} d_j \leq C, \forall i \in [1, \ldots, k]$. Each route corresponds to a set of customers serviced by one of the k vehicles. To simplify the presentation, let us define some additional notation.

By associating a variable with each edge in the graph, we obtain the following formulation of this ILP [41]:

$$\min \sum_{e \in E} c_e x_e,$$

$$x(\delta(\{0\})) = 2k, \tag{4.56}$$

$$x(\delta(\{v\})) = 2 \qquad \forall\, v \in V, \tag{4.57}$$

$$x(\delta(S)) \geq 2b(S) \quad \forall\, S \subseteq V,\ |S| > 1, \tag{4.58}$$

$$x_e \in \{0, 1\} \quad \forall\, e \in E(V), \tag{4.59}$$

$$x_e \in \{0, 1, 2\} \,\forall\, e \in \delta(0). \tag{4.60}$$

Here, $b(S)$ represents a lower bound on the number of vehicles required to service the set of customers S. Inequalities (4.56) ensure that exactly k vehicles depart from and returning to the depot, while inequalities (4.57) require that each customer must be serviced by exactly one vehicle. Inequalities (4.58), known as the *generalized subtour elimination constraints* (GSECs) can be viewed as a generalization of the subtour elimination constraints from TSP, and enforce connectivity of the solution, as well as ensuring that no route has total demand exceeding capacity C. For ease of computation, we can define $b(S) = \lceil (\Sigma_{i \in S} d_i)/C \rceil$, a trivial lower bound on the number of vehicles required to service the set of customers S.

The set of feasible solutions to the VRP is

$$\mathcal{F} = \{x \in \mathbb{R}^E \mid x \text{ satisfies } (4.56) - (4.60)\}$$

and we call $\mathcal{P} = \text{conv}(\mathcal{F})$ the *VRP polyhedron*. Many classes of valid inequalities for the VRP polyhedron have been reported in the literature (see [51] for a survey). Significant effort has been devoted to developing efficient algorithms for separating an arbitrary fractional point using these classes of inequalities (see [38] for recent results).

We concentrate here on the separation of GSECs. The separation problem for GSECs was shown to be $\mathcal{NP}$-complete by Harche and Rinaldi (see [5]), even when $b(S)$ is taken to be $\lceil (\Sigma_{i \in S} d_i)/C \rceil$. In [38], Lysgaard, et al. review heuristic procedures for generating violated GSECs. Although GSECs are part of the formulation presented above, there are exponentially many of them, so we generate them dynamically. We discuss three relaxations of the VRP: the *Multiple Traveling Salesman Problem*, the *Perfect b-Matching Problem*, and the *Minimum Degree-Constrained k-Tree Problem*. For each of these alternatives, violation of GSECs by solutions to the relaxation can be easily discerned.

Perfect b-Matching Problem. With respect to the graph G, the *Perfect b-Matching Problem* is to find a minimum weight subgraph of G such that $x(\delta(v)) = b_v \; \forall v \in V$. This problem can be formulated by dropping the GSECs from the VRP formulation, resulting in the feasible set

$$\mathcal{F}' = \{x \in \mathbb{R}^E \mid x \text{ satisfies } (4.56), (4.57), (4.59), (4.60)\}.$$

In [50], Müller-Hannemann and Schwartz, present several fast polynomial algorithms for solving b-Matching. The polyhedron $\mathcal{P}_O$ consists of the GSECs (4.58) in this case.

In [49], Miller uses the b-Matching relaxation to solve the VRP by branch, relax, and cut. He suggests generating GSECS violated by b-matchings as follows. Consider a member s of $\mathcal{F}'$ and its support graph G_s (a b-Matching). If G_s is disconnected, then each component immediately induces a violated GSEC. On the other hand, if G_s is connected, we first remove the edges incident to the depot vertex and find the connected components, which comprise the routes described earlier. To identify a violated GSEC, we compute the total demand of each route, checking whether it exceeds capacity. If not, the solution is feasible for the original ILP and does not violate any GSECs. If so, the set S of customers on any route whose total demand exceeds capacity induces a violated GSEC. This separation routine runs in $O(|V| + |E|)$ time and can be used in any of the integrated decomposition methods previously described. Figure 4.20a shows an example vector that could arise during execution of either price and cut or decompose and cut, along with a decomposition into a convex combination of two b-Matchings, shown

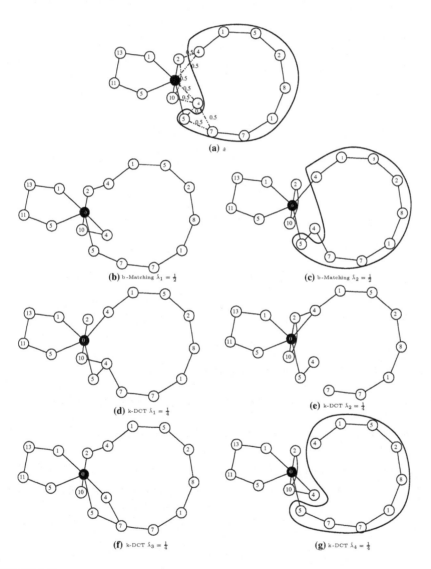

FIGURE 4.20
Example of a decomposition into b-Matchings and k-DCTs.

in Figure 4.20b and c. In this example, the capacity $C = 35$ and by inspection we find a violated GSEC in the second b-Matching (c) with S equal to the marked component. This inequality is also violated by the optimal fractional solution, since $\hat{x}(\delta(S)) = 3.0 < 4.0 = 2b(S)$.

Minimum Degree-Constrained k-Tree Problem. A *k-tree* is defined as a spanning subgraph of G that has $|V| + k$ edges (recall that G has $|V| + 1$ vertices). A *degree-constrained k-tree* (k-DCT), as defined by Fisher in [23], is a k-tree with degree $2k$ at vertex 0. The *Minimum k-DCT Problem* is that of finding a minimum cost k-DCT, where the cost of a k-DCT is the sum of the costs on the edges present in the k-DCT. Fisher [23] introduced this relaxation as part of a Lagrangian relaxation-based algorithm for solving the VRP.

The k-DCT polyhedron is obtained by first adding the redundant constraint

$$x(E) = |V| + k, \tag{4.61}$$

then deleting the degree constraints (4.57), and finally, relaxing the capacity to $C = \Sigma_{i \in s} d_i$. Relaxing the capacity constraints gives $b(S) = 1$ for all $S \subseteq V$, and replaces the set of constraints (4.58) with

$$\sum_{e \in \delta(S)} x_e \geq 2, \forall S \subseteq V, \ |S| > 1. \tag{4.62}$$

The feasible region of the Minimum k-DCT Problem is then

$$\mathcal{F}' = \{x \in \mathbb{R}^E \mid x \text{ satisfies (4.56), (4.58), (4.59), (4.61)}\}.$$

This time, the polyhedron $\mathcal{P}_O$ is comprised of the constraints (4.57) and the GSECs (4.58). Since the constraints (4.57) can be represented explicitly, we focus again on generation of violated GSECs. In [62], Wei and Yu give a polynomial algorithm for solving the Minimum k-DCT Problem that runs in $O(|V|^2 \log |V|)$ time. In [48], Martinhon et al. study the use of the k-DCT relaxation for the VRP in the context branch, relax, and cut. Again, consider separating a member s of $\mathcal{F}'$ from the polyhedron defined by all GSECS. It is easy to see that for GSECs, an algorithm identical to that described above can be applied. Figure 4.20a also shows a vector that could arise during the execution of either the price and cut or decompose-and-cut algorithms, along with a decomposition into a convex combination of four k-DCTs (Figure 4.20d to g). Removing the depot edges, and checking each components demand, we easily identify the violated GSEC in k-DCT (g).

Multiple Traveling Salesman Problem. The *Multiple Traveling Salesman Problem* (k-TSP) is an uncapacitated version of the VRP obtained by adding the degree constraints to the k-DCT polyhedron. The feasible region of the k-TSP is

$$\mathcal{F}' = \{x \in \mathbb{R}^E \mid x \text{ satisfies (4.56), (4.57), (4.59), (4.60), (4.62)}\}.$$

Although the k-TSP is an $\mathcal{N}P$-hard optimization problem, small instances can be solved effectively by transformation into an equivalent TSP obtained by adjoining to the graph $k - 1$ additional copies of vertex 0 and its incident edges. In this case, the polyhedron $\mathcal{P}_O$ is again comprised solely of the GSECs (4.58). In [57], Ralphs et al. report on an implementation of branch, decompose and cut using the k-TSP as a relaxation.

4.8.2 Three-Index Assignment Problem

The *Three-Index Assignment Problem* (3AP) is that of finding a minimum-weight clique cover of the complete tri-partite graph $K_{n,n,n}$. Let I, J, and K be three disjoint sets with $|I| = |J| = |K| = n$ and set $H = I \times J \times K$. 3AP can be formulated as the following binary integer program:

$$\min \sum_{(i,j,k) \in H} c_{ijk} x_{ijk},$$

$$\sum_{(j,k) \in J \times K} x_{ijk} = 1 \quad \forall\, i \in I, \tag{4.63}$$

$$\sum_{(i,k) \in I \times K} x_{ijk} = 1 \quad \forall\, j \in J, \tag{4.64}$$

$$\sum_{(i,j) \in I \times J} x_{ijk} = 1 \quad \forall\, k \in K, \tag{4.65}$$

$$x_{ijk} \in \{0, 1\} \quad \forall\, (i, j, k) \in H. \tag{4.66}$$

A number of applications of 3AP can be found in the literature (see Piersjalla [18,19]). 3AP is known to be $\mathcal{N}P$-hard [26]. As before, the set of feasible solutions to 3AP is noted as

$$\mathcal{F} = \{x \in \mathbb{R}^H \mid x \text{ satisfies } (4.63) - (4.66)\}$$

and we set $\mathcal{P} = \text{conv}(\mathcal{F})$.

In [7], Balas and Saltzman study the polyhedral structure of $\mathcal{P}$ and introduce several classes of facet-inducing inequalities. Let $u, v \in H$ and define $|u \cap v|$ to be the numbers of coordinates for which the vectors u and v have the same value. Let $C(u) = \{w \in H \mid |u \cap w| = 2\}$ and $C(u, v) = \{w \in H \mid |u \cap w| = 1, |w \cap v| = 2\}$. We consider two classes of facet-inducing inequalities $Q_1(u)$ and $P_1(u, v)$ for $\mathcal{P}$,

$$x_u + \sum_{w \in C(u)} x_w \leq 1 \quad \forall\, u \in H, \tag{4.67}$$

$$x_u + \sum_{w \in C(u,v)} x_w \leq 1 \quad \forall\, u, v \in H, |u \cap v| = 0. \tag{4.68}$$

Note that these include all the clique facets of the intersection graph of $K_{n,n,n}$ [7]. In [6], Balas and Qi describe algorithms that solve the separation problem for the polyhedra defined by the inequalities $Q_1(u)$ and $P_1(u, v)$ in $O(n^3)$ time.

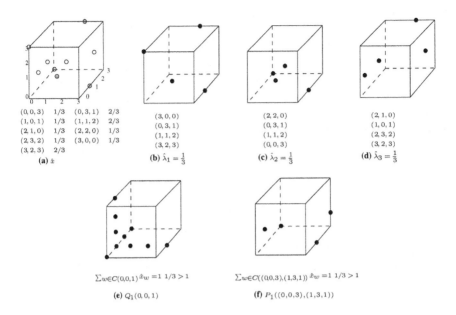

(0, 0, 3)	1/3	(0, 3, 1)	2/3
(1, 0, 1)	1/3	(1, 1, 2)	2/3
(2, 1, 0)	1/3	(2, 2, 0)	1/3
(2, 3, 2)	1/3	(3, 0, 0)	1/3
(3, 2, 3)	2/3		

(a) $\hat{x}$

(3, 0, 0)
(0, 3, 1)
(1, 1, 2)
(3, 2, 3)

(b) $\hat{\lambda}_1 = \frac{1}{3}$

(2, 2, 0)
(0, 3, 1)
(1, 1, 2)
(0, 0, 3)

(c) $\hat{\lambda}_2 = \frac{1}{3}$

(2, 1, 0)
(1, 0, 1)
(2, 3, 2)
(3, 2, 3)

(d) $\hat{\lambda}_3 = \frac{1}{3}$

$\sum_{w \in C(0,0,1)} \hat{x}_w = 1 \, 1/3 > 1$

(e) $Q_1(0, 0, 1)$

$\sum_{w \in C((0,0,3),(1,3,1))} \hat{x}_w = 1 \, 1/3 > 1$

(f) $P_1((0,0,3),(1,3,1))$

FIGURE 4.21
Example of a decomposition into assignments.

Balas and Saltzman consider the use of the classical *Assignment Problem* (AP) as a relaxation of 3AP in an early implementation of branch, relax, and cut [8]. The feasible region of the AP is

$$\mathcal{F}' = \{x \in \mathbb{R}^H \mid x \text{ satisfies } (4.64) - (4.66)\}.$$

The AP can be solved in $O(n^{5/2} \log(nC))$ time where $C = \max_{w \in H} c_w$, by the cost-scaling algorithm [2]. The polyhedron $\mathcal{P}_O$ is here described by constraints (4.63), the constraints $Q_1(u)$ for all $u \in H$, and the constrains $P_1(u, v)$ for all $u, v \in H$. Consider generating a constraint of the form $Q_1(u)$ for some $u \in H$ violated by a given $s \in \mathcal{F}'$. Let $L(s)$ be the set of n triplets corresponding to the nonzero components of s (the assignment from J to K). It is easy to see that if there exist $u, v \in L(s)$ such that $u = (i_0, j_0, k_0)$ and $v = (i_0, j_1, k_1)$, i.e., the assignment *overcovers* the set I, then both $Q(i_0, j_0, k_1)$ and $Q(i_0, j_1, k_0)$ are violated by s. Figure 4.21 shows the decomposition of a vector $\hat{x}$ (a) that could arise during the execution of either the price-and-cut or decompose-and-cut algorithms, along with a decomposition of $\hat{x}$ into a convex combination of assignments (b to d). The pair of triplets (0, 3, 1) and (0, 0, 3) satisfies the condition just discussed and identifies two valid inequalities, $Q_1(0, 3, 3)$ and $Q_1(0, 0, 1)$, that are violated by the second assignment, shown in (c). The latter also violates $\hat{x}$ and is illustrated in (e). This separation routine runs in $O(n)$ time.

Now consider generation of a constraint of the form $P_1(u, v)$ for some $u, v \in H$ violated by $s \in \mathcal{F}'$. As above, for any pair of assignments that correspond to nonzero components of s and have the form (i_0, j_0, k_0), (i_0, j_1, k_1), we know s violates $P_1((i_0, j_0, k_0), (i, j_1, k_1))$, $\forall i \neq i_0$ and $P_1((i_0, j_1, k_1), (i, j_0, k_0))$, $\forall i \neq i_0$. The inequality $P_1((0, 0, 3), (1, 3, 1))$ is violated by the second assignment, shown in Figure 4.21(c). This inequality is also violated by $\hat{x}$ and is illustrated in (f). Once again, this separation routine runs in $O(n)$ time.

4.8.3   Steiner Tree Problem

Let $G = (V, E)$ be a complete undirected graph with vertex set $V = \{1, \dots, |V|\}$, edge set E and a positive weight c_e associated with each edge $e \in E$. Let $T \subset V$ define the set of *terminals*. The *Steiner Tree Problem* (STP), which is $\mathcal{N}P$-hard, is that of finding a subgraph that spans T (called a *Steiner tree*) and has minimum edge cost. In [13], Beasley formulated the STP as a side constrained *Minimum Spanning Tree Problem* (MSTP) as follows. Let $r \in T$ be some terminal and define an artificial vertex 0. Now, construct the augmented graph $\bar{G} = (\bar{V}, \bar{E})$ where $\bar{V} = V \cup \{0\}$ and $\bar{E} = E \cup \{\{i, 0\} \mid i \in (V \setminus T) \cup \{r\}\}$. Let $c_{i0} = 0$ for all $i \in (V \setminus T) \cup \{r\}$. Then, the STP is equivalent to finding a minimum spanning tree (MST) in $\bar{G}$ subject to the additional restriction that any vertex $i \in (V \setminus T)$ connected by edge $\{i, 0\} \in \bar{E}$ must have degree one.

By associating a binary variable x_e with each edge $e \in \bar{E}$, indicating whether or not the edge is selected, we can then formulate the STP as the following integer program:

$$\min \sum_{e \in E} c_e x_e,$$

$$x(\bar{E}) = |\bar{V}| - 1, \tag{4.69}$$
$$x(E(S)) \leq |S| - 1 \quad \forall S \subseteq \bar{V}, \tag{4.70}$$
$$x_{i0} + x_e \leq 1 \qquad \forall e \in \delta(i), i \in (V \setminus T), \tag{4.71}$$
$$x_e \in \{0, 1\} \qquad \forall e \in \bar{E}. \tag{4.72}$$

Inequalities (4.69) and (4.70) ensure that the solution forms a spanning tree on $\bar{G}$. Inequalities (4.70) are subtour elimination constraints (similar to those used in the TSP). Inequalities (4.71) are the side constraints that ensure the solution can be converted to a Steiner tree by dropping the edges in $\bar{E} \setminus E$.

The set of feasible solutions to the STP is

$$\mathcal{F} = \{x \in \mathbb{R}^{\bar{E}} \mid x \text{ satisfies (4.69)} - \text{(4.72)}\}.$$

We set $\mathcal{P} = \text{conv}(\mathcal{F})$ as before. We consider two classes of valid inequalities for $\mathcal{P}$ that are lifted versions of the subtour elimination constraints (SEC).

$$x(E(S)) + x(E(S \setminus T \mid \{0\})) \leq |S| - 1 \,\forall S \subseteq V, S \cap T \neq \emptyset, \tag{4.73}$$
$$x(E(S)) + x(E(S \setminus \{v\} \mid \{0\})) \leq |S| - 1 \,\forall S \subseteq V, S \cap T = \emptyset, v \in S. \tag{4.74}$$

The class of valid inequalities (4.73) were independently introduced by Goemans [27], Lucena [44] and Margot, Prodon, and Liebling [47], for another extended formulation of STP. The inequalities (4.74) were introduced in [27,47]. The separation problem for inequalities (4.73) and (4.74) can be solved in $O(|V|^4)$ time through a series of max-flow computations.

In [45], Lucena considers the use of MSTP as a relaxation of STP in the context of a branch, relax, and cut algorithm. The feasible region of the MSTP is

$$\mathcal{F}' = \{x \in \mathbb{R}^{\hat{E}} \mid x \text{ satisfies } (4.69), (4.70), (4.72)\}.$$

The MSTP can be solved in $O(|E|\log|V|)$ time using Prim's algorithm [55]. The polyhedron $\mathcal{P}_O$ is described by the constraints (4.71), (4.73), and (4.74). Constraints (4.71) can be represented explicitly, but we must dynamically generate constraints (4.73) and (4.74). In order to identify an inequality of the form (4.73) or (4.74) violated by a given $s \in \mathcal{F}'$, we remove the artificial vertex 0 and find the connected components on the resulting subgraph. Any component of size greater than 1 that does not contain r and does contain a terminal, defines a violated SEC (4.73). In addition, if the component does not contain any terminals, then each vertex in the component that was not connected to the artificial vertex defines a violated SEC (4.74).

Figure 4.22 gives an example of a vector (a) that might have arisen during execution of either the price and cut or decompose-and-cut algorithms, along with a decomposition into a convex combination of two MSTs (b,c). In this figure, the artificial vertex is black, the terminals are gray and $r = 3$. By removing the artificial vertex, we easily find a violated SEC in the second spanning tree (c) with S equal to the marked component. This inequality is also violated by the optimal fractional solution, since $\hat{x}(E(S)) + \hat{x}(E(S \setminus T \mid \{0\})) = 3.5 > 3 = |S| - 1$. It should also be noted that the first spanning tree (b), in this case, is in fact feasible for the original problem.

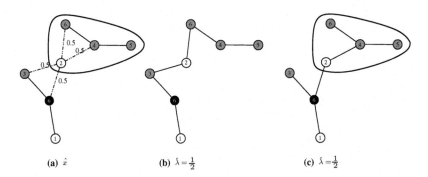

(a) $\hat{x}$ (b) $\hat{\lambda} = \frac{1}{2}$ (c) $\hat{\lambda} = \frac{1}{2}$

FIGURE 4.22
Example of a decomposition into minimum spanning trees.

4.9 Conclusions and Future Work

In this chapter, we presented a framework for integrating dynamic cut generation (outer methods) and traditional decomposition methods (inner methods) to yield new integrated methods that may produce bounds that are improved over those yielded by either technique alone. We showed the relationships between the various methods and how they can be viewed in terms of polyhedral intersection. We have also introduced the concept of structured separation and a related paradigm for the generation of improving inequalities based on decomposition and the separation of solutions to a relaxation. The next step in this research is to complete a computational study using the software framework introduced in Section 4.7 that will allow practitioners to make intelligent choices between the many possible variants we have discussed.

References

1. Aardal, K. and van Hoesel, S., Polyhedral techniques in combinatorial optimization, *Statistica Neerlandica*, 50, 3, 1996.
2. Ahuja, R., Magnanti, T., and Orlin, J., *Network Flows: Theory, Algorithms, and Applications*, Prentice Hall, Englewood Cliffs, NJ, 1993.
3. Applegate, D., Bixby, R., Chvátal, V., and Cook, W., TSP cuts which do not conform to the template paradigm, in *Computational Combinatorial Optimization*, pages 261–303, Springer, 2001.
4. Applegate, D., Bixby, R., Chvátal, V., and Cook, W., Implementing the Dantzig-Fulkerson-Johnson algorithm for large traveling salesman problems, Mathematical Programming, 97, 91, 2003.
5. Augerat, P. et al., Computational Results with a Branch and Cut Code for the Capacitated Vehicle Routing Problem, Technical Report 949-M, Université Joseph Fourier, Grenoble, France, 1995.
6. Balas, E. and Qi, L., Linear-time separation algorithms for the three-index assignment polytope, *Discrete Applied Mathematics*, 43, 1, 1993.
7. Balas, E. and Saltzman, M., Facets of the three-index assignment polytope, *Discrete Applied Mathematics*, 23, 201, 1989.
8. Balas, E. and Saltzman, M., An algorithm for the three-index assignment problem, *Operations Research*, 39, 150, 1991.
9. Barahona, F. and Anbil, R., The volume algorithm: Producing primal solutions with a subgradient method, *Mathematical Programming*, 87, 385, 2000.
10. Barahona, F. and Jensen, D., Plant location with minimum inventory, *Mathematical Programming*, 83, 101, 1998.
11. Barnhart, C., Hane, C. A., and Vance, P. H., Using branch-and-price-and-cut to solve origin-destination integer multi-commodity flow problems, *Operations Research*, 48, 318, 2000.
12. Barnhart, C., Johnson, E. L., Nemhauser, G. L., Savelsbergh, M. W. P., and Vance, P. H., Branch and price: Column generation for solving huge integer programs, *Operations Research*, 46, 316, 1998.

13. Beasley, J., A SST-based algorithm for the steiner problem in graphs, *Networks*, 19, 1, 1989.
14. Beasley, J., Lagrangean relaxation, in *Modern Heuristic Techniques for Combinatorial Optimization*, edited by Reeves, C., Wiley, 1993.
15. Belloni, A. and Lucena, A., A Lagrangian heuristic for the linear ordering problem, in *Metaheuristics: Computer Decision-Making*, edited by Resende, M. and de Sousa, J. P., pages 123–151, Kluwer Academic, 2003.
16. Calheiros, F. C., Lucena, A., and de Souza, C., Optimal rectangular partitions, *Networks*, 41, 51, 2003.
17. Caprara, A., Fischetti, M., and Toth, P., Algorithms for the set covering problem, *Annals of Operations Research*, 98, 353, 2000.
18. Carraresi, P., Frangioni, A., and Nonato, M., Applying bundle methods to optimization of polyhedral functions: An applications-oriented development, Technical Report TR-96-17, Universitá di Pisa, 1996.
19. Dantzig, G. and Wolfe, P., Decomposition principle for linear programs, *Operations Research*, 8, 101, 1960.
20. Danzig, G. and Ramser, R., The truck dispatching problem, *Management Science*, 6, 80, 1959.
21. de Aragão, M. P. and Uchoa, E., Integer program reformulation for robust branch-and-cut-and-price, Working paper, Pontifíca Universidade Católica do Rio de Janeiro, 2004, Available from http://www.inf.puc-rio.br/~uchoa/doc/rbcp-a.pdf.
22. Fisher, M., The Lagrangian relaxation method for solving integer programming problems, *Management Science*, 27, 1, 1981.
23. Fisher, M., Optimal solution of vehicle routing problems using minimum k-trees, *Operations Research*, 42, 626, 1994.
24. Fukasawa, R., de Aragão, M. P., Reis, M., and Uchoa, E., Robust branch-and-cut-and-price for the capacitated minimum spanning tree problem, in *Proceedings of the International Network Optimization Conference*, pages 231–236, Evry, France, 2003.
25. Galati, M., DECOMP user's manual, Technical report, Lehigh University Industrial and Systems Engineering, 2005.
26. Garey, M. R. and Johnson, D. S., *Computers and Intractability: A Guide to the Theory of NP-Completeness*, W. H. Freeman and Company, New York, 1979.
27. Goemans, M., The steiner tree polytope and related polyhedra, *Mathematical Programming*, 63, 157, 1994.
28. Goffin, J. L. and Vial, J. P., Convex nondifferentiable optimization: A survey focused on the analytic center cutting plane method, Technical Report 99.02, Logilab, Geneva, Switzerland, 1999.
29. Gondzio, J. and Sarkissian, R., Column generation with a primal-dual method, Technical report, University of Geneva, Logilab, HEC Geneva, 1996.
30. Grötschel, M. and Padberg, M., On the symmetric travelling salesman problem I: Inequalities, *Mathematical Programming*, 16, 265, 1979.
31. Grötschel, M. and Padberg, M., On the symmetric travelling salesman problem II: Lifting theorems and facets, *Mathematical Programming*, 16, 281, 1979.
32. Grötschel, M. and Lovász, L. and Schrijver, A., The ellipsoid method and its consequences in combinatorial optimization, *Combinatorica*, 1, 169, 1981.
33. Guignard, M., Efficient cuts in Lagrangean "relax-and-cut" schemes, *European Journal of Operational Research*, 105, 216, 1998.
34. Guignard, M., Lagrangean relaxation, *Top*, 11, 151, 2003.

35. Held, M. and Karp, R. M., The traveling salesman problem and minimum spanning trees, *Operations Research*, 18, 1138, 1970.
36. Huisman, D., Jans, R., Peeters, M., and Wagelmans, A., Combining column generation and Lagrangian relaxation, Technical report, Erasmus Research Institute of Management, Rotterdamn, The Netherlands, 2003.
37. Hunting, M., Faigle, U., and Kern, W., A Lagrangian relaxation approach to the edge-weighted clique problem, *European Journal of Operational Research*, 131, 119, 2001.
38. J. Lysgaard, A. L. and Eglese, R., A new branch-and-cut algorithm for the capacitated vehicle routing problem, *Mathematical Programming*, 100, 423, 2004.
39. Kohl, N., Desrosiers, J., Madsen, O., Solomon, M., and Soumis, F., 2-path cuts for the vehicle routing problem with time windows, *Transportation Science*, 33, 101, 1999.
40. Kopman, L., *A New Generic Separation Routine and Its Application In a Branch and Cut Algorithm for the Capacitated Vehicle Routing Problem*, PhD thesis, Cornell University, 1999.
41. Laporte, G., Nobert, Y., and Desrouchers, M., Optimal routing with capacity and distance restrictions, *Operations Research*, 33, 1050, 1985.
42. Linderoth, J. and Galati, M., Knapsack constrained circuit problem, Unpublished working paper, 2004.
43. Lübbecke, M. E. and Desrosiers, J., Selected topics in column generation, Technical Report 008-2004, Technische Universität Berlin, 2004.
44. Lucena, A., Tight bounds for the Steiner problem in graphs, Talk given at TIMS-XXX-SOBRAPO XXIII Joint International Meeting, Rio de Janeiro, 1991.
45. Lucena, A., Steiner problem in graphs: Lagrangian relaxation and cutting planes, COAL Bulletin, 28, 2, 1992.
46. Lucena, A., Nondelayed relax-and-cut algorithms, Working paper, Departmaneto de Administraç ao, Universidade Federal do Rio de Janeiro, 2004.
47. Margot, F., Prodon, A., and Liebling, T., Tree polyhedron on 2-tree, *Mathematical Programming*, 63, 183, 1994.
48. Martinhon, C., Lucena, A., and Maculan, N., Stronger k-tree relaxations for the vehicle routing problem, *European Journal of Operational Research*, 158, 56, 2004.
49. Miller, D., A matching based exact algorithm for capacitated vehicle routing problems, *ORSA Journal on Computing*, 7, 1, 1995.
50. Müller-Hannemann, M. and Schwartz, A., Implementing weighted b-matching algorithms: Toward a flexible software design, in *Proceedings of the Workshop on Algorithm Engineering and Experimentation (ALENEX99)*, volume 1619 of *Lecture notes in Computer Science*, pages 18–36, Baltimore, MD, 1999, Springer-Verlag.
51. Naddef, D. and Rinaldi, G., Branch-and-cut algorithms for the capacitated VRP, in *The Vehicle Routing Problem*, edited by Toth, P. and Vigo, D., pages 53–84, SIAM, 2002.
52. Nemhauser, G. and Wolsey, L., *Integer and Combinatorial Optimization*, Wiley, New York, 1988.
53. Padberg, M. and Rinaldi, G., An efficient algorithm for the minimum capacity cut problem, *Mathematical Programming*, 47, 19, 1990.
54. Pigatti, A., *Modelos e algoritmos para o problema de aloção generalizada e aplicações*, PhD thesis, Pontifícia Universidade Católica do Rio de Janeiro, 2003.
55. Prim, R., Shortest connection networks and some generalizations, *Bell System Technical Journal*, 36, 1389, 1957.

56. Ralphs, T., *Parallel Branch and Cut for Vehicle Routing*, PhD thesis, Cornell University, 1995.
57. Ralphs, T., Kopman, L., Pulleyblank, W., and Trotter Jr., L., On the capacitated vehicle routing problem, *Mathematical Programming*, 94, 343, 2003.
58. Rousseau, L. M., Gendreau, M., and Feillet, D., Interior point stabilization for column generation, Working paper, Cahier du Gerad, 2003, Available from http://www.lia.univ-avignon.fr/fich_art/380-IPS.pdf.
59. V. Chvátal, *Linear Programming*, W.H. Freeman and Company, 1983.
60. van den Akker, J., Hurkens, C., and Savelsbergh, M., Time-indexed formulations for machine scheduling problems: Column generation, *INFORMS Journal on Computing*, 12, 111, 2000.
61. Vanderbeck, F., Lot-sizing with start-up times, *Management Science*, 44, 1409, 1998.
62. Wei, G. and Yu, G., An improved $O(n^2 \log n)$ algorithm for the degree-constrained minimum k-tree problem, Technical report, The University of Texas at Austin, Center for Management of Operations and Logistics, 1995.

5

Airline Scheduling Models and Solution Algorithms for the Temporary Closure of Airports

Shangyao Yan and Chung-Gee Lin

CONTENTS

5.1   Introduction

The temporary closure of airports usually results in significant perturbations of flight schedules. For example, typhoons or periods of heavy rain happen from time to time in Taiwan each year. The airports usually have to be temporarily closed, and many flights are then canceled or delayed. The poor scheduling of flights, or an entire fleet, may result in a substantial loss of profit and decreased levels of service for airline carriers. Thus, effective and efficient flight/fleet incident management is important for carriers in order to regain their normal services as soon as possible following a schedule perturbation.

The current process for handling the schedule perturbations from a temporary closure of airports for the carriers in Taiwan is inefficient and ineffective from a system perspective, especially for large flight networks. The typical process is as follows. When an airport is closed temporarily, a scheduling group in the System Operations Control department (SOC) starts by choosing a suitable strategy, the expected time for the airport to open, the regular planned schedule, projected (with booked) demand on all flights and the allocation of the currently available airplanes. The detailed flight/fleet schedule is then adjusted by a trial-and-error process until a feasible solution is found. For example, if a flight cannot be served because the airport is closed, then it may be delayed until the airport is reopened, or it may be canceled if it is delayed too long. For another example, if a flight cannot be served by its scheduled aircraft (probably due to the closure of an upstream airport), then the flight may be suitably delayed so that a holding aircraft (or an incoming aircraft) can be rescheduled to serve this flight; however, if there is no holding aircraft (or no incoming aircraft), then the flight may be canceled, or served by an aircraft obtained from another station using a ferry flight. Such a process generally involves a series of local adjustments (usually by hand) of the aircraft routes and related flights. When the network size grows, this process becomes inefficient for finding the optimal solution. Typically, only a feasible solution is obtained under the real time constraint. The drafted schedule is then shuttled to other groups in the SOC for the application of other constraints (for example crew constraints or maintenance constraints). The schedule will be executed if proves feasible; otherwise, it will be returned to the scheduling group for schedule revisions. The process is repeated until all groups are satisfied with the revised schedule.

Due to deregulation, the flight networks for carriers in Taiwan have recently grown. In particular, to enhance operations, carriers have been purchasing different types of aircraft suitable for different flight mileages to form multi-fleet operations, in order to improve their profit. If the schedule is perturbed following accidents, in a multi-fleet operation airplanes of different types can support each other through new routing with a temporary flight schedule. For example; a) some idle larger-sized aircraft can serve flights scheduled for smaller-sized aircraft, b) some idle smaller-sized aircraft can serve flights scheduled for larger-sized aircraft, if passengers not able to board the smaller-sized aircraft are allowed to cancel, or can be reaccommodated on suitable flights or alternate modes to the same destinations. For another example, some flights could be delayed so that other-sized aircraft can be rescheduled to serve these flights if this is profitable from the system aspect. Since a single-fleet scheduling model may not be adequate to evaluate such schedule perturbations, it would be helpful for carriers to have multi-fleet models to handle schedule perturbations both efficiently and effectively and thus to reduce losses resulting from airport closures.

As a result of growing size of fleet, it is increasingly difficult for the traditional approach to adequately handle schedule incidents. It would be helpful for carriers to have systematic computerized models to handle schedule perturbations efficiently and effectively, so as to reduce losses resulting from airport closures. The past studies on airline schedule perturbations can typically be classified into two topics: one, the shortage of aircraft and the other, the closure of airports (Yan and Lin [22]).

For schedule perturbations on aircraft shortages, relatively more research has been done. In particular, Deckwitz [5] introduced "positioning flights" and "sliding flights". The former denotes a relocation of some reserved airplanes to serve certain flights. The latter indicates the delay of a flight's departure time within a slot time to better adjust that flight's connections. Jedlinsky [11] applied a minimum cost network flow problem to solve schedule perturbation using the out-of-kilter algorithm. Etschmaier and Rothstein [6] built a simulation model for evaluating flight punctuality. Teodorovic and Guberinic [17] developed a nonlinear integer model by setting the objective to be a minimization of the total passenger delay. Gershkoff [9] used a time-space framework to formulate fleet routing and applied a successive shortest path method to find arc chains for the cancellation of a series of flights because of a shortage of aircraft. Teodorovic and Stojkovic [18] developed a greedy heuristic for solving a goal programming problem. Given some perturbations in the flight schedule, the heuristic finds the new set of aircraft routings that first minimizes the number of cancellations, and then minimizes the overall passenger delays. Jarrah et al. [10] proposed two minimum cost network flow models to systematically adjust aircraft routing and flight scheduling in real time, to minimize the total cost incurred from a shortage of aircraft. The "delay model" applied flight delays and aircraft swaps at a station, and the "cancellation model" applied flight cancellations and aircraft swaps among stations to handle aircraft routing and flight scheduling. Yan and Yang [24] have incorporated flight cancellations, delays and ferry flights into four time-space network models in order to resolve schedule recovery problems arising from a temporary shortage of aircraft, for single-fleet and nonstop flight operations. Yan and Tu [23] have extended Yan and Yang's models to multiple-fleet and multiple-stop flight operations. Thengvall et al. [19] have extended the work of Yan et al. [22–24] to the building of recovery schedules by the inclusion of an incentive in the formulation that will minimize any deviation from the original aircraft routings.

For schedule perturbations on airport closures, Jedlinsky [11] has introduced a network using the local modification of links to handle schedule perturbations resulting from the closure of an airport. The modified network was formulated as a minimum cost network flow problem which was solved using the out-of-kilter algorithm. Teodorovic [15] has developed a model for designing the most meteorologically reliable airline schedule, which however, was not focused on schedule perturbations to execute a planned airline schedule. Yan and Lin [22] have used time-space networks to develop eight scheduling models for single-fleet and multiple-stop operations dealing with the temporary closure of airports. Thengvall et al. [20] have incorporated flight cancellations, delays, ferry flights and substitutions between fleets and sub-fleets, to formulate three multi-commodity network models for the large carriers that have to deal with a hub closure. Compared to the aircraft shortage literature, not much research has been done on schedule perturbations resulting from the temporary closure of airports. Moreover, most of the former research on airline schedule perturbations was focused on the operations of a single fleet.

To improve the traditional scheduling process, this research modifies the process by solving the scheduling problem using several strategic models which combine the traditional scheduling rules in a systematic scheme.

This research aims to develop a set of models to help carriers handle schedule perturbations resulting from the temporary closure of airports. A basic model is first constructed as a multiple time-space network, from which several perturbed network models are developed for rescheduling following incidents. The scope of this research is confined to the operations of a multiple fleet as well as one-stop and nonstop flights. The operations involved in other types of multi-stop flights and the schedule perturbation caused by other types of incidents are left to be addressed in the future. In addition, this research focuses on a case where an airport is temporarily closed. Although the model developed in this research needs to be modified to handle cases where more than one airport is closed at the same time, we believe modification to be easy, a subject also left for future research.

Although there is a significant interdependence between the airline schedule design process and aircraft maintenance, as well as the crew scheduling processes, these processes are usually separated in order to facilitate problem solving (Teodorovic, [16]). This research thus excludes the constraints of aircraft maintenance and crew scheduling in the modeling. In practice, the fleet routes and flight schedules obtained from these models can serve as a good initial solution for the minor modification of these constraints. However, actually incorporating these constraints into a complicated framework would be a topic for future research.

The rest of this chapter is organized as follows: first, we formulate our models, including a basic model and several strategic models. These models are formulated as integer programs and their solutions are developed hereafter. A case study is then performed to test the models in the real world. Finally, we offer conclusions.

5.2 The Basic Model

In practice, most carriers' primary objective for adjusting schedule perturbation following incidents is to resume their normal services as soon as possible in order to maintain their levels of service. Given the time length of a schedule perturbation, carriers typically aim to adjust a schedule to minimize the loss of system profit. Thus, the basic model is designed to minimize the schedule-perturbed period after an incident and to obtain the most profitable schedule given the schedule-perturbed period.

The basic model contains several fleet networks, each formulating a single fleet operation in the schedule-perturbed period. This research suggests using a time-space network, as shown in Figure 5.1, to formulate a single fleet operation in the basic mode, because it is natural to represent conveyance routings in the dimensions of time and space (see, for example, Yan et al. [21]).

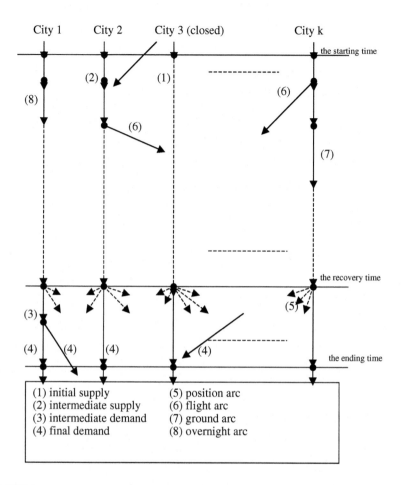

FIGURE 5.1
The single-fleet basic model.

To develop strategic models based on the basic model, only a few nodes, links, or additional side constraints need to be modified, without changing the original network structure. In Figure 5.1, the horizontal axis represents airport locations, while the vertical axis represents the time duration. Each node represents a specific airport at a specific time. Each arc represents an activity for an airplane. There are four types of arcs described below:

Flight arc: representing a nonstop flight or a one-stop flight. The time window for a nonstop flight is calculated as a block of time, from the time when an airplane begins to prepare for this flight to the time when the airplane finishes the flight and is ready for the next flight service or holding. If the flight is a multiple-stop flight, then the time block is from the beginning of the first leg to the end of the last leg. The time block includes the time for investigating aircraft before departure, fueling, passenger/baggage boarding and deplaning, and flight time in the air. The arc cost equals the flight cost minus passenger revenues. The arc flow upper bound is one, meaning that at most one airplane can serve this flight. The arc flow lower bound is zero, indicating that this flight can be canceled. Additional charges for canceling flights can be incurred; for example, to reaccommodate the passengers of canceled flights on suitable flights or alternate modes to the same destinations. To reflect such charges, we set the cost function for a flight from i to j as $Cc_{ij}^n + Cd_{ij}^n * X_{ij}^n$, where X_{ij}^n is the flow on arc (i, j) in the nth fleet network, which is equal to zero or one, Cc_{ij}^n is the cancellation cost and Cd_{ij}^n equals the flight arc cost minus the cancellation cost, $C_{ij}^n - Cc_{ij}^n$. Therefore, if the flight (i, j) is served (that is, $X_{ij}^n = 1$), then the cost is C_{ij}^n; otherwise (that is, $X_{ij}^n = 0$), the cost is Cc_{ij}^n. Note that for flight arcs connected to the closed airport within the unavailable period, their upper bounds are set as zero, meaning that these flights are canceled unless other strategies are applied to modify the flights to be served.

Ground arc: representing the holding of airplanes at an airport in a time window. The time window lies between two adjacent flight event times (landing/taking-off) at an airport. The arc cost represents the expenses incurred by holding an airplane at the airport in the time window, including the airport tax, airport holding charges and gate use charges. The arc flow upper bound is equal to the apron capacity, representing the greatest number of airplanes that can be held at this airport during this time window. The arc flow lower bound is zero, meaning that a minimum of zero airplanes are held at this airport in this time window.

Overnight arc: representing the holding of airplanes overnight at an airport. The time window is set for the overnight duration between two consecutive days. The arc cost is the cost for an airplane held overnight, which is similar to that of a ground arc with an additional overnight charge. The arc flow upper bound and lower bound are set to be the same as those of the ground arcs.

Position arc: representing a ferry flight between two airports. This is similar to a flight arc; except that the arc cost excludes passenger revenue, and the block time excludes passenger boarding/deplaning time and package handling time, and the arc flow upper bound is equal to the airport departure capacity.

The network contains a perturbed period from a starting time to an ending time. The recovery time indicates the time that the closed airport is opened. The starting time is when an airport is closed. At the starting time all airplanes located at airports or in the air, are set as initial or intermediate node supplies. The ending time is when the fleet resumes its normal operations. The ending time is determined as follows. In order to ensure a feasible solution in the designed network and to minimize the perturbed period, position arcs are added from the nodes when the closed airport is opened. Through these position arcs, the airplanes can be at least relocated to suitable airports by ferry flights. Thus the fleet can resume its regular service. In order to ensure that all airplanes are relocated in time to the destination of such ferry flights, we set the ending time to be the latter of the following two times: the last arrival time of all the position arcs and the last arrival time of flights which depart earlier than the times their corresponding ferry flights arrive. Final or intermediate node demands are set according to the normal fleet allocation at airports, or in the air, at the ending time. Note that if more than one airport is closed at the same time, the ending time can be set using the same method, except it is the last one calculated among them all. This research however, assumes only one closed airport at a time for simplification. It should also be noted that the airport authority might possibly update estimates of an airport's opening time. For this reason, the models should be dynamically applied. In particular, a new network should be updated in such a case, a new starting time and a new ending time. After solving the new network, the schedule should be revised again.

To formulate the basic model involving multi-fleet operations, we use a multiple time-space network (as shown in Figure 5.2) assuming there are three fleets in the operation. Each fleet network in Figure 5.2 is designed the same as that in Figure 5.1. Note that the ending time of the system is the last ending time calculated in each fleet network.

The integer program formulating the basic model is shown below:

$$\text{Min } Z = \sum_{n \in M} \left(\sum_{i,j \in A^n \setminus F^n} C_{ij}^n X_{ij}^n + \sum_{i,j \in F^n} C_{dij}^n X_{ij}^n + \sum_{i,j \in F^n} C_{cij}^n \right) \quad (5.1)$$

$$\text{s.t.} \sum_{j \in O^n(i)} X_{ij}^n - \sum_{k \in I^n(i)} X_{ki}^n = b_i^n, \quad \forall i \in N, \forall n \in M \quad (5.2)$$

$$0 \le X_{ij}^n \le U_{ij}^n, \quad \forall (i, j) \in A, \quad \forall n \in M \quad (5.3)$$

$$X_{ij}^n \in I, \quad \forall (i, j) \in A, \quad \forall n \in M \quad (5.4)$$

where:

n :	the nth fleet
M :	the set of all fleets
N^n :	the set of all nodes in the nth fleet network
A^n :	the set of all arcs in the nth fleet network
F^n :	the set of all flight arcs in the nth fleet network
$O^n(i)$:	the set of head nodes for arcs emanating from node i in the nth fleet network
$I^n(i)$:	the set of tail nodes for arcs pointing into node i in the nth fleet network
b_i^n :	the node supply/demand of node i in the nth fleet network
$C_{ij}^n, X_{ij}^n, U_{ij}^n$:	arc (i, j) cost, flow and flow upper bound, respectively in the nth fleet network
C_{cij}^n :	the cancellation cost for flight (i, j) in the nth fleet network
C_{dij}^n :	equal to $C_{ij}^n - C_{cij}^n$

The objective of this model is to "flow" all node supplies to all node demands in each network at a minimum cost. Since passenger revenues are formulated as negative costs, the objective is equivalent to a maximization of

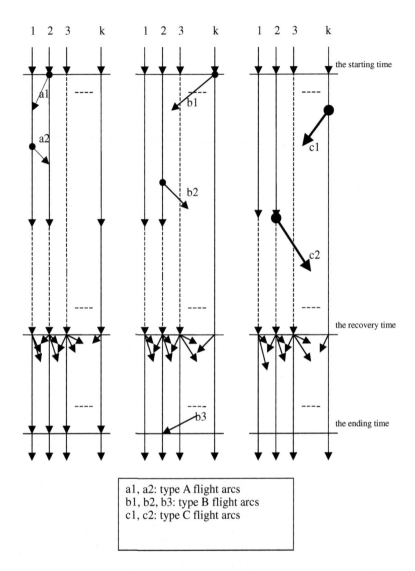

al, a2: type A flight arcs
b1, b2, b3: type B flight arcs
c1, c2: type C flight arcs

FIGURE 5.2
The multi-fleet basic model.

the total system profit. Constraint (5.1) is the flow conservation constraint, at every node, in each fleet network. Constraint (5.2) ensures that all arc flows are within their upper and lower bounds in each fleet network, and constraint (5.3) ensures that all arc flows are integers in each fleet network. Note that the basic model can be used to evaluate the cancellation of flights. When $n = 1$, the integer program formulates a single-fleet scheduling model. As $n \geq 2$, the integer program describes a multi-fleet scheduling model. Since any fleet network is independent of the other fleet networks, the basic model can be characterized as a single-fleet scheduling model.

5.3 The Strategic Models

To make the basic model more useful, four practical scheduling rules are incorporated to develop strategic models, in particular, (a) the swap of aircraft types (b) flight delays, (c) the modification of one-stop flights, and (d) the ferrying of idle aircraft. Note that the first rule can only be applied to multi-fleet scheduling. Although the last three rules are designed for single-fleet scheduling, with the combination of the first rule, the last three rules can also be applied to multi-fleet scheduling. Carriers may create different models based on the basic model and the selected scheduling rules, and then choose the best one for application. Modifications for these on the network are described below.

5.3.1 Rule (a) The Swap of Aircraft Types

An example of the modifications for this rule is shown in Figure 5.3; assume that there are three types of aircraft in operation where the capacity of type A is the smallest and that of type C is the largest. Since larger aircraft can serve smaller-type flights, smaller-type flight arcs can be added into larger-type fleet networks. For example, as shown in Figure 5.3, type A flight arcs (e.g., "a1", "a2", and "a3") can be added into the type B and type C networks. Type B flight arcs (e.g., "b1", "b2", and "b3") can be added into the type C network. Some of the type B flight arcs (e.g., "b1") may be added into the type A network. If passengers not able to get on the type A aircraft (because of fewer seats), they can be reaccommodated on other suitable flights or alternate modes to the same destinations. Similarly, some type C flight arcs (e.g., "c1" and "c2") may be added into the type A or type B network. Note that, similar to flight cancellations of Yan and Lin [22], additional charges may be incurred for reaccommodating passengers who are unable to get on the smaller-sized aircraft, on suitable flights or alternate modes to the same destinations. Also note that, when using another type of aircraft to serve a flight, this type of aircraft should be feasible in terms of the flight mileage and the associated airport facilities.

It should be mentioned that, if flight swaps happen when using another type of aircraft to serve flights, then swap costs (for example, the additional costs of switching gates or switching crew members) should be included when modeling. After all, the smaller-type flight arc cost in the larger-type fleet network (e.g., "a1" in the type C fleet network) equals the larger-type of aircraft's flight expenses, plus the swap cost, minus the on-board passenger revenue. Similarly, the larger-type flight arc cost in the smaller-type fleet network (e.g., "b1" in the type A fleet network) equals the smaller-type of aircraft's flight expenses, plus the swap cost, minus the on-board passenger revenue, plus the additional charges acquired for passengers not getting on the smaller-type aircraft.

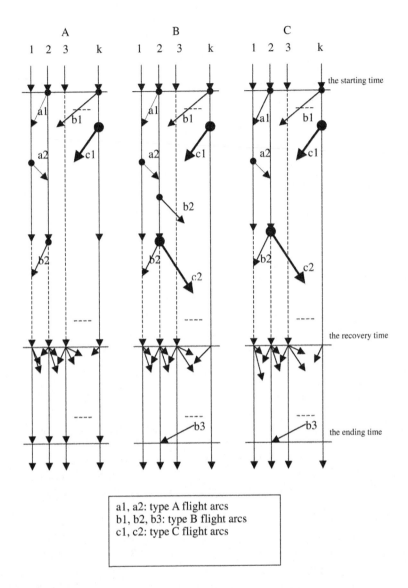

al, a2: type A flight arcs
bl, b2, b3: type B flight arcs
cl, c2: type C flight arcs

FIGURE 5.3
Network modifications for the swap of aircraft types.

5.3.2 Rule (b) Flight Delays

For the evaluation of flight delays, several alternate flight arcs (also called sliding arcs in Deckwitz [5]) are added, as shown in Figure 5.4, with respect to a flight arc, to formulate a choice of delays. Parameters for these sliding arcs are set similar to those of flight arcs with additional delay costs and lost passenger revenues. For example, as shown in Figure 5.4, "r" denotes the set of the rth flight arc and its alternate flight arcs (two dashed arcs). If one dashed

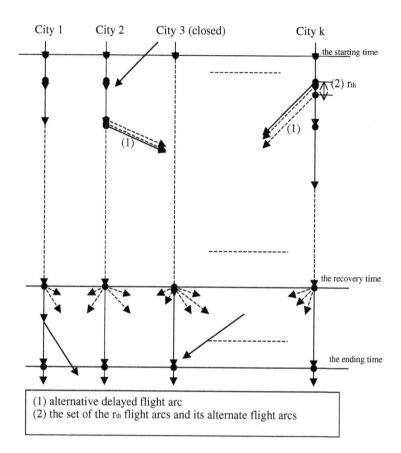

City 1   City 2   City 3 (closed)   City k

the starting time

(2) r_th

(1)

(1)

the recovery time

the ending time

(1) alternative delayed flight arc
(2) the set of the r_th flight arcs and its alternate flight arcs

FIGURE 5.4
Network modifications for flight delays.

arc contains a flow at the optimal solution, it means that the departure time of the flight should be slid to a later time. As at most one departure time is assigned for a flight, a side constraint should be introduced for a flight arc and its sliding arcs. Assuming that X_{ij}^n is the flow in the flight arc (i, j) in the nth fleet network and that D_r^n is the rth set of bundle arcs including the rth flight arc and its sliding arcs in the nth fleet network, then all of the side constraints for the fleet are $\sum_n \sum_{i, j \in D_r} X_{ij}^n \leq 1, \forall r$. Note that basically users can set full sliding arcs within the slot time for every flight. However, based on both needs and experience, for example, considering the level of service and the constraints on slot times or solution efficiency, a number (representing the maximum allowable delay) may be chosen for setting sliding arcs for each flight or partial flights. Certainly, the more sliding arcs that are added, the better the results that are expected, while the larger the problem size grows. A tradeoff between system profit and model computation time may be made in practice.

5.3.3 Rule (c) Modification of Multi-Stop Flights

Some uneconomic segments in a multi-stop flight may be suitably canceled to improve system profits when the aircraft are not enough for service, due to a schedule perturbation. To evaluate the cancellation of one-stop flight segments, this research suggests an extension of the modeling technique by Simpson [14]. As shown in Figure 5.5, X_{ij}^n, X_{jk}^n, and X_{ik}^n stand for a nonstop flight from i to j, from j to k, or from i to k respectively in the nth fleet network. X_{ijk}^n stands for a one-stop flight from i to j to k in the nth fleet network. In the technique proposed by Simpson [14], X_{ik}^n is an addition. As introduced by Simpson [14], two nonstop segments (X_{ij}^n and X_{jk}^n) are bridged by a one-stop arc (X_{ijk}^n). A dummy node (a or b) is placed on flight (i, j) and flight (j, k). This construction allows either (i, j) or (j, k) segments to be operated independently as nonstop flights. Any previous arrivals at j cannot use the attractive "one-stop" arc joining the two flights. There are two cost arcs created for use if the flights operate independently. If the flights are connected, the X_{ij}^n flow directly transfers to the X_{jk}^n flow and receives an additional benefit of R_{ik}^n for doing so. This model, however, does not consider the passenger demand from i to k.

To evaluate the decomposition of a one-stop flight into complete nonstop flights, this research introduces another nonstop segment, X_{ik}^n, and a side constraint. Firstly, since the flight segment from i to k can be chosen at most once, that is, the sum of X_{ik}^n and X_{ijk}^n can be at most one, an additional constraint, $\sum_n (X_{ik}^n + X_{ijk}^n) \leq 1$, should be added.

If X_{ik}^n has flows (should be 1), then X_{ijk}^n should be zero, meaning that the one-stop flight from i to j to k is not served. Similarly, if X_{ijk}^n has flows (should be 1), then X_{ik}^n should be zero, meaning that the passenger demand from i to k

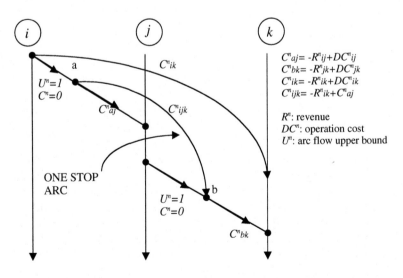

FIGURE 5.5
Network modifications for a one-stop flight arc.

has been served by the one-stop flight i to j to k, and there is no need for providing the nonstop flight from i to k. Note that the arc cost of X_{ik}^n is the flight cost minus the passenger revenue obtained on the nonstop flight from i to k; all other arc costs are the same as those mentioned in Simpson [14]. Similar to rule (b), users can evaluate all or part of the one-stop flights for their needs in practice. To reduce the model's complexity, this research focuses on the scheduling of nonstop and one-stop flights. Modifications for two-stop or more-stop flights can be considered in future research.

5.3.4 Rule (d) Ferrying of Idle Aircraft

This rule uses ferry flights to relocate idle airplanes of the same type to where and when the system needs them for the best routing. As shown in Figure 5.6, position arcs with ferry flight costs are systematically added. For example, if there is a flow in the arc (b, s), it means that an aircraft should be relocated,

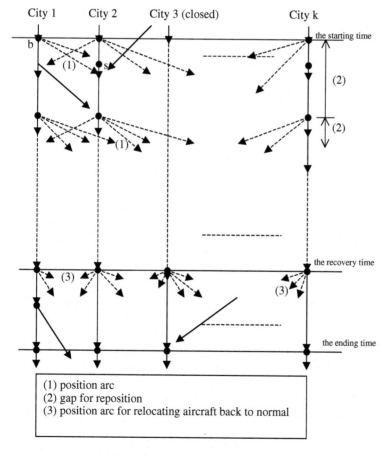

FIGURE 5.6
Network modifications for the ferrying of idle aircraft.

via a ferry flight from city 1 to city 2, when the perturbation begins. Note that ferry arcs should not be connected to the airport when it is closed. As in rules (b) and (c), users can add full position arcs into the network, or, they may add position arcs whenever and wherever they feel it to be suitable for the possible relocation of airplanes. The model will then determine the optimal solution. Obviously, the more position arcs that are added, the better the effects of the ferry flights expected, while the larger the network size grows. A tradeoff between system profit and model computation time may be made in practice. No additional side constraints, but additional arcs are generated in this modification.

Although modifications of the basic model for rules (b), (c), and (d) in each fleet network are referred to in Yan and Lin [22], if rule (a) is applied together with rule (b), then, because at most one departure time is assigned for a flight, a side constraint should be introduced for a flight arc and its alternate flight arcs among all associated fleet networks to ensure that at most one flight is served. Similarly, if rule (a) is applied together with rule (c), then the additional side constraint mentioned in Yan and Lin [22] should be extended across all the associated fleet networks. Besides, to ensure that each nonstop flight (which is modified from a one-stop flight) is served at most once, a side constraint should be introduced for each nonstop flight across all the associated fleet networks.

5.4 Solution Methods

The aforementioned models are formulated as pure network flow problems, network flow problems with side constraints, or multiple commodity network flow problems. As in the basic model, a strategic model, not including rule (a), can be characterized as a single-fleet scheduling model, because any fleet network is independent of the other fleet networks. Thus, the basic model, incorporating rules (b), (c), or (d), can be formulated as pure network flow problems or network flow problems with side constraints. Moreover, the basic model incorporated rule (a) and any of rules (b), (c), and (d) can be formulated as multi-commodity network flow problems. These models are designed to have optimal solutions, so that in actual applications they do not generate infeasible or unbounded solutions (Yan and Lin [22]). Numerous methods are capable of solving the pure network flow problem, including the out-of-kilter algorithm, the network simplex method, the cost scaling algorithm and the right-hand-side scaling algorithm (Ahuja et al. [1]). This research suggests using the network simplex method to solve pure network flow problems due to its demonstrated efficiency, see Ahuja et al. [1] or Kennington and Helgason [12].

Since the network flow problems with side constraints and the multiple commodity network flow problems are characterized as NP-hard, (see Garey and Johnson [8]), traditional exact integer solution techniques are applicable, for example, the branch-and-bound or the cutting planes. This research, how-ever, suggests using a Lagrangian relaxation with subgradient methods (LRS) for an approximation of the near-optimal solutions in order to avoid a long computation time for solving realistically large problems.

LRS is known for its fast convergence and efficient allocation of memory space (Fisher [7]). The LRS process is addressed as follows: By dualizing the side constraints or the bundle constraints with a Lagrangian multiplier, we produce a Lagrangian problem that is easy to solve and whose optimal objective is a lower bound (for minimization problems) on the optimal value of the original problem. We then use a Lagrangian heuristic to find a feasible solution whose objective is an upper bound (for minimization problems) on the optimal value of the original problem. To reduce the gap between the upper bound and the lower bound, we modify the Lagrangian multiplier using a subgradient method, then solve for another lower and upper bound. Better bounds are updated. The process is repeated until the gap is converged to within 0.1% of error.

Since the proposed network flow problem with side constraints is a special case of the proposed multiple commodity network flow problem, we intro-duce the LRS on the basis of the latter problem. The detailed application of the LRS in this research is summarized below in three parts.

(5.1) A lower bound for each iteration

The side constraints are relaxed by a nonnegative Lagrangian multiplier and are added to the objective function, resulting in a Lagrangian program which contains several independent pure network flow problems. Each pure net-work flow problem can be solved using the network simplex method and the sum of the optimal objective values can be proven to be the lower bound of the original problem (Fisher [7]).

(5.2) An upper bound for each iteration

A Lagrangian heuristic is developed to find a feasible solution (an upper bound of the optimal solution), from the lower bound solution (typically an infeasible solution), for each iteration. If a certain side constraint is violated, there should be at least two flight arcs in this set of side arcs with positive flow, which is equal to one. We choose the arc from among them with the largest arc cost (after being modified by LR), to reduce its flow from one to zero. In order to maintain the flow conservation on the arc's tail and head, we find the least cost flow augmenting path from this arc's tail to the arc's head in the associated fleet network by a modified label correcting algorithm (Powell [13]), and augment a unit of flow through the path; thus, we reduce it by one unit of bundle flow, surely increasing the objective value. Any arc flow involved with side constraints cannot be increased during the flow augmentation to prevent the violation of other side constraints. If the side constraint is not yet satisfied, we choose another arc with the next highest cost from this set to reduce by another unit of bundle flow until this side constraint is satisfied. If all the side constraints are scanned and modified to become feasible, then we have found a feasible flow. Otherwise, we discard the upper bound from this iteration. Such a situation, however, was not found in the case study presented in Section 5.5. Note that since the networks are designed to have feasible solutions, we can find an initial upper bound (corresponding to the initial feasible solution) at the beginning of the algorithm.

(5.3) Solution process:

The solution steps are listed below:

Step 0: Set the initial Lagrangian multipliers.
Step 1: Solve the Lagrangian problem optimally using the network simplex method to get a lower bound. Update the lower bound.
Step 2: Apply the Lagrangian heuristic to find an upper bound and update the upper bound.
Step 3: If the gap between the lower bound and the upper bound is within a 0.1% error, stop the algorithm.
Step 4: Adjust the Lagrangian multipliers using the subgradient method (Camerini et al. [2]).
Step 5: Set $k = k + 1$. Go to Step 1.

Because the solutions mentioned above are fleet flows and are incapable of expressing the route of each airplane, we suggest using a flow decomposition method (Yan and Lin [22]) to decompose the link flows in each fleet network into arc chains, with each arc chain representing an airplane's route with the associated aircraft type in the perturbed period. If some airplanes' routes are different from their original ones, flight swaps of the same aircraft type should be done for these aircraft. Note that according to a major Taiwan airline carrier, the swap cost for aircraft of the same type in its operations is small and can be omitted. Thus a suitable aircraft swap does not cause deterioration in the system performance. Also note that the arc chains may not be unique. When applying the models, the SOC scheduling group may solve several arc chain patterns, then send them to other groups for the application of aircraft maintenance and crew scheduling constraints. Hence, there will be more options for the SOC to choose the best routing and a satisfactory solution for all groups could be relatively easy to find. Even if all the patterns are infeasible, then instead of revising only one schedule, as is done currently, several alternate patterns with the same best profit would provide more flexibility for schedule revisions. Thus, both the scheduling process and the quality of a schedule could be improved. However, more practical arc chains can be studied in the future in order to meet the regulations for aircraft maintenance or crew scheduling.

5.5 Case Study

To test how well the models may be applied in real world, we demonstrate a case study based on data from a major Taiwan airline's international operations (China Airlines [3]). There are 24 cities involved in its operations.

The flight timetable, used for testing, is rotated once a week and includes 273 flights. About 20 percent of them are one-stop flights; the others are non-stop flights. There are several types of aircraft involved in its operations, including B737's, AB3's, AB6's, MD11's, B74L's, B744's and B747's. For simplicity, we had three types of aircraft in this case study; type A indicates B737's (3 airplanes) with 120 seats, type B includes AB3's, AB6's, MD11's, and B74L's (17 airplanes) with an average of 269 seats, and type C includes B744's and B747's (6 airplanes) with an average of 403 seats.

For ease of testing, as calculated in Yan and Lin [22], all the cost parameters were set according to the airline's reports and the Taiwan government regulations, with reasonable simplifications, refer to China Airlines [3] and Civil Aeronautics Administration [4]. Note that according to the airline the swap cost for different aircraft types in its operations is very small compared to the flight cost. For simplicity, the swap cost between different fleets is assumed to be zero. Besides, since the airline has contracts with other airlines for transporting passengers under irregular operations, without additional charges, for simplicity, we assume that the additional charges, for reaccommodating passengers not getting on the smaller-sized aircraft for suitable flights to the same destinations, are zero. Note that the cost typically affects the test results. Because the case study is only for demonstration purposes at the current stage, the evaluation of the application of the models to actual operations is left to future work.

For the simplification of strategy (b), we only add two alternate flight arcs after each flight arc, each alternate flight denoting a delay of 30 minutes (in other words, we do not allow a delay of more than one hour in this test). For strategy (d), we add position arcs between every OD pair every 6 hours from the starting time to the recovery time. We assume that an airport (Taipei) is closed at 3:00 AM on Wednesday and will be reopened 8 hours later. Consequently, the starting time is 3:00 AM on Wednesday; the recovery time is 11:00 AM on Wednesday and the ending time by definition is 6:20 AM on Friday.

Eighteen scenarios were done in this test. Scenario 1 (indicated as "Normal") denotes a normal operation. Referred to Gershkoff [9], Scenario 2 ("SSP") applies the successive shortest paths to randomly finding a series of aircraft routes from the starting time to the ending time. The number of canceled flights can thus be calculated after fleet assignment. To assure that the fleet can resume its normal operations after the ending time, ferry flights could be used in the assignment. Scenario 3 through Scenario 18 are associated with our strategic models. For example, Scenario "B" indicates the basic model (flight cancellations); "Ba" denotes the combined model for flight cancellations and the swap of aircraft types; "Bb" denotes the combined model for flight cancellations and flight delays; "Bc" denotes the combined model for flight cancellations and the modification of multi-stop flights; "Bd" denotes the combined model for flight cancellations and the ferrying of idle aircraft; and "Babcd" denotes the combined model for flight cancellations, the swap of aircraft types, flight delays, the modification of multi-stop flights, and the ferrying of idle aircraft. Models "B" and "Bd" are pure network flow problems and the other models are network flow problems with side constraints or multi-commodity network flow problems. We used the network simplex method to solve the pure network flow problems and applied the Lagrangian relaxation-based algorithm to solve the network flow problems with side constraints or multi-commodity network flow problems. The convergence gap was set as 0.1% of the error. The same flow decomposition method (Yan and Lin [22]) was used to decompose the link flows into arc chains in each fleet network, with each arc chain representing an airplane's route in the perturbed period.

We note that the SSP method is easy to implement using our networks. In particular, the label correcting algorithm (Ahuja et al. [1]) is applicable for solving the fleet assignment. The SSP method used here is for the preliminary evaluation of our models. Several C programs were developed for; (1) the analysis of raw data, (2) the building of the basic model, (3) the development and solution of the strategic models, and (4) the output of data. The case study was implemented on an HP735 workstation. Sixteen strategic models were tested with problem sizes of up to 7,515 nodes and 29,355 arcs. All of the results show that the models could be useful for actual operations. The results are summarized in Table 5.1 and specifically analyzed as follows.

(1) The algorithms performed very well, indicating they could be useful in practice. In particular, models "B" and "Bd" were optimally solved in two seconds of CPU time. The other models converged within 0.1% of the error in at most 250 seconds of CPU time. Note that the CPU times for network generation and data output in each scenario are relatively short compared with model solutions and can be neglected. This shows that the network simplex algorithm should be efficient for solving pure network problems, like "B" and "Bd", and this research indicates that the proposed Lagrangian relaxation-based algorithm could also be efficient for solving the multi-commodity network flow problems. Compared with the efficiency of the traditional approach, the algorithms are superior.

(2) From column (10) in Table 5.1, we find that the multi-fleet scheduling models yield a higher profit than the single-fleet scheduling models. In particular, all eight strategic models containing strategy "a" suggest that a fleet can temporarily serve other types of flights in scheduling, so as to improve the system profit. For example, three type C flights in Model "Ba" are served by the type B fleet. The result for Model "Ba" (−72,896,930) is better than that for the basic model "B" (−72,691,067) which is a single-fleet scheduling model, an improvement of NT$ 205,863 (about 0.28%). For another example, six type B flights and three type C flights in Model "Babcd" are served by the type A fleet, and the type B fleet respectively. The result for Model "Babcd" (−81,028,533) is better than that for Model "Bbcd" (−79,730,287), an improvement of NT$ 1,298,246 (about 1.63%). Similarly, Models "Bab", "Bac", "Bad", "Babc", "Babd", and "Bacd" yield higher profits than "Bb", "Bc", "Bd", "Bbc", "Bbd", and "Bcd" respectively. The improvements in profit are NT$ 223,466 (0.28%), 407,149 (0.56%), 205,663 (0.28%), 1,602,025 (2.04%), 221,264 (0.28%), and 1,111,457 (1.53%) respectively.

TABLE 5.1

Results of All Scenarios

Scenario	Computation Time (sec)	# Iteration	Objective Value (NT$)	Converged Gap %	# Nodes A	# Nodes B	# Nodes C	# Links A	# Links B	# Links C	# Side Constraints
Normal											
SSP	—	—	− 87,674,685	—	—	—	—	—	—	—	—
B	0.65	—	− 47,906,444	—	2,473	2,473	2,473	5,480	5,529	5,483	—
Ba	1.09	1	− 72,691,067	0	2,473	2,473	2,473	5,480	5,529	5,483	0
Bb	124.88	117	− 72,896,930	0.1	2,473	2,473	2,473	5,548	5,548	5,548	76
Bc	14.29	11	− 78,720,690	0.04	2,473	2,473	2,473	5,548	5,585	5,494	76
Bd	91.17	76	− 72,691,067	0.1	2,473	2,473	2,473	5,480	5,641	5,483	108
Bab	1.86	1	− 72,835,603	0	2,473	2,473	2,473	9,530	9,579	9,533	0
Bac	39.66	31	− 78,944,156	0.1	2,473	2,473	2,473	5,623	5,623	5,623	76
Bad	250.2	193	− 73,098,216	0.1	2,505	2,505	2,473	5,660	5,660	5,660	108
Bbc	159.55	78	− 73,041,266	0.1	2,473	2,473	2,473	9,598	9,598	9,598	76
Bbd	41.86	31	− 78,720,690	0.02	2,473	2,505	2,473	5,488	5,697	5,494	108
Bcd	62.84	25	− 78,924,326	0.03	2,473	2,473	2,473	9,538	9,535	9,544	76
Babc	84.14	37	− 72,835,603	0	2,473	2,505	2,473	9,530	9,691	9,533	108
Babd	61.14	45	− 80,322,715	0.08	2,505	2,505	2,505	5,735	5,735	5,735	108
Bacd	90.36	35	− 79,145,591	0.01	2,473	2,473	2,473	9,673	9,673	9,673	108
Bbcd	130.56	55	− 73,947,060	0.03	2,505	2,505	2,505	9,710	9,710	9,710	76
Bbcd	95.2	37	− 79,730,287	0	2,473	2,505	2,473	9,538	9,747	9,544	108
Babcd	159.71	59	− 81,028,533	0.07	2,505	2,505	2,505	9,785	9,785	9,785	108

(continued)

TABLE 5.1 (Continued)

Results of All Scenarios

	# Original Flights			# Flights Served by Other Fleet		# Canceled Flights			# Delayed Flights			# Modified Multi-stop Flights			# Ferry Flights			# Canceled Flights and Flight Segments Total
Scenario	A	B	C	B/A	C/B	A	B	C	A	B	C	A	B	C	A	B	C	Total
Normal	8	57	11	—	—	—	—	—	—	—	—	—	—	—	—	—	—	—
SSP	8	57	11	—	—	3	25	3	—	—	—	—	—	—	1	6	1	31
B	8	57	11	0	0	2	14	2	0	0	0	0	0	0	0	0	0	18
Ba	8	57	11	0	3	2	15	1	0	0	0	0	0	0	0	1	0	18
Bb	8	57	11	0	0	2	8	2	0	4	0	0	0	0	0	0	0	12
Bc	8	57	11	0	0	2	14	2	0	0	0	0	0	0	0	0	0	18
Bd	8	57	11	0	0	2	14	2	0	0	0	0	0	0	0	2	0	18
Bab	8	57	11	2	3	2	9	1	0	4	0	0	0	0	0	1	0	12
Bac	8	57	11	0	3	4	13	1	0	0	0	0	0	0	1	1	0	18
Bad	8	57	11	0	3	2	15	1	0	0	0	0	0	0	0	3	0	18
Bbc	8	57	11	0	0	2	8	2	0	4	0	0	0	0	0	0	0	12
Bbd	8	57	11	0	0	2	8	2	0	3	0	0	0	0	0	5	0	12
Bcd	8	57	11	0	0	2	14	2	0	0	0	0	0	0	0	2	0	18
Babc	8	57	11	6	2	4	1	2	0	9	0	0	3	0	1	1	0	10
Babd	8	57	11	2	3	2	9	1	0	3	0	0	0	0	4	2	0	12
Bacd	8	57	11	3	3	2	13	1	0	0	0	0	0	0	4	1	0	16
Bbcd	8	57	11	0	0	2	5	2	0	7	0	0	2	0	0	4	0	11
Babcd	8	57	11	6	3	5	1	1	0	10	0	0	1	0	4	3	0	8

(3) All of our models yield a higher profit than the SSP approach does. The best result (−81,028,533), for Model "Babcd", is closest to the profit achieved in normal operations (−87,674,685). Model "Babcd" causes much less profit loss (NT$ 6,646,152) than using the SSP approach (NT$ 39,768,241). Though the basic model, "B", is the worst among the strategic models, its objective (−72,691,067) is significantly better than that of the SSP method (−47,906,444). We note that if more rules are incorporated to develop a strategic model, then the model will be more flexible for optimizing a temporary schedule, so that system profits can be improved more. For example, the result for "Babcd" is better than that for "Babc", which is better than "Bab". The reader can see other examples in Table 5.1. The reason that our models outperform the SSP method could be that although the SSP method cancels a series of uneconomic flights (3 type A, 25 type B, and 3 type C flights), it does not consider the swapping of aircraft, the delaying of flights, the modification of multi-stop flights, the ferrying of idle aircraft, or a combination of these in adjusting the schedule. Besides, its randomness for finding shortest paths neglects the combinatorial complexity of systematic routings, possibly resulting in an inferior combination of aircraft routes. Through the effective adjustment of flight schedules or fleet routes, fewer flights are canceled (for example, only 7 (5 type A, 1 type B, and 1 type C) flights are canceled in "Babcd". Thus, higher profits are achieved. We also find that the strategic models containing strategy "b", for example, "Bb", "Bab", "Bbc", "Bbd", "Babc", "Babd", "Bbcd", or "Babcd" yield significantly higher profits (more than 78,000,000) than other models (less than 74,000,000). This implies that a flight delay is an important and effective strategy in schedule adjustment.

(4) Other than the profit considerations, the degree of schedule perturbation may be a criterion for carriers to evaluate levels of service for all strategic models. Typically the more flights perturbed, the higher the chance that the level of service is affected. Furthermore, the number of canceled flights (including one-stop flight segments) and delayed flights in the case study may serve as an index of schedule perturbation. Although the influence on the level of service for flight cancellations and flight delays could be different, carriers may evaluate them separately, or use the technique of multiple criteria decision making. In this study, since a flight has at most a delay of one hour, its influence on the level of service could be omitted when compared to a canceled flight. Thus, in terms of the level of service, "Babcd" could be the best (7 canceled flights and 1 canceled one-stop flight segment, including 3 departure flights at Taipei during the closed period) and SSP the worst (31 canceled flights), as shown in Figure 5.2. Note that strategic models containing "b" canceled less flights (less than or equal to 12 flights and one-stop flight segments) than other models (more than 15). Consequently, considering both profit and level of service, strategic models containing "b" perform better than others. In particular, Model "Babcd" is the best of them all.

It should be noted that if a solution obtained from a strategic model, which is associated with a large degree of schedule perturbation, might not be accepted by carriers under real operating constraints (for example, crew availability or

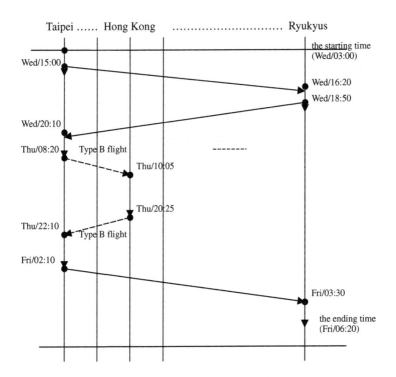

FIGURE 5.7
An example of a type A aircraft route of multi-fleet case (Scenario "Babcd").

aircraft maintenance rule), then carriers may choose another model to find a solution with less perturbation of a flight schedule, or they may make minor modifications of the perturbed schedule to satisfy the operating constraints. From this, a DSS might be helpful for the application of these strategic models in real time operations.

(5) The flow decomposition method was applied to trace the path of each airplane. In particular, fleet flows are decomposed into several arc chains, each representing an airplane's route. An example of an airplane path, Taipei-Ryukyus-Taipei-Hong Kong-Taipei-Ryukyus, in the perturbed period, (from Wednesday 3:00 a.m. to Friday 6:20 a.m.), is shown in Figure 5.7. Note that the two flights between Taipei and Hong Kong, which were originally served by the type B fleet, are served by the type A aircraft. Other aircraft routes can be similarly traced and are not reported on here.

5.6 Conclusions

Efficient and effective management to assist carriers to quickly resume their normal services after a schedule perturbation is important in today's competitive airline markets. This research applied network flow techniques to develop several systematic strategic models, combining various practical scheduling rules, to help carriers handle schedule perturbations resulting from the temporary closure of airports. These models were based on a basic model formulated as a time-space network from which various strategic models were developed. These models minimize the schedule-perturbed time after incidents so that carriers can resume their normal services as soon as possible in order to maintain their levels of service. In addition, these models combine the swapping of aircraft, flight cancellations, flight delays, the modification of multi-stop flights, and the ferrying of idle aircraft in a combined and systematic framework to effectively adjust a schedule following incidents, so that a carrier can maintain its profitability.

Mathematically, these strategic models were formulated as, pure network flow problems, network flow problems with side constraints, or multiple commodity network flow problems. We used the network simplex method to solve the pure network flow problems and developed a Lagrangian relaxation-based solution algorithm to solve the network flow problems with side constraints and multiple commodity network flow problems.

A case study of a major Taiwan airline's operations was presented to test the models. Sixteen strategic models and the SSP approach were tested, with substantial problem sizes of up to 7515 nodes and 29355 arcs. Several C programs were developed to apply these models on an HP735 workstation. The algorithms performed very well. In particular, models "B" and "Bd" were solved optimally in two seconds of CPU time; other models converged within 0.1% of the error in at most 250 seconds of CPU time. We found that the multi-fleet scheduling models yield a higher profit than the single-fleet scheduling models. The improvements in profit for all multi-fleet strategic models were significant, between NT\$ 205,663 and 1,602,025. The improved ratios fell between 0.28 and 2.04%. This shows that a fleet can temporarily serve other types of flights in scheduling, so as to improve the system profit. We also found that all of our models yielded a higher profit than the SSP approach does. All of these results show that the models could be useful for actual operations. Since the case study is only for demonstration at the current stage, the evaluation of impact on the application of the models to actual operations is left to future work.

Although this research is motivated by Taiwan airline operations, the models developed here are not specific and could be applicable for operations in other areas. It might be possible that more than one airport is temporarily closed at the same time. To resolve this problem, the networks developed in this research should be modified. In particular, the flow upper bounds of the flight arcs connected to these closed airports during the closed period are set to zero and the ending time of the network can be set as the last one calculated among all the closed airports. In addition, how to combine several smaller-type flights to form a larger-type flight is not considered in this research. Though it may make the modeling more complicated, it could be an effective strategy in multi-fleet operations and could be a direction for future research. Besides, the incorporation of other routing factors (for example, different or multiple objectives involved in actual operations) and systematic models for handling schedule perturbation caused by other types of incidents can be performed as well in future research. A computerized decision support system useful for users, to apply these strategic models in real time operations, could also be another direction of future research.

Finally, we note that the techniques and ideas used in this research could be similarly applied to more general daily airline operations control problems, for example, a temporary limitation on airport operations from the FAA because of congestion, or other types of incidents (for example, temporary aircraft shortage). Particularly, the time-space network could be used with modifications on the starting time/the initial supply, the ending time/the final demand, and the nodes/arcs according to the planned schedule and the incident. The practical scheduling rules could also be added into the basic network to develop various strategic models which should be formulated as pure network flow problems, network flow problems with side constraints, or multiple commodity network flow problems. Then, the solution techniques used in this research, with suitable modifications, may be applicable for solving the problems. All of these could be topics for future research.

Acknowledgments

This research was supported by a grant from the National Science Council of Taiwan. We thank China Airlines for providing the test problems.

References

1. Ahuja, R. K., Magnanti, T. L., and Orlin, J. B., *Network Flows, Theory, Algorithms, and Applications*, Prentice Hall, Englewood Cliffs, 1993.
2. Camerini, K., Fratta L., and Maffioli, F., On improving relaxation methods by modified gradient techniques, *Mathematical Programming Study*, 3, 6, 1975.
3. China Airlines, *China Airlines Annual Report 1992*, Taipei, R.O.C., 1993.
4. Civil Aeronautics Administration, *CAA's Fare Regulation for Using Airport Buildings, Land, and Other Relative Equipment*, Ministry of Transportation and Communication, R.O.C ,1984.
5. Deckwitz, T. A., *Interactive Dynamic Airplane Scheduling*, Flight Transportation Laboratory Report R-84-5, Massachusetts Institute of Technology, 1984.
6. Etschmaier, M. M. and Rothstein, M., *Estimating the Punctuality Rate Inherent in an Airline Schedule*, Technical Report 19, Department of Industrial Engineering, University of Pittsburgh, 1973.
7. Fisher, M. L., The Lagrangian relaxation method for solving integer programming problems, *Management Science*, 27, 1, 1981.
8. Garey, M. R. and Johnson, D. S., *Computers and Intractability: A Guide to the Theory of NP-completeness*, W. H. Freeman & Company, San Francisco, 1979.
9. Gershkoff, I., *Aircraft Shortage Evaluator*, presented at ORSA/TIMS Joint National Meeting, St. Louis, MO, 1987.

10. Jarrah, A.I., et al., A Decision Support Framework for Airline Flight Cancellations and Delays, *Transportation Science*, 27, 266, 1993.
11. Jedlinsky, D. C., *The Effect of Interruptions on Airline Schedule Control*, Masters thesis, Massachusetts Institute of Technology, 1967.
12. Kennington, J. L. and Helgason, R. V., *Algorithms for Network Programming*, John Wiley & Sons, New York, 1980.
13. Powell, W. B., A review of sensitivity results for linear networks and a new approximation to reduce the effects of degeneracy, *Transportation Science*, 23, 231, 1989.
14. Simpson, R. W., *Scheduling and Routing Models for Airline Systems*, Flight Transportation Laboratory Report R-68-3, Massachusetts Institute of Technology, 1969.
15. Teodorovic, D., A model for designing the meteorologically most reliable airline schedule, *European Journal of Operational Research*, 21, 156, 1985.
16. Teodorovic, D., *Airline Operations Research*, Gordon and Breach Science Publishers, New York, 1988.
17. Teodorovic, D. and Guberinic, S., Optimal dispatching strategy on an airline network after a schedule perturbation, *European Journal of Operational Research*, 15, 178, 1984.
18. Teodorovic, D. and Stojkovic, G., Model for operational daily airline scheduling, *Transportation Planning and Technology*, 14, 273, 1990.
19. Thengvall, B. G., Bard, J. F., and Yu, G., Balancing user preferences for aircraft schedule recovery during airline irregular operations, *IIE Transactions on Operations Engineering*, 32, 181, 2000.
20. Thengvall, B. G., Yu, G., and Bard, J. F. Multiple fleet aircraft schedule recovery following hub closure, *Transportation Research*, 35A, 289, 2001.
21. Yan, S., Bernstein, D., and Sheffi, Y., Intermodal pricing using network flow techniques, *Transportation Research*, 29B, 171, 1995.
22. Yan, S. and Lin, C., Airline scheduling for the temporary closure of airports, *Transportation Science*, 31, 72, 1997.
23. Yan, S. and Tu, Y., Multi-fleet routing and multi-stop flight scheduling for schedule perturbation, *European Journal of Operational Research*, 103, 155, 1997.
24. Yan, S. and Yang, D., A decision support framework for handling schedule perturbation, *Transportation Research*, 30B, 405, 1996.

6

Determining an Optimal Fleet Mix and Schedules: Part I — Single Source and Destination

Hanif D. Sherali and Salem M. Al-Yakoob

CONTENTS

6.1 Introduction: Problem Description

Efficient scheduling of oceanic transportation vessels presents serious challenges to concerned decision-makers due to the complexity of the operation and the potential savings that can be attained. A vessel usually costs tens of millions of U.S. dollars with daily operational costs amounting to tens of thousands of U.S. dollars. Chartering and spot-chartering of vessels is also very expensive and should be avoided to the extent possible. In general, large-scale vessel scheduling is very intricate and requires repeated revamping to accommodate changes in demand, market conditions, and weather effects. Manual approaches are often inefficient and expensive, and hence, it is imperative to utilize quantitative methods to generate and update vessel schedules efficiently and in a timely fashion.

The problem that is studied in this chapter is concerned with the transportation of a product from a single source to a single destination. The next chapter examines the case of multiple sources and destinations, along with the option of leasing transshipment depots. For example, we might be interested in shipping crude oil from a refinery to a storage facility that has a known demand structure defined by different daily consumption rates. The daily export from the source depends on the availability of the product at the source, the availability of vessels, and the current storage level at the destination's storage facility. Consumption rates at the destination might not be fixed during the entire time horizon and might vary based on, for example, seasonal considerations. The level of the product at the destination's storage facility is desired to lie within certain lower and upper bounds, and hence, daily penalties are imposed based on limited shortage or excess quantities with respect to these bounds.

Typically the organization that handles the transportation of the product from the source to the destination owns vessels of different types, where each type is characterized by the vessel size, speed, loading and unloading times, etc. Moreover, there are vessels that are available for chartering during all or part of the time horizon. Spot-chartering is also available if desired, whereby a vessel is chartered for a particular voyage or voyages, not for a period of time. In this case, all operating costs are undertaken by the vessel owner. (See for example, Rana and Vickson [24].)

The organization aims to satisfy demand requirements at a minimum overall cost that is comprised of operational expenses, penalties resulting from exceeding lower and upper storage levels, and chartering expenses. This requires an efficient utilization of the organization's self-owned vessels, and a minimal reliance on chartered and spot-chartered vessels that tend to have high associated costs. The combinatorial nature of this problem makes the manual scheduling of vessels inefficient and costly because of unduly high resulting chartering and penalty expenses.

In this chapter, we develop mathematical models that generate cost-effective vessel schedules in a timely fashion. The problem described above is faced by many oil companies such as the Kuwait Petroleum Corporation (KPC), which is required by agreed upon contracts to meet specified demands of crude oil that vary based on daily consumption rates at a given destination. The single source-destination vessel scheduling operation arises in many situations, for example, when large quantities of crude oil need to be transported from Kuwait to a destination located in Europe, North America, or Asia. In this case, the destination might be a single customer or a collection of customers that are located within close proximities of each other. Hence, the single source and destination problem considered herein, although a special case of the multiple source and destination problem addressed elsewhere [31], is important in its own right and is frequently faced in practice.

6.2   Related Literature

Transportation routing and scheduling problems have been widely researched in the literature, with a bulk of the published work dealing with vehicle routing and scheduling problems. Vessel scheduling, in particular, has attracted the least attention among all transportation modes. Ronen [26, 27] highlighted a noticeable scarcity of published research dealing with designing, planning, and managing sea-borne transportation systems, and gave a number of impediments associated with vessel-scheduling problems such as the complexity and the high uncertainties associated with such problems. However, recently, there has been increasing interest in research related to maritime transportation as evidenced, for example, by a special issue devoted to this area in Transportation Science (Psaraftis [23]). Also, the book by Perakis [20] gives an overview of models for a number of problems related to fleet operations and deployment. For a comprehensive review on ship routing and scheduling research conducted over the last decade, the reader may refer to Christiansen et al. [8].

In order to accurately model vessel routing and scheduling problems, novel mathematical formulations are often needed. Moreover, models for large-scale vessel scheduling problems are rather hard to solve due to complex constraint structures that result from the operational intricacies of such problems, and their inherent combinatorial characteristics arising from an explicit consideration of the different components of the operation. Vessel scheduling can therefore be an arduous and time-consuming task, and is often complicated by the dynamic nature of such problems. Yet, many organizations still largely handle their vessel scheduling problems only manually.

Typically there are three modes of operation in seaborne shipping: liner, tramp, and industrial (see, for example, Christiansen et al. [8] and Lawrence [17]). Liners operate based on a fixed published itinerary and schedule similar to bus or passenger train lines. Tramp ships operate like a taxi by following the available cargos. In industrial shipping, the cargo owner controls the fleet of ships. Accordingly, vessel routing and scheduling problems can be partitioned into four categories: a) liner, b) tramp, c) industrial, and d) other related models. The first three are not sharply defined or mutually exclusive, nor collectively exhaustive (Lawrence [17]); hence, the fourth category of problems captures applications that cannot be clearly classified as liner, tramp, or industrial. It is worth mentioning that water transportation routing and scheduling models mainly deal with the transport and delivery of cargo.

Considerably fewer models have been developed to tackle vessel routing and scheduling problems in the context of transporting passengers. This is a result of the fact that most vessels move cargo around the world, while passengers mainly travel by air or land.

The literature on modeling techniques and approaches for the liner shipping is fairly limited; however, in recent years an increased activity in this area has become evident (see, for example, Christiansen et al. [8] and Lane et al. [16]). Powell and Perakis [22] formulated an integer programming model for a fleet deployment problem. The objective of the model was to minimize the operating costs for a fleet of liner ships involved in various routes. It was shown, based on real liner shipping data, that substantial savings could be achieved via using the proposed modeling approach. Note that the work of Powell and Perakis [22] was an extension to the work of Perakis and Jaramillo [19] and Jaramillo and Perakis [13]. In the latter two papers, a linear programming approach was used to solve a fleet deployment problem. Cho and Perakis [6] investigated a fleet size and design of a liner routing problem for a container shipping company. The problem was formulated by generating a subset of candidate routes for the different ships *a priori*, and was then solved as a linear program, where the columns represent the candidate routes. This model was also extended to a mixed-integer program that additionally incorporates investment alternatives for expanding fleet capacity. Xinlian et al. [31] formulated a fleet planning model for a problem similar to that investigated by Cho and Perakis [6]. The model aims to determine the ship types to add to the existing fleet of ships as well as an optimal fleet deployment plan.

Datz et al. [9] developed a simple calculative method for scheduling a liner and suggested some techniques for evaluating the financial results of such a schedule. Nemhauser and Yu [18] studied a model for rail service that can be used for a liner problem. Dynamic programming was used to find the optimal frequency of services that maximizes profit over the planning horizon. Demand for service was a function of two variables, namely service frequency and timing. Rana and Vickson [24] developed a deterministic mathematical programming model for optimally routing a chartered container vessel. The formulation involves nonlinearities, which were handled by converting the nonlinear problem into a number of mixed-integer programs. Benders' decomposition was applied to the resulting mixed-integer programs, wherein the integer network subprograms were solved by a specialized algorithm. Later, Rana and Vickson [25] extended their work in [24] by allowing multiple ships. They formulated a mathematical programming model for a container-ship routing problem that determines the following: 1) an optimal sequence of ports of call for each vessel, 2) the number of trips each vessel makes in the planning horizon, and 3) the amount of cargo delivered between any two ports by each ship. The problem was solved by Lagrangian relaxation by decomposing it into several sub-problems, one for each vessel. Each sub-problem was further decomposed into a number of mixed-integer programs. For other related literature on liner models, the reader may refer to Christiansen et al. [8] and Al-Yakoob [1].

Very little research has been conducted on the allocation, routing, and scheduling of tramp shipping (see Appelgren [2, 3]). Appelgren [2] discussed a ship scheduling problem obtained from a Swedish ship-owning company. In this problem, a ship-owning company was engaged in world-wide operations involving a large number of vessels. A set of cargos were provided for the planning period, which covered 2 to 4 months. The author designed an algorithm that used the Dantzig-Wolf decomposition method for linear programming, where the subprograms were modeled as network flow problems and were solved by dynamic programming. The master program in the decomposition algorithm was modeled as a linear program having only zero-one elements in the matrix and on the right-hand-side. This algorithm was tested on instances involving about 40 ships and 50 cargos.

Later, Appelgren [3] utilized integer programming methods to solve a vessel scheduling problem. The problem was to determine an optimal sequence of cargos for each vessel in a given fleet during a specified time period. This paper was an attempt to deal with some of the shortcomings associated with the technique used by the author in [2], where a decomposition algorithm was used, which, however, produced nonintegral solutions that could not be interpreted as valid schedules. To avoid fractional solutions, a branch-and-bound algorithm was developed, where the branching was performed on one of the "essential" noninteger variables and the bounds were computed by the decomposition algorithm. A decision support system for both tramp and industrial shipping was described in Fagerholt [10], where a heuristic hybrid search algorithm was employed for solving such ship scheduling problems.

Since the problem under consideration can be classified as an industrial shipping problem, we present next some industrial scheduling literature that is most closely related to our problem.

A vessel scheduling problem concerned with transporting crude oil from the Middle East to Europe and North America was considered by Brown et al. [5], who formulated a mixed-integer partitioning model and utilized column generation techniques to solve this problem. The demand structure was specified by a sequence of cargos that needed to be delivered during the planning horizon. The approach adopted by Brown et al. [5] attempted to generate a partial set of complete feasible schedules (in a column generation framework), along with the generated schedules' costs, and then utilized an elastic set partitioning programming model to derive a prescribed solution. This same problem was investigated by Perakis and Bremer [21] who also applied a set partitioning approach to solve the problem. The work of Brown et al. [5] was extended in Bausch et al. [4] by allowing a shipload to have up to five products. A similar set partitioning approach to that of Brown et al. [5] was used by Bausch et al. [4] to solve the problem.

Sherali et al. [30] presented mixed-integer programming models and specialized rolling horizon algorithms for an oil tanker routing and scheduling problem to ship various products from one source to different destinations. The demand structure that was investigated by Sherali et al. [30] was determined by the total demand for each product at each destination, along with the respective sub-demand that was required to be satisfied within agreed-upon time intervals. It was assumed that some vessels could carry more than one product. The modeling approach adopted in Sherali et al. [30] as well as in this chapter differs from the modeling approach of Brown et al. [5] in that it combines the process of constructing and selecting feasible schedules at once.

Ronen [28] proposed a mixed-integer programming model and a cost-based heuristic procedure for a vessel transportation problem that is similar to the one considered in this chapter. The approach adopted by Ronen [28] separates the solution of the problem into two stages. The first stage determines the slate of shipments to be made, and the second stage generates vessel schedules. In contrast, the models developed here decide upon the fleet size mix as well as the detailed vessel schedules all together. Another application within the oil industry, which involves the transport of refined oil products from a refinery to several depots, was investigated by Scott [29]. A Lagrangian relaxation approach was adopted to generate a set of potentially good schedules, and an enhanced version of Benders' decomposition was used to decide upon an optimal schedule from within the generated set of schedules. Fagerholt and Christiansen [11] studied a multi-product scheduling problem similar to the one presented by Bausch et al. [4]. However, in this paper, the authors assumed that each ship in the fleet is equipped with a flexible cargo hold that can be partitioned into many holds in a specified number of ways. A set partitioning approach was utilized by the authors for this problem and a detailed algorithm for finding optimal schedules for the individual ships was described in Fagerholt and Christiansen [12]. An inventory-routing problem was studied by Christiansen [7], which attempts to attain a degree of balance in the ammonia supply at all producing and/or consuming company-owned plants around the world.

A ship scheduling problem that does not clearly fall into liner, tramp, or industrial shipping was investigated by Koenigsberg and Lam [14] who studied queuing aspects for a small system of liquid gas tankers operating in closed routes between a small number of terminals. For any particular system, their model provides the expected number of ships at each stage, the expected number waiting in each stage, and most importantly, the expected waiting time at ports. Exponential service time distributions were used; however, a series of parallel simulation computations were also employed to analyze the impact of other distributions. Later, Koenigsberg and Meyers [15] extended the work of Koenigsberg and Lam [14] by developing an analytical model of a system having two independent fleets that share a common loading port. Exponential distributions of service times were used in all queuing stages. The authors used a simulation program to investigate the behavior of the system when the service time distributions were not exponential and demonstrated a good level of conformity between the simulation and the analytical results for exponential distributions.

For further details on vessel routing and scheduling problems and models, the reader may refer to Al-Yakoob [1], Christiansen [8], Perakis [20], and Ronen [26, 27].

Although many vessel scheduling problems and models have been discussed in the aforementioned literature, none of these fully address the peculiar aspects of the problem considered in this chapter. The specific nature of the demand structure and the consideration of different vessel types and cost components require novel mathematical formulations and solution methods to accurately model the problem and to derive solutions in a manner that can be practically implemented. The remainder of this chapter is organized as follows. The next section provides a detailed description of the problem. A mixed-integer programming model is then developed in Section 6.4, and subsequently, an aggregate reformulation of this model is derived in Section 6.5. Solution algorithms and computational results are presented in Section 6.6, along with a comparison with the *ad-hoc* scheduling procedure that is currently in use at KPC. Finally, Section 6.7 provides a summary and some concluding remarks.

6.3 Modeling Preliminaries

This section introduces aspects of the problem that will be used to formulate integer programming models in Section 6.4 and Section 6.5. In particular, Section 6.3.1 presents notation and assumptions, and Section 6.3.2 describes a function for computing penalties associated with exceeding certain lower and upper allowable storage levels. For the reader's convenience, a glossary of notation (sequenced in order of appearance) is provided in Appendix A.

6.3.1 Problem Notation and Assumptions

Let $h = 1, \ldots, H$ index the days of the time horizon under consideration. Typically, a time horizon is associated with a contract to deliver the product from the source to the destination based on some given consumption rates as discussed in Section 6.1. Therefore, the terms "time horizon" and "contract horizon" will henceforth be used interchangeably. Note that a contract may be signed, and then another contract may be signed prior to the end of the first contract horizon. Suppose that there are T vessel types, indexed by $t = 1, \ldots, T$. It is assumed that vessels of the same type have similar features such as capacity, speed, loading and unloading times, etc. The capacity of a vessel of type t is denoted by Ω_t and the total number of vessels of this type that are available for use during all or part of the time-horizon is given by M_t (note that M_t is composed of self-owned vessels in addition to vessels that are available for chartering). For a vessel type $t \in \{1, \ldots, T\}$, let $n = 1, \ldots, M_t$ index all vessels of this type, and let O_t and $CH_t = M_t - O_t$ respectively denote the number of self-owned vessels and the number of available vessels of this type that can be possibly chartered. Accordingly, let $n = 1, \ldots, O_t$ and $n = O_t + 1, \ldots, O_t + CH_t \equiv M_t$ respectively index self-owned vessels and vessels that are available for chartering of type t. Let us also denote $O = \sum_{t=1}^{T} O_t$ and $CH = \sum_{t=1}^{T} CH_t$. Let $\$_{t,n}$ be the cost (in U.S. dollars) of chartering a vessel n of type t, for $n = O_t + 1, \ldots, O_t + CH_t = M_t$, and for each $t \in \{1, \ldots, T\}$. Let $UT_{t,n}$ be the maximum number of days vessel n of type t can be used during the time horizon. This time restriction is typically needed for maintenance purposes.

Let T_t represent the time required to load a vessel of type t at the source, plus the time this vessel takes to travel from the source to the destination, unload time at the destination, and then travel back from the destination to the source. We assume that there is a unique prescribed route from the source to the destination, and likewise, from the destination to the source. Let $T_t = T_{1,t} + T_{2,t}$, where $T_{1,t}$ is the time required to load a vessel of type t at the source plus and then travel to the destination, and $T_{2,t}$ is the time required to unload at the destination and then travel back to the source. Vessels of the same type are assumed to have equal values of T_t (and their splits), and this duration is also assumed to be independent of h; i.e., T_t is independent of the day the leg starts (weather effects are neglected). The values of T_t and its splits are derived from the design speed of the vessels of type t. Let $DC_{t,n}$ denote the daily operational cost of vessel n of type t so that the total operational cost of vessel n of type t is given by $C_{t,n} = T_t(DC_{t,n})$, which covers the round-trip expenses from the source to the destination, and is independent of the day the trip starts.

Let Q denote a production capacity or certain imposed quota of the product at the source. At the beginning of the time horizon, the storage level at the destination is given by w. This level may represent a single storage facility or a collection of storage facilities at the destination; however, for the sake of modeling, we only deal with a combined aggregate storage level. Let SL_1 and SL_2 denote the minimum and maximum desired levels, respectively, at the destination's storage facility, which should be maintained to the extent possible in order to avoid penalties. Accordingly, let π denote the daily penalty for each shortage or excess unit at the destination. The permitted shortage and excess quantities at the destination with respect to the desired levels SL_1 and SL_2 to the extent given by A_1 and A_2, respectively. Let $b_1 = SL_1 - A_1$ and $b_2 = SL_2 + A_2$. Let $UB > b_2$ be a sufficiently large upper bound on the maximum allowable storage level on any given day of the time horizon. Shortages levels falling below b_1 or in excess of b_2 (up to UB), while permitted, are highly undesirable, and incur a significantly greater penalty $\lambda > \pi$ per unit.

Let R_j denote the expected consumption rate at the destination on day j, for $j = 1, \ldots, H$. The different daily consumption rates arise from possible seasonal changes during the time-horizon, as well as from client-specific considerations. Thus, the total cumulative consumption at the destination over the days $j = 1, \ldots, h$ is given by $TC_h = \Sigma_{j=1}^{h} R_j$.

6.3.2 Penalty Functions

The daily storage levels determine the overall penalty over the time horizon, being given by the summation of all daily penalties as described below. Let S_h be the storage level on day h. Define Type I and Type II penalty functions as follows:

Type I penalty: $P_I(S_h) = \pi \text{ maximum } \{0, (SL_1 - S_h), (S_h - SL_2)\}$ if $S_h \in [b_1, b_2]$

and

$$\textbf{Type II penalty:} P_{II}(S_h) = \begin{cases} \pi A_1 + \lambda(b_1 - S_h) & \text{if } S_h \in (0, b_1), \\ \pi A_2 + \lambda(S_h - b_2) & \text{if } S_h \in (b_2, UB), \end{cases}$$

where π and $\lambda > \pi$ are as defined above. Note that if $S_h \in [SL_1, SL_2]$, then the storage level lies within the desired bounds and no penalty is induced. If $S_h \in [b_1, SL_1) \cup (SL_2, b_2]$, then a penalty is incurred based on the respective shortage or excess quantity. On the other hand, if $S_h \in [0, b_1) \cup (b_2, UB)$, then a sufficiently large additional penalty rate is imposed continuously beyond that of $P_I(.)$ to indicate the undesirability of such a storage level on any given day of the time horizon.

PROPOSITION 6.1

$$\text{Let } S_h = S_{1,h} - S_{2,h} - S_{3,h} + S_{4,h} + S_{5,h}, \tag{6.1}$$

where

$$SL_1 \leq S_{1,h} \leq SL_2, 0 \leq S_{2,h} \leq A_1, 0 \leq S_{3,h} \leq b_1,$$
$$0 \leq S_{4,h} \leq A_2, \text{ and } 0 \leq S_{5,h} \leq UB - b_2. \tag{6.2}$$

Define $P(S_h) : [0, UB] \rightarrow [0, \infty)$ *as the linear penalty function:*
$P(S_h) = \pi(S_{2,h} + S_{4,h}) + \lambda(S_{3,h} + S_{5,h})$. *Then any minimization objective formulation that incorporates the term* $P(S_h)$ *defined above along with (1.1) and (1.2) will automatically enforce the sum of the type I and type II penalties* $P_I(S_h) + P_{II}(S_h)$.

PROOF Noting that $0 < \pi < \lambda$ we have that for any $S_h \in [0, UB)$, the corresponding representation of S_h in terms of $S_{1,h}, S_{2,h}, S_{3,h}, S_{4,h}$, and $S_{5,h}$ is determined as follows, where in each case, the remaining (unspecified) variables from this list have zero values. If $S_h \in [SL_1, SL_2]$, then $S_h = S_{1,h}$. If $S_h \in [b_1, SL_1)$, then $S_h = S_{1,h} - S_{2,h}$, where $S_{1,h} = SL_1$ and $S_{2,h} = SL_1 - S_h$, while if $S_h \in (SL_2, b_2]$, then $S_h = S_{1,h} + S_{4,h}$, where $S_{1,h} = SL_2$ and $S_{4,h} = S_h - SL_2$. Likewise, if $S_h \in [0, b_1)$, then $S_h = S_{1,h} - S_{2,h} - S_{3,h}$, where $S_{1,h} = SL_1$, $S_{2,h} = A_1$, and $S_{3,h} = b_1 - S_h$, while if $S_h \in (b_2, UB)$, then $S_h = S_{1,h} + S_{4,h} + S_{5,h}$, where $S_{1,h} = SL_2$, $S_{4,h} = A_2$, and $S_{5,h} = S_h - b_2$. In each case, the function $P(S_h)$ is readily verified to impose the required Type I plus Type II penalties as described above, and this completes the proof. ∎

6.4 Model Development

In this section, we formulate a mixed-integer programming model for the prescribed vessel scheduling problem. The variables and constraints of the model are respectively presented in Section 6.4.1 and Section 6.4.2, and the objective function and the proposed mathematical model are respectively given in Section 6.4.3 and Section 6.4.4.

6.4.1 Model Variables

Define the following sets of binary decision variables. Let

$$X_{h,t,n} = \begin{cases} 1 & \text{if vessel } n \text{ of type } t \text{ departs the source toward the} \\ & \text{destination on day } h, \\ 0 & \text{otherwise,} \end{cases}$$

Since a vessel cannot be dispatched from the source on day h unless it is available on this day, another set of binary variables is defined as follows. Let

$$Y_{h,t,n} = \begin{cases} 1 & \text{if vessel } n \text{ of type } t \text{ is available at the source on day } h, \\ 0 & \text{otherwise.} \end{cases}$$

Finally, in order to represent the chartering decisions, let

$$Z_{t,n} = \begin{cases} 1 & \text{if vessel } n \text{ of type } t \text{ is selected for chartering during} \\ & \text{(all or only part of) the time horizon,} \\ 0 & \text{otherwise.} \end{cases}$$

A vessel may be chartered for the entire duration of the time horizon or for only a specified subset of it, depending on its availability. The chartering expense, denoted by $\$_{t,n}$ for a vessel n of type t is incurred as a fixed-cost based on the availability duration of this vessel, whenever it is chartered during a specified interval of the time horizon, regardless of its usage during this time interval. The reason for this is that selected chartered vessels will be under the control of the (leased-to) company (in our case, KPC) for the specified time interval, and the (leased-to) company is free to make any related dispatching decisions during this interval.

REMARK 6.1
Note that subsets of these binary decisions variables are *a priori* known to be effectively zero, i.e., inadmissible. The following are examples of such zero variables. A vessel n of type t might not be available for use from the first day of the time horizon. This occurs if this vessel is a self-owned vessel that is involved in a trip from a previously signed demand contract that will terminate sometime during the current time horizon. In other words, previous contracts might have committed certain vessels over durations concurrent with the present contract horizon. Accordingly, suitable sets of variables should be defined to be zero for the present contract horizon problem to signify the unavailability of such vessels. This might also happen if this vessel is a chartered vessel that will become available sometime after the first day of the time-horizon. In either case, the vessel availability variable $Y_{h,t,n}$ is set to zero until this vessel becomes available at the source. Also, for a chartered vessel n of type t, the leasing conditions may specify some last operational day for this vessel so that sufficient time is allowed for the leasing company to transfer this vessel to another organization or perform a scheduled maintenance.

In this case, the availability variable $Y_{h,t,n}$ is set to zero whenever $h + T_t$ is greater than that specified day. Note that naturally whenever $Y_{h,t,n} = 0$, then $X_{h,t,n} = 0$ because a vessel is dispatched on a given day only if it is available on that day. Let ϕ_X and ϕ_Y denote the index sets of X and Y variables, respectively, that are restricted to be fixed at specified binary values by virtue of such considerations.

6.4.2 Model Constraints

The various constraints of the model are formulated as described in turn below.

(A) Representation of the destination's storage level

The daily storage level of the product at the destination must remain within $[b_1, b_2]$ to the extent possible as discussed in the previous section, and appropriate daily penalties are imposed based on the specific levels of the storage. Representation of the storage level is given by the following constraints.

$$(C_1) \quad S_h = w + \sum_t \sum_n \sum_{h_1:h_1+T_{1,t}\in\{1,\dots,h\}} \Omega_t X_{h_1,t,n} - TC_h, \forall h \in \{1, \dots, H\},$$

$$(C_2) \quad S_h = S_{1,h} - S_{2,h} - S_{3,h} + S_{4,h} + S_{5,h}, \forall h \in \{1, \dots, H\},$$

where $SL_1 \leq S_{1,h} \leq SL_2, 0 \leq S_{2,h} \leq A_1, 0 \leq S_{3,h} \leq b_1, 0 \leq S_{4,h} \leq A_2$, and $0 \leq S_{5,h} \leq UB - b_2$.

Constraint (C_1) gives the storage level on day h based on the daily consumption rates and the shipments of the product that are delivered on or before day h. Constraint (C_2) represents S_h in terms of $S_{1,h}, S_{2,h}, S_{3,h}, S_{4,h}$, and $S_{5,h}$ as described in (1.1) and (1.2) so that appropriate penalties would be incurred in the objective function based on this representation as stated in Proposition 6.1.

(B) Availabilities of vessels

The vessel availability constraints are given as follows:

$$(C_3) \quad Y_{h,t,n} = Y_{h-1,t,n} - X_{h-1,t,n} + \sum_{h_1:h_1+T_t=h} X_{h_1,t,n}, \forall h \geq 2, t, n,$$

$$(C_4) \quad X_{h,t,n} \leq Y_{h,t,n}, \forall h, t, n.$$

A vessel n of type t can be dispatched from the source to the destination on day h only if it is available at the source on that day. This vessel is available at the source on day h if either the vessel was available at the source on the previous day and it was not dispatched, or this vessel was not available there during the previous day but it arrived on the current day. On the other hand, this vessel is unavailable on day h at the source if it was available there on the previous day and it was dispatched on that day (assuming that $T_t \geq 2$ for all vessel-types t), or it was unavailable on the previous day and it did not arrive on the current day. Constraint (C_3) examines the availability of vessel n of type t at the source on day h by incorporating these cases, and then Constraint (C_4) permits dispatching of vessels conditioned upon this availability.

(C) Chartering of vessel

The following constraint examines if a vessel $n \in \{O_t + 1, \ldots, M_t\}$ of type t is selected for chartering or not.

$$(C_5) \quad Y_{h,t,n} \leq Z_{t,n}, \forall h, t, n \in \{O_t + 1, \ldots, M_t\}.$$

Hence, a vessel $n \in \{O_t + 1, \ldots, M_t\}$ that is available for chartering is used only if $Z_{t,n} = 1$.

(D) Capacity restrictions and maintenance requirements

A production capacity or certain imposed daily quota might restrict the maximum amount of the product that can be shipped to the destination on any day of the time horizon. This restriction is represented by the following constraint.

$$(C_6) \quad \sum_t \sum_n \Omega_t X_{h,t,n} \leq Q, \forall h.$$

Furthermore, the following constraint enforces that any vessel n of type t can be used for at most $UT_{t,n}$ days during the time horizon since different age vessels might have different usage allowances. This restriction might be needed for maintenance purposes and is enforced by the following constraint.

$$(C_7) \quad \sum_h T_t X_{h,t,n} \leq UT_{t,n}, \forall t, n.$$

Other forms of scheduled maintenance restrictions can be also accommodated in the model by setting certain X-variables to zero.

6.4.3 Objective Function

The objective function is composed of the following terms.

(a) Operational costs (both for self-owned and chartered vessels) given by

$$\sum_h \sum_t \sum_n C_{t,n} X_{h,t,n}.$$

(b) Penalty costs resulting from shortage or excess levels at the destination's storage are given by the following term based on the representation stated in Proposition 1 above.

$$\sum_h \pi [S_{2,h} + S_{4,h}] + \sum_h \lambda [S_{3,h} + S_{5,h}].$$

(c) The chartering expenses are given by $\sum_t \sum_{n=O_t+1}^{M_t} \$_{t,n} Z_{t,n}.$

6.4.4 Overall Model Formulation

The objective function terms (a), (b), and (c) together with the constraints yield the following model for the Vessel Scheduling Problem (VSP). (All indices are assumed to take on only their respective relevant values.)

VSP:

$$\text{Minimize} \sum_{h} \sum_{t} \sum_{n} C_{t,n} X_{h,t,n} + \sum_{h} \pi [S_{2,h} + S_{4,h}]$$

$$+ \sum_{h} \lambda [S_{3,h} + S_{5,h}] + \sum_{t} \sum_{n=O_t+1}^{M_t} \$_{t,n} Z_{t,n},$$

subject to

(C_1) $\quad S_h = w + \sum_{t} \sum_{n} \sum_{\substack{h_1: \\ h_1+T_{1,t} \in [1,\dots,h]}} \Omega_t X_{h_1,t,n} - TC_h, \forall h,$

(C_2) $\quad S_h = S_{1,h} - S_{2,h} - S_{3,h} + S_{4,h} + S_{5,h}, \forall h,$

(C_3) $\quad Y_{h,t,n} = Y_{h-1,t,n} - X_{h-1,t,n} + \sum_{\substack{h_1: \\ h_1+T_t=h}} X_{h_1,t,n}, \forall h \geq 2, t, n,$

(C_4) $\quad X_{h,t,n} \leq Y_{h,t,n}, \forall h, t, n,$

(C_5) $\quad Y_{h,t,n} \leq Z_{t,n}, \forall h, t, n \in \{O_t+1, \dots, M_t\},$

(C_6) $\quad \sum_{t} \sum_{n} \Omega_t X_{h,t,n} \leq Q, \forall h,$

(C_7) $\quad \sum_{h} \sum_{t} X_{h,t,n} T_t \leq UT_{t,n}, \forall t, n,$

$X_{h,t,n} \in \{0,1\}, \forall h, t, n,$ if $X_{h,t,n} \notin \phi_X$, and fixed at zero or one otherwise, $Y_{1,t,n} \in \{0,1\}, \forall t, n, 0 \leq Y_{h,t,n} \leq 1, \forall h \geq 2, t, n,$ if $Y_{h,t,n} \notin \phi_Y$, and fixed at zero or one otherwise, $0 \leq Z_{t,n} \leq 1, \quad \forall t, n = O_t+1, \dots, O_t+CH_t, S_h \geq 0, SL_1 \leq S_{1,h} \leq SL_2, 0 \leq S_{2,h} \leq A_1, 0 \leq S_{3,h} \leq b_1, 0 \leq S_{4,h} \leq A_2,$ and $0 \leq S_{5,h} \leq UB-b_2, \forall h.$

REMARK 6.2
Note that by Constraint (C_3), the integrality of the Y-variables is guaranteed once the integrality of the X-variables and the Y-variables corresponding to the first day of the time horizon is enforced. The integrality of the Z-variables is then automatically enforced by Constraint (C_5) along with the fourth term of the objective function. This holds true since if a vessel $n \in \{O_t+1, \dots, O_t+CH_t\}$ of type t is selected for chartering, then this vessel is used in at least one trip from the source to the destination, in which case $Y_{h,t,n} = 1$ for some day h, and hence, the corresponding Constraint (C_5) then enforces $Z_{t,n}$ to also take on a value of one. On the other hand, if this vessel is

not selected for chartering, then $Y_{h,t,n} = 0$ for all days of the time-horizon, in which case the most attractive value for $Z_{t,n}$ is zero based on the fourth term of the objective function.

In the next section, we derive an aggregated formulation of Model VSP, and in Section 6.6, we present a problem size analysis for both formulations.

6.5   An Aggregate Formulation AVSP for Model VSP

The vessel scheduling problem at hand can be alternatively formulated by deciding upon the number of vessels of each type that are needed to be dispatched every day instead of having to make dispatching decisions about individual vessels. In this section, we derive an aggregated version of Model VSP that retains the essential characteristics of the operation while being far more computationally tractable than Model VSP. This formulation can be ideally used for problem instances having relatively longer time horizons in order to deal with the ensuing large number of binary variables.

6.5.1   Formulation of Model AVSP

Assume that the daily operational costs of vessels of a given type are the same, and that the chartering expenses of all such vessels are identical. In case the operational costs of vessels of the same type are not identical, then we take the average of all such costs, and likewise for the chartering expenses. However, if there are significant differences between these daily operational costs or between the chartering expenses, then we may accordingly partition a vessel-type into various sub-types so that the assumed cost representation is adequate. Hence, let DC_t denote the average daily operational cost of a vessel of type t and let $C_t = T_t (DC_t)$.

Define $x_{h,t}$ as an integer variable that represents the number of vessels of type t that are dispatched from the source on day h. Define $y_{h,t}$ to be an integer decision variable that represents the maximum number of vessels of type t that are available for dispatching from the source on day h. As mentioned in Remark 6.1, vessels might become available for use at different days of the time horizon due to, for example, their involvement in trips from previous demand contracts that will terminate sometime during the current time horizon. Hence, we let $O_{h,t}$ be the number of self-owned vessels of type t that will become available for use for the first time at the source on day h of the time horizon, and we let $CH_{h,t}$ be the number of vessels of type t that will become available for chartering on day h of the time horizon. Let $\alpha_{h,t} = O_{h,t} + CH_{h,t}$ and note that $O_t = \Sigma_h O_{h,t}$ and $CH_t = \Sigma_h CH_{h,t}$. Accordingly, we let $z_{h,t}$ be the integer variable that denotes the number of vessels of type t that are actually selected for chartering on day h of the time horizon and let $\$_{h,t}$ denote the average chartering cost of a vessel of type t that will become available for use on day h of the time horizon.

Let $A_{h,t}$ be the subset of indices for vessels of type t (both self-owned and vessels available for chartering) that will become available for use at the source for the first time on day h of the time horizon. Hence, for a given day h and vessel type t, we let $UT_{h,t} = \frac{\sum_{n \in A_{h,t}} UT_{t,n}}{\alpha_{h,t}}$, which basically gives the average usage allowance for a vessel of type t that will become available for use for the first time on day h of the time horizon. Accordingly, $UT_t = \sum_h UT_{h,t}$ gives the average usage allowance for a vessel of type t.

Note that in this aggregated model, the variable $y_{h,t}$ represents the maximum number of vessels of type t that could be consigned on day h as necessary; the actual number used, and in particular the chartering decisions, are governed in this model via the dispatching variables $x_{h,t}$. Similar to the index sets ϕ_X and ϕ_Y, we let ϕ_x and ϕ_y denote the index sets of x-and y-variables, respectively, that are *a priori* restricted to be zero, or fixed at some known positive integer values.

The aggregated model is formulated as stated below, where the S-variables are defined as for Model VSP.

AVSP:

Minimize $\displaystyle\sum_h \sum_t C_t x_{h,t} + \sum_h \pi [S_{2,h} + S_{4,h}] + \sum_h \lambda [S_{3,h} + S_{5,h}] + \sum_h \sum_t \$_{h,t} z_{h,t},$

subject to

(AC_1) $\quad S_h = w + \displaystyle\sum_t \sum_{\substack{h_1: \\ h_1 + T_{1,t} \in \{1, \dots, h\}}} \Omega_t x_{h_1,t} - TC_h, \ \forall h,$

(AC_2) $\quad S_h = S_{1,h} - S_{2,h} - S_{3,h} + S_{4,h} + S_{5,h}, \ \forall h,$

(AC_3) $\quad y_{h,t} = y_{h-1,t} - x_{h-1,t} + \displaystyle\sum_{\substack{h_1: \\ h_1 + T_t = h}} x_{h_1,t} + O_{h,t} + z_{h,t}, \ \forall h \geq 2, t,$

(AC_4) $\quad x_{h,t} \leq y_{h,t}, \ \forall h, t,$

(AC_5) $\quad y_{1,t} = O_{1,t} + z_{1,t}, \ \forall t,$

(AC_6) $\quad \displaystyle\sum_t \Omega_t x_{h,t} \leq Q, \forall h,$

$(AC_{7.1})$ $\quad z_t = \displaystyle\sum_h z_{h,t}, \ \forall h,$

$(AC_{7.2})$ $\quad \displaystyle\sum_h T_t \, x_{h,t} \leq UT_t \, (O_t + z_t), \ \forall t,$

$x_{h,t} \in \{0, 1, \dots, M_t\}, \ \forall h, t,$ if $x_{h,t} \notin \phi_x$ and fixed at zero or one otherwise,
$0 \leq y_{h,t} \leq M_t, \ \forall h, t,$ if $y_{h,t} \notin \phi_y$ and fixed at zero or one otherwise,

$\quad z_{h,t} \in \{1, \dots, CH_{h,t}\}, \ \forall h, t,$
$\quad S_h \geq 0, \ SL_1 \leq S_{1,h} \leq SL_2, \ 0 \leq S_{2,h} \leq A_1, \ 0 \leq S_{3,h} \leq b_1,$
$\quad 0 \leq S_{4,h} \leq A_2, \quad \text{and} \quad 0 \leq S_{5,h} \leq UB - b_2, \ \forall h.$

6.5.1.1   Objective Function and Constraints

The objective function of Model AVSP is similar to that of Model VSP, which represents the total operational costs for both the self-owned vessels and the chartered vessels, the penalties resulting from shortage or excess levels at the destination, and the chartering expenses. Constraints (AC_1), (AC_2), (AC_4), (AC_6), and $(AC_{7.2})$ used are basically representations of Constraints (C_1), (C_2), (C_4), (C_6), and (C_7), respectively, in an aggregate sense. Note that Constraint $(AC_{7.1})$ is a definitional constraint that computes the total number of vessels of type t that are actually selected for chartering. Constraint (AC_3) is a representation of (C_3) in an aggregate sense, however, the right-hand-side of (AC_3) also accounts for the first-time availabilities of the self-owned and chartered vessels. Note that Constraints (AC_5) in concert with Constraints (AC_3) and (AC_4) are sufficient to account for the chartered vessels.

Note that the parameters $O_{h,t}$ and $CH_{h,t}$ used in this model are necessary because, unlike as in Model VSP, we no longer maintain a track of individual vessels. The above reformulation adopts an aggregated approach as indicated by the integer decision variables that represent the number of dispatched vessels (without any individual vessel identity) over the time horizon, in lieu of using the previous binary variables. This formulation of the problem is more compact, however, at the expense of having to relax the individual vessel's total usage and downtime restrictions (C_7) to $(AC_{7.1})$, which now represents an aggregate usage constraint for vessels of type t, because we no longer specifically account for each individual vessel's activity. Likewise, for the commissioned chartered vessels, we assume that these vessels are available for the duration of use as prescribed by the model solution. This relaxed constraint needs to be dealt with separately while implementing the model-based decision. Note that the integrality of the y-variables is automatically enforced based on a similar reasoning as discussed in Remark 6.2 above.

REMARK 6.3
Consider the following constraint.

$$(\mathbf{AC_{5.1}})\ z_t \geq \left(x_{h,t} + \sum_{\substack{h_1 < h: \\ (h_1+T_t)>h}} x_{h_1,t} - O_{h,t} \right), \forall h, t.$$

Note that the right-hand-side of the Constraint $(AC_{5.1})$ yields the number of vessels of type t that are being used on day h beyond the self-owned number of vessels $O_{h,t}$. Thus, this constraint can be used to tighten the continuous relaxation of Model AVSP.

6.5.1.2   Extracting Schedules for Individual Vessels

A feasible schedule for individual vessels can be extracted from the solution of Model AVSP as follows. Having solved Model AVSP the variable $z_{h,t}$ at optimality specifies the number of vessels of type t that are needed for chartering on day h, and hence, we can determine the fleet size mix. Then, we can begin dispatching different vessels of type t on each day h based on the values of the x-variables. In this process, the downtime unavailabilities of various vessels and the balancing of days for which the different vessels are put into service could be incorporated. It is worth mentioning that, in practice, there is some flexibility in scheduling maintenance and in vessel usage constraints. Hence, this facilitates the conversion of the model solution to one that can be implemented on an individual ship basis without perturbing the overall solution and its associated cost and while satisfying ($C_{7.2}$) to the extent possible.

6.6 Computational Results and Rolling Horizon Heuristics

In this section, we present computational results related to solving Models VSP and AVSP based on ten test problems that represent various operational scenarios. The test problems are given in Appendix B. Section 6.6.1 provides computational experience for solving Models VSP and AVSP directly by the CPLEX package (version 7.5). In Section 6.6.2, we develop two rolling horizon heuristics to solve problem instances that cannot be solved directly via Model AVSP. Finally, in Section 6.6.3, we present an *ad-hoc* procedure that is intended to simulate a manual scheduling procedure used by KPC and the schedules obtained via this procedure are compared against those derived via the proposed modeling approach.

Notationally, we will let $\overline{P}$ denote the linear relaxation of any model P. The optimal objective function value of model P will be denoted by $v(P)$. The best upper bound and lower bound found for model P will be respectively denoted by $v_{UB}(P)$ and $v_{LB}(P)$. All runs below are made on a Pentium 4, CPU 1.70 GHz computer having 512 MB of RAM using CPLEX-7.5, with coding in Java. The test problem instances are labeled as defined in Appendix B: I_i for $i = 1, \ldots,$ 10. The symbol "•" will be used to indicate that no meaningful solution of a given model was obtained using CPLEX due to out-of-memory difficulties.

6.6.1 Computational Experience in Solving Models VSP and AVSP

Table 6.1 and Table 6.2 report computational results related to solving Models VSP and AVSP using CPLEX-MIP-7.5 based on the ten test problems.

REMARK 6.4
The number of constraints and variables in Model AVSP are substantially less than the respective number of constraints and variables in Model VSP, as observed from Table 6.1 and Table 6.2. For example, the number of constraints and variables for Model VSP based on test instance I_{10} are respectively given by 10,451 and 11,126, while those for Model AVSP for this test instance are 2,366 and 4,242, respectively. Note that these numbers are obtained from CPLEX after performing necessary preprocessing and elimination steps.

TABLE 6.1

Linear Relaxation of Model VSP

I_i	Rows	Columns	Nonzero Entries	$v(\overline{\text{VSP}})$ ($)	CPU Time (seconds)
I_1	493	598	4,460	390,000	0.01
I_2	503	727	7,787	615,000	0.01
I_3	1,636	1,811	24,601	615,000	0.08
I_4	1,614	1,958	35,859	1,290,000	0.11
I_5	5,176	5,519	146,415	3,930,000	0.61
I_6	4,399	4,728	114,657	4,290,000	0.50
I_7	5,082	5,487	173,115	7,400,000	0.69
I_8	10,590	11,106	441,532	6,630,000	1.61
I_9	8,812	9,407	421,789	7,980,000	1.53
I_{10}	10,451	11,126	556,298	14,999,166	13.89

This will motivate the utilization of Model AVSP in concert with rolling horizon algorithms that are designed below to solve problem instances having relatively long time horizons.

REMARK 6.5

Solutions for the linear relaxations of both Models VSP and AVSP were readily obtained for all the test problems. We were unable to obtain meaningful solutions for Model VSP for any of the test problems due to out-of-memory difficulties. For Model AVSP, we were able to solve it directly only for test problems $I_1, \ldots, I_4$, while for test problems $I_5, \ldots, I_{10}$, we encountered out-of-memory difficulties before reaching meaningful solutions.

6.6.2   Rolling Horizon Algorithms

In this section, we present rolling horizon algorithms similar to those proposed in [30] to facilitate the derivation of good quality solutions with reasonable effort for the test problem instances that cannot be solved directly

TABLE 6.2

Statistics Related to Solving Model AVSP

I_i	Rows	Columns	Nonzero Entries	$v(\overline{\text{AVSP}})$($)	CPU Time (seconds)	$v(\text{AVSP})$($)	CPU Time (seconds)
I_1	105	235	688	390,000	0.00	420,000	0.01
I_2	205	477	2,273	615,000	0.02	660,000	6,363.16
I_3	342	682	3,990	615,000	0.01	660,000	0.02
I_4	519	1,039	8,696	1,459,166	0.03	1,500,000	0.06
I_5	698	1,394	15,198	4,520,000	0.03	?	14,002.11
I_6	1,172	2,108	27,405	10,125,000	0.06	?	310,593.03
I_7	1,173	2,199	33,700	7,765,000	0.08	?	318,887.36
I_8	1,406	2,622	45,954	7,190,000	0.13	?	105,854.48
I_9	2,061	3,689	81,638	9,260,000	0.25	?	771,486.74
I_{10}	2,366	4,242	107,672	16,157,333	0.45	?	39,559.61

via Model AVSP. These algorithms are based on a sequential fixing of integer variables and are presented below along with related computational results. Note that since optimal solutions of Model AVSP based on test problems $I_1, \ldots, I_4$ were obtained using CPLEX, we only apply the proposed rolling horizon algorithms for test problem instances $I_5, \ldots, I_{10}$.

Let AVSP($\overline{H}$, opt‑gap) denote the relaxation of Model AVSP for which integrality is enforced only on the pertinent x-variables that correspond to day 1 through day $\overline{H}$ of the time horizon, and an optimal solution is required to be found within a tolerance "opt-gap" of optimality. Note that $v(\text{AVSP}(\overline{H}, \text{opt-gap}))$ provides a lower bound for problem AVSP, where $v(\overline{\text{AVSP}}) \leq v(\text{AVSP}(\overline{H}, \text{opt-gap}))$. Furthermore, increasing $\overline{H}$ and decreasing opt-gap will tighten this lower bound. Using the default CPLEX optimality gap, we solved AVSP($\overline{H}$, opt-gap) for the largest possible integer n, where $\overline{H} \equiv 10n \leq H$, such that a meaningful solution is obtained within a maximum of 7 hours of run time. Let $v_{LB} = v(\text{AVSP}(\overline{H}, \text{opt-gap}))$ for the largest computationally feasible $\overline{H}$ as discussed above. Table 6.3 displays the results obtained.

(A) Rolling Horizon Algorithm RHA1

In Model AVSP, let the vector x be partitioned as $(x_1, x_2, \ldots, x_H)$, where x_h denotes the vector of x-variables associated with the hth day of the time horizon. Let H_1 be the length of the horizon interval for which the corresponding x-variables are restricted to be integer valued, and the remaining variables are declared to be continuous. Accordingly, in a rolling-horizon framework, let H_2 be the duration of the initial subset of this interval for which the determined decisions are permanently fixed. Let $KK = \lceil ((H - H_1)/H_2) + 1 \rceil$ and let AVSP(H_1, H_2, opt-gap, k), for $k = 1, \ldots, KK$, denote Model AVSP having the following characteristics:

(a) x_h is enforced to be integer valued for $h \leq H_1 + (k - 1)H_2$, and is relaxed to be continuous otherwise, $\forall h = 1, \ldots, H$.

(b) x_h for $h \leq (k - 1)H_2$ is fixed at the values found from the solution to Model AVSP(H_1, H_2, opt-gap, $(k - 1)$).

(c) The optimality gap tolerance for fathoming is set at opt-gap.

The rolling-horizon algorithm (RHA1) then proceeds as follows.

TABLE 6.3

Statistics Related to Solving $v(\text{AVSP}(\overline{H}, \text{opt-gap}))$

I_i	v_{LB} (\$)	$\overline{H}$ (days)	CPU Time (seconds)
I_5	4,580,000	80	0.28
I_6	11,069,999	80	2,501.30
I_7	7,840,000	80	0.61
I_8	7,265,000	90	0.73
I_9	9,335,000	90	1.61
I_{10}	16,265,333	90	2.44

Initialization: Let H_1 be some integer number less than or equal to H, $H_2 \leq (H_1/2)$, and $k = 1$. Let opt-gap be some selected optimality gap criterion. Solve Model AVSP(H_1, H_2, opt-gap, 1).

Main Step: If $k = KK$; then terminate the algorithm; the proposed solution is that obtained from solving Model AVSP(H_1, H_2, opt-gap, KK). Otherwise, increment k by one and solve the Model AVSP(H_1, H_2, opt-gap, k). Repeat the Main Step.

Table 6.4 reports the results obtained by using RHA1 for some fixed judicious values of H_1 and H_2 as determined via some computational experimentation, where opt-gap is set to the CPLEX default optimality tolerance. Here, v_{RHA1} gives the solution value obtained by the rolling-horizon algorithm RHA1, and perct_opt(v_{RHA1}) $\equiv 100 \left(1 - \frac{v_{\text{RHA1}} - v_{LB}}{v_{\text{RHA1}}}\right)$ gives the percentage of optimality of v_{RHA1} with respect to the lower bound v_{LB}.

(B) Rolling Horizon Algorithm RHA

Motivated by the fact that more stringent values of H_1 and opt-gap can enhance the quality of the solution obtained, we devised another rolling horizon algorithm as follows.

Divide the time horizon into $\omega \geq 1$ disjoint partitions, HP_1, HP_2, $\ldots$, HP_ω that together span the time horizon $\{1, \ldots, H\}$, where each partition covers at least two time periods. We will denote each such HP_i as $\{HI^i, \ldots, HF^i\}$ for $i = 1, \ldots, \omega$, where $HF^i \geq HI^i + 1$, $HI^1 = 1$, $HF^\omega = H$, and $HI^i = HF^{i-1} + 1$, $\forall i$.

Accordingly, let $\text{AVSP}_1 = \text{AVSP}(HP_1, \text{opt-gap})$, and for $i = 2, \ldots, \omega$, let $\text{AVSP}_i = \text{AVSP}(HP_i, \text{opt-gap})$ where the values of the x-variables that correspond to the days in $\cup_{j=1}^{i-1} HP_j$ are fixed as obtained from $\text{AVSP}_1, \ldots, \text{AVSP}_{i-1}$.

The rolling horizon algorithm RHA2 then proceeds as follows.

Initialization: Let ω be some positive integer. Let $i = 1$ and let opt-gap be some selected optimality gap criterion. Solve AVSP_1.

Main Step: If $i = \omega$, then terminate the algorithm; the proposed solution is that obtained from solving Model AVSP_ω. Otherwise, increment i by one and solve Model AVSP_i. Repeat the Main Step.

TABLE 6.4

Statistics Related to Solving AVSP Using RHA1

| I_i | $H_1 = 30, H_2 = 10$ (days) | | perct_opt (v_{RHA1}) (Percentage Optimality for RHA1) |
	$v_{RHA1}(\$)$	CPU Time (seconds)	
I_5	5,008,541	30.22	91.44
I_6	13,309,002	256.56	83.17
I_7	9,620,070	15.78	81.49
I_8	9,253,200	128.90	75.51
I_9	11,213,700	33.98	83.24
I_{10}	20,395,863	20.19	79.74

TABLE 6.5

Statistics Related to Solving AVSP Using RHA2

I_i	$v(\text{AVSP}_1)$ ($)	HF^1 (days)	CPU Time (seconds)	$v(\text{AVSP}_2)$ ($)	HF^2 (days)	CPU Time (seconds)	$v(\text{AVSP}_3)$ ($)	HF^3 (days)	CPU Time (seconds)
I_5	4,580,000	80	0.28	4,680,000	120	47.41	N/A		
I_6	11,069,999	80	2,501.30	11,400,000	120	0.52	N/A		
I_7	7,840,000	80	0.61	7,900,000	120	0.83	8,000,000	150	22.14
I_8	7,265,000	90	0.73	7,310,000	150	860.27	7,400,000	180	1.64
I_9	9,335,000	90	1.61	9,339,999	170	49.89	9,360,000	210	0.95
I_{10}	16,265,333	90	2.44	16,293,333	170	34.52	16,450,000	240	5.77

Let v_{RHA2} denote the solution value obtained by the rolling horizon algorithm RHA2, and let $\text{perct\_opt}(v_{\text{RHA2}}) = 100(1 - \frac{v_{\text{RHA2}} - v_{LB}}{v_{\text{RHA2}}})$, which gives the percentage of optimality of v_{RHA2} with respect to the lower bound v_{LB}. Moreover, let $\text{perct\_imp}(v_{\text{RHA2}}, v_{\text{RHA1}}) = 100(\frac{v_{\text{RHA1}} - v_{\text{RHA2}}}{v_{\text{RHA1}}})$, which gives the percentage of improvement in total cost of algorithm RHA2 over algorithm RHA1. Table 6.5 and Table 6.6 report statistics related to algorithm RHA2, where opt-gap is set to the default CPLEX optimality gap.

Note that the sets of test problems $\{I_5, I_6\}$, and $\{I_7, \ldots, I_{10}\}$ were solved using two and three partitions, respectively. Furthermore, algorithm RHA2 produced better results than RHA1 for all the test problems under consideration. Note that the performance of these algorithms (particularly the latter) may be enhanced by using higher-grade computers that would permit fixing or relaxing fewer variables at each step.

6.6.3 Comparison with an *ad-hoc* Scheduling Procedure

In this section, we compare the performance of the proposed modeling approach vs. an *ad-hoc* scheduling procedure that represents the process adopted by Kuwait Petroleum Corporation. Since chartering expenses are large relative to operational costs and penalties imposed for undesirable

TABLE 6.6

Statistics Related to Solving AVSP Using RHA2

I_i	v_{RHA2}($)	Total CPU Time (seconds)	perct_opt (v_{RHA2}) (Percentage Optimality for RHA2)	perct_imp ($v_{\text{RHA2}}, v_{\text{RHA1}}$) (Percentage Improvement of RHA2 over RHA1)
I_5	4,680,000	47.69	97.86	6.56
I_6	11,400,000	2,501.82	97.10	14.34
I_7	8,000,000	23.58	98.00	16.84
I_8	7,400,000	862.64	98.17	20.02
I_9	9,360,000	52.45	99.73	16.53
I_{10}	16,450,000	42.73	98.87	19.34

storage levels, this *ad-hoc* procedure attempts to fully utilize company-owned vessels before resorting to chartered vessels, and operates as follows:

(A) Examine the first day of the time horizon, say h, on which delivery of shipments is possible without exceeding SL_2, and dispatch company-owned vessels for delivery on day h so that the total storage level on that day will not exceed SL_2. This involves exploring various feasible departure days, and the usage of different combinations of vessels. This process is repeated for the company-owned vessels, taking into account the availability of vessels during the days of the time horizon, until it becomes impossible to do so.

(B) Repeat Step A, however, without exceeding b_2, i.e., allowing for the Type I penalty when the storage level lies within $(SL_2, b_2]$.

(C) Repeat Step A, however, without exceeding UB, i.e., allowing for the Type I and Type II penalties when the storage level lies within $(SL_2, UB]$.

(D) Now, if the storage level on any given day of the horizon is not below zero, then we are done, and hence, there is no need for chartered vessels. Otherwise, let hh be a day when the storage level becomes negative, and select the smallest sized vessel that is available for chartering to be dispatched for delivery on a day in $(hh - \delta)$, where δ is to be determined by the scheduler. Repeat this step as necessary, however, using the already selected chartered vessel(s). At each pass through this process, if the storage level on any given day of the time horizon is nonnegative, then we are done. On the other hand, if there exist no more chartered vessels while for some day we still have a storage level below zero, then no feasible solution is found.

Let AH denote the foregoing *ad-hoc* procedure. Table 6.7 presents some computational statistics for comparing schedules obtained via Procedure AH with those generated by the proposed modeling approach. Here, v_{AH} represents the total cost of the solution obtained via AH and v_{min} gives the objective value of the best solution obtained via the proposed modeling approach. Also, let perct_imp$(v_{min}, v_{AH}) = 100[(\frac{v_{AH}-v_{min}}{v_{AH}})]$, which gives the percentage improvement in the total cost of the schedules generated by the proposed modeling approach over those obtained via the *ad-hoc* procedure.

Observe that for each of the test problems $I_1, \ldots, I_{10}$, the overall cost obtained via the proposed modeling approach is often substantially better than the overall cost obtained via the *ad-hoc* procedure AH. For example, in test problem I_{10}, the improvement in total cost obtained via the modeling approach over that obtained via the existing AH procedure is $47,090,000. Notice also that there is a large variance in the performance of the *ad-hoc* procedure AH vs. the proposed approach. The reason for this is two-fold. First, the AH procedure makes myopic decisions and is unable to recognize complex compromises that sometimes need to be made for attaining an overall efficient

TABLE 6.7

Comparison of the *ad-hoc* Procedure AH vs. the Proposed
Modeling Approach

I_i	v_{min} ($)	v_{AH} ($)	(Percentage Improvement of the Proposed Approach over the *Ad-hoc* Approach) perct_imp (v_{min}, v_{AH})
I_1	420,000	720,000	41.66
I_2	660,000	720,000	8.33
I_3	660,000	2,480,000	73.38
I_4	1,500,000	3,940,000	61.93
I_5	4,680,000	6,720,000	30.35
I_6	11,400,000	42,178,000	72.97
I_7	8,000,000	10,880,000	26.47
I_8	7,400,000	8,282,000	10.65
I_9	9,360,000	56,450,000	83.42
I_{10}	16,450,000	24,140,000	31.85

solution. Second, because of such myopic decisions that unwisely use the self-owned vessel resources, the *ad-hoc* procedure often needs to resort to unnecessary chartering of vessels, which is an expensive venture. On the other hand, the proposed modeling approach makes more effective and robust decisions.

6.7 Summary, Conclusions, and Future Research

In this chapter, we presented mixed-integer programming models for determining an optimal mix of vessels of different types that are needed to transport a product from a source to a destination based on a stream of consumption rates at the destination's facility. Various cost components such as daily operational costs of vessels, chartering expenses, and penalties associated with undesirable storage levels are incorporated in the models. Such single source-destination vessel scheduling problems are faced by oil companies, for example, in which the product is crude oil, the source is a refinery facility, and the destination is a storage location that belongs to a client. Problems of this type also arise in practice where large quantities of crude oil need to be shipped from a country such as Kuwait to specific aggregated clusters of locations in Europe, North America, or Asia.

Due to the combinatorial nature of the problem, a manual scheduling of vessels is often expensive and requires an inordinate amount of effort for constructing and revamping vessel schedules. Therefore, it is imperative to utilize modeling approaches to advantageously compromise between the various cost components in order to avoid unduly high vessel chartering and penalty expenses. The proposed modeling approach enables the transporting organization to generate and revamp vessels' schedules as frequently as necessary and in a timely fashion. This approach also allows the organization to contemplate long-term plans regarding the size of the self-owned fleet and the need for chartered vessels.

The efficiency of the proposed modeling approach is assessed by comparing it against an *ad-hoc* scheduling procedure that represents the actual scheduling of vessels in a related case study concerning Kuwait Petroleum Corporation (KPC). Using a set of ten realistic test problems, the results indicate that the proposed approach substantially improves upon the manual procedure, resulting in savings ranging from $60,000 to $47,090,000.

The current manual practice at KPC suffers from a lack of robustness because of the myopic nature of decisions made. Often, such decisions encumber the self-owned vessels ineffectively, resulting in relatively large expenses for chartering additional vessels. On the other hand, the developed procedure determines near-optimal solutions more robustly within 97 to 99 percent of optimality.

This work can be extended to examine the cost effectiveness of simultaneously investigating multiple sources and destinations instead of associating specific vessels with designated source-destination combinations. We can also explore the impact of leasing temporary transshipment storage depots on the overall chartering and penalty costs, whereby fewer chartered vessels might need to be acquired. These two extensions are the subject of a companion follow-on paper [31].

Acknowledgments

This research work was supported by Kuwait University under Research Grant No. SM-06/02 and the National Science Foundation under Research Grant No. DMI-0094462. Special thanks to Mrs. Lulwa Al-Shebeeb for her contribution in the computational implementation of the solution algorithms.

References

1. Al-Yakoob, S.M., Mixed-integer mathematical programming optimization models and algorithms for an oil tanker routing and scheduling problem, PhD. dissertation, Department of Mathematics, Virginia Polytechnic Institute and State University, Blacksburg, 1997.
2. Appelgren, L.H., A column generation algorithm for a ship scheduling problem, *Transportation Science*, 3, 53, 1969.
3. Appelgren, L.H., Integer programming methods for a vessel scheduling problem, *Transportation Science*, 5, 64, 1971.
4. Bausch, D.O., Brown, G.G., and Ronen, D., Scheduling short-term marine transport of bulk products, *Maritime Policy and Management*, 25(4), 335, 1998.
5. Brown, G.G., Graves, G.W., and Ronen, D., Scheduling ocean transportation of crude oil, *Management Science*, 32, 335, 1983.

6. Cho, S.C. and Perakis, A.N., Optimal liner fleet routing strategies, *Maritime Policy and Management*, 23(3), 249, 1996.
7. Christiansen, M., Decomposition of a combined inventory and time constrained ship routing problem, *Transportation Science*, 3(1), 3, 1999.
8. Christiansen, M., Fagerholt, K., and Ronen, D., Ship routing and scheduling: status and prospective, *Transportation Science*, 38(1), 1, 2004.
9. Datz, I.M., Fixman, C.M., Friedberg, A.W., and Lewinson, V.A., A description of the maritime administration mathematical simulation of ship operations, *Trans. SNAME*, 493, 1964.
10. Fagerholt, K., A computer-based decision support system for vessel fleet scheduling — Experience and future research, *Decision Support Systems*, 37(1), 35, 2004.
11. Fagerholt, K. and Christiansen, M., A combined ship scheduling and allocation problem, *Journal of the Operational Research Society*, 51(7), 834, 2000a.
12. Fagerholt, K. and Christiansen, M., A traveling salesman problem with allocation time window and precedence constraints — an application to ship scheduling, *International Transactions in Operational Research*, 7(3), 231, 2000b.
13. Jaramillo, D.I. and Perakis, A.N., Fleet deployment optimization for liner shipping, Part 2: Implementation and results, *Maritime Policy and Management*, 18(4), 235, 1991.
14. Koenigsberg, E. and Lam, R.C., Cyclic queue models of fleet operations, *Operations Research*, 24(3), 516, 1976.
15. Koenigsberg, E. and Meyers, D.A., An interacting cyclic queue model of fleet operations, *The Logistics and Transportation Review*, 16, 59, 1980.
16. Lane, D.E., Heaver, T.D., and Uyeno, D., Planning and scheduling for efficiency in liner shipping, *Maritime Policy and Management*, 14(2), 109, 1987.
17. Lawrence, S.A., *International Sea Transport: the Years Ahead*, Lexington Books, Lexington, 1972.
18. Nemhauser, G.L. and Yu, P.L., A problem in bulk service scheduling, *Operations Research*, 20, 813, 1972.
19. Perakis, A.N. and Jaramillo, D.I., Fleet deployment optimization for liner shipping, part 1: Background, problem formulation and solution approaches, *Maritime Policy and Management*, 18(3), 183, 1991.
20. Perakis, A.N., Fleet operations optimization and fleet deployment, in *The Hand Book of Maritime Economics and Business*, Grammenos, C.T., Ed., Lloyds of London Publications, London, 2002, 580.
21. Perakis, A.N. and Bremer, W.M., An operational tanker scheduling optimization system: background, current practice and model formulation, *Maritime Policy and Management*, 19(3), 177, 1992.
22. Powell, B.J. and Perakis, A.N., Fleet deployment optimization for liner shipping: an integer programming model, *Maritime Policy and Management*, 24(2), 183, 1997.
23. Psaraftis, H.N., Foreword to the focused issue on maritime transportation, *Transportation Science*, 33(1), 1, 1999.
24. Rana, K. and Vickson, R.G., A model and solution algorithm for optimal routing of a time-chartered containership, *Transportation Science*, 22(2), 83, 1988.
25. Rana, K. and Vickson, R. G., Routing container ships using Lagrangian relaxation and decomposition. *Transportation Science*, 25(3), 201, 1991.
26. Ronen, D., Cargo ships routing and scheduling: survey of models and problems, *European Journal of Operational Research*, 12, 119, 1983.

27. Ronen, D., Ship scheduling: the last decade, *European Journal of Operational Research*, 71, 325, 1993.

28. Ronen, D., Marine inventory routing: shipment planning, *Journal of the Operational Research Society*, 53, 108, 2002.

29. Scott, J. L., A transportation model, its development and application to a ship scheduling problem, *Asia-Pacific Journal of Operational Research*, 12, 111, 1995.

30. Sherali, H.D., Al-Yakoob, S.M., and Merza, H., Fleet management models and algorithms for an oil-tanker routing and scheduling problem, *IIE Transactions*, 31, 395, 1999.

31. Sherali, H.D., Al-Yakoob, S.M. Determining an optimal fleet mix and schedules: Part II—multiple sources and destinations, and the option of leasing transshipment depots. Department of Mathematics and Computer Science, Kuwait University, 2004. (Part I of the sequence is the present chapter.)

32. Xinlian, X., Tangfei, W., and Daisong, C., A dynamic model and algorithm for fleet planning, *Maritime Policy and Management*, 27(1), 53, 2000.

Appendix A: Glossary of Notation

- $h = 1, \ldots, H$: Indices of the days of the time horizon.
- $t = 1, \ldots, T$: Indices for the types of vessels.
- Ω_t : Capacity (in barrels) of a vessel of type t.
- M_t : Number of vessels of type t.
- $n = 1, \ldots, M_t$: Indices of all vessels of type t.
- O_t and $CH_t = M_t - O_t$: Respectively, the number of self-owned vessels and the number of available vessels of this type that can be possibly chartered.
- $n = 1, \ldots, O_t$: Indices for the self-owned vessels.
- $n = O_t + 1, \ldots, O_t + CH_t \equiv M_t$: Indices for the self-owned vessels of type t that are available for chartering.

- $O = \displaystyle\sum_{t=1}^{T} O_t.$

- $CH = \displaystyle\sum_{t=1}^{T} CH_t.$

- $\$_{t,n}$: Cost (in U.S. dollars) of chartering a vessel n of type t.
- $UT_{t,n}$: Maximum number of days vessel n of type t can be used during the time horizon.
- T_t : Time (in days) required to load a vessel of type t at the source, plus the time this vessel takes to travel from the source to the destination, unload time at the destination, and then travel back from the destination to the source.

- $T_t = T_{1,t} + T_{2,t}$.
- $T_{1,t}$: Time (in days) required to load a vessel of type t at the source and then travel to the destination.
- $T_{2,t}$: Time (in days) required to unload a vessel of type t at the destination and then travel back to the source.
- $DC_{t,n}$: Daily operational cost (in U.S. dollars) of vessel n of type t.
- $C_{t,n} = T_{1,t} (DC_{t,n})$.
- Q : A production capacity (in barrels) or certain imposed quota of the product at the source.
- w : Storage level (in barrels) at the destination at the beginning of the time horizon.
- SL_1 and SL_2 : The minimum and maximum desired levels (in barrels), respectively, at the destination's storage facility, which should be maintained to the extent possible in order to avoid penalties.
- π : Daily penalty (in U.S. dollars) for each shortage or excess unit at the destination.
- A_1 and A_2 : Permitted shortage and excess quantities (in barrels) at the destination with respect to the desired levels SL_1 and SL_2, respectively.
- $b_1 = SL_1 - A_1$.
- $b_2 = SL_2 + A_2$.
- UB : Upper bound on the maximum storage level (in barrels) at the destination.
- λ : Type II penalty (in U.S. dollars).
- R_j : Expected consumption rate (in barrels) at the destination on day j, for $j = 1, \ldots, H$.
- $TC_h = \sum_{j=1}^{h} R_j$.
- S_h : A continuous variable representing the storage level on day h.
- $P_I(S_h) = \pi \text{ maximum } \{0, (SL_1 - S_h), (S_h - SL_2)\}$ if $S_h \in [b_1, b_2]$.
- $P_{II}(S_h) = \begin{cases} \pi A_1 + \lambda (b_1 - S_h) & \text{if } S_h \in (0, b_1) \\ \pi A_2 + \lambda (S_h - b_2) & \text{if } S_h \in (b_2, UB). \end{cases}$
- $S_h = S_{1,h} - S_{2,h} - S_{3,h} + S_{4,h} + S_{5,h}$,

  where $SL_1 \leq S_{1,h} \leq SL_2, 0 \leq S_{2,h} \leq A_1, 0 \leq S_{3,h} \leq b_1,$
  $0 \leq S_{4,h} \leq A_2,$ and $0 \leq S_{5,h} \leq UB - b_2.$
- $P(S_h) = \pi (S_{2,h} + S_{4,h}) + \lambda (S_{3,h} + S_{5,h}).$
- $X_{h,t,n} = \begin{cases} 1 & \text{if vessel } n \text{ of type } t \text{ departs the source toward} \\ & \text{the destination on day } h, \\ 0 & \text{otherwise.} \end{cases}$

- $Y_{h,t,n} = \begin{cases} 1 & \text{if vessel } n \text{ of type } t \text{ is available at the source on day } h, \\ 0 & \text{otherwise.} \end{cases}$

- $Z_{t,n} = \begin{cases} 1 & \text{if vessel } n \text{ of type } t \text{ is selected for chartering during} \\ & \text{(all or only part of) the time horizon,} \\ 0 & \text{otherwise.} \end{cases}$

- ϕ_X and ϕ_Y : Sets of X and Y variables, respectively, that are restricted to be fixed at specified binary values by virtue of such considerations.

- DC_t : Average daily operational cost (in U.S. dollars) of a vessel of type t.

- $C_t = T_t\,(DC_t)$.

- $x_{h,t}$: An integer variable that represents the number of vessels of type t that are dispatched from the source on day h.

- $y_{h,t}$: An integer decision variable that represents the maximum number of vessels of type t that are available for dispatching from the source on day h.

- $O_{h,t}$: Number of vessels of type t that will become available for use at the source for the first time on day h of the time horizons.

- $CH_{h,t}$: Number of vessels of type t that will become available for chartering on day h of the time horizon.

- $\alpha_{h,t} = O_{h,t} + CH_{h,t}$: Number of vessels of type t that will become available for use on day h of the time horizon.

- $O_t = \sum_{h} O_{h,t}$.

- $CH_t = \sum_{h} CH_{h,t}$.

- $z_{h,t}$: An integer variable that denotes the number of vessels of type t that are actually selected for chartering on day h of the time horizon.

- $\$_{h,t}$: Average chartering cost (in U.S. dollars) of a vessel of type t that will become available for use on day h of the time horizon.

- $A_{h,t}$: A subset of indices for the vessels of type t (both self-owned and vessels available for chartering) that will become available for use at the source for the first time on day h of the time horizon.

- $UT_{h,t} = \dfrac{\sum_{n \in A_{h,t}} UT_{t,n}}{\alpha_{h,t}}$, which basically gives the average usage allowance (in days) for a vessel of type t that will become available for use for the first time on day h of the time horizon.

- ϕ_x and ϕ_y : Sets of x- and y-variables, respectively, that are a priori restricted to be zero, or fixed at some known positive integer value.

- $\overline{P}$: Linear relaxation of any model P.
- $v(P)$: Optimal objective function value of model P.
- $v_{UB}(P)$: Best upper bound (solution) found for model P.
- $v_{LB}(P)$: Best lower bound found for model P.
- I_i : Test problem i, for $i = 1, \ldots, 10$.
- "$\cdot$": Indicate that no meaningful solution of a given model is obtained by a direct application of CPLEX due to out-of-memory difficulties.
- AVSP($\overline{H}$, opt-gap) : Relaxation of Model AVSP for which integrality is enforced only on the pertinent x-variables that correspond to day 1 through day $\overline{H}$ of the time horizon, and an optimal solution is required to be found within a tolerance "opt-gap" of optimality.
- RHA1 and RHA2 : Rolling horizon algorithms.
- v_{RHA1} : Solution value (in U.S. dollars) obtained by RHA1.
- $\text{perct\_opt}(v_{RHA1}) \equiv 100 \left(1 - \frac{v_{RHA1} - v_{LB}}{v_{RHA1}} \right)$
- v_{RHA2} : Solution value (in U.S. dollars) obtained by RHA2.
- $\text{perct\_opt}(v_{RHA2}) = 100 \left(1 - \frac{v_{RHA2} - v_{LB}}{v_{RHA2}} \right)$.
- $\text{perct\_imp}(v_{RHA2}, v_{RHA1}) = 100 \left(\frac{v_{RHA1} - v_{RHA2}}{v_{RHA1}} \right)$.
- AH: The *ad-hoc* procedure.
- v_{AH} : Total cost (in U.S. dollars) obtained via AH.
- v_{min} : Objective value (in U.S. dollars) of the best solution obtained via the proposed modeling approach.
- $\text{perct\_imp}(v_{min}, v_{AH}) = 100 \left[\left(\frac{v_{AH} - v_{min}}{v_{AH}} \right) \right]$.

Appendix B: Test Problems

This appendix presents statistics related to ten test problems. Assume that there are four vessel types. Note that the models described in this chapter

TABLE B.1

Vessel Types

t	Ω_t (barrels)	$\$_t(\$)$	$DC_t(\$)$
1	400,000	3,000,000	6,000
2	600,000	4,000,000	8,000
3	800,000	5,500,000	10,000
4	1,200,000	7,000,000	13,500

TABLE B.2

Test Problems

I_i	H (days)	T (vessel types)	(O_1, O_2, O_3, O_4) (vessels)	(CH_1, CH_2, CH_3, CH_4) (vessels)	(T_1, T_2, T_3, T_4) (days)	w (barrels)
I_1	30	1	(4,0,0,0)	(5,0,0,0)	(10,0,0,0)	4,000,000
I_2	60	1	(2,0,0,0)	(2,0,0,0)	(10,0,0,0)	7,000,000
I_3	60	2	(2,3,0,0)	(2,7,0,0)	(10,20,0,0)	7,000,000
I_4	90	2	(2,2,0,0)	(2,3,0,0)	(10,20,0,0)	7,000,000
I_5	120	2	(3,4,0,0)	(8,7,0,0)	(20,30,0,0)	7,000,000
I_6	120	4	(1,1,1,1)	(3,4,2,6)	(20,30,40,50)	7,000,000
I_7	150	3	(2,5,2,0)	(1,5,2,0)	(20,30,40,0)	4,000,000
I_8	180	3	(5,1,6,0)	(2,7,9,0)	(20,30,40,0)	7,000,000
I_9	210	4	(4,2,1,4)	(3,2,1,4)	(20,30,40,50)	7,000,000
I_{10}	240	4	(5,2,2,2)	(1,2,2,6)	(30,40,50,60)	7,000,000

consider a single source-destination operation. We assume that the daily consumption rate for a given destination is the same for the duration of a given time-horizon (denoted by R_1), and that the daily operational costs of vessels of a given type are also the same. The Type I penalty is fixed at \$3 and the Type II penalty is fixed at \$500. Specific details for the test problems are presented in Table B.1 and Table B.2, where Table B.2 describes our set of ten test problems. Note that $R_1 = 150,000$, $SL_1 = 2,250,000$, $SL_2 = 10,000,000$, and $A_1 = A_2 = 1,000,000$ for all the test problems. Also, note that $T_i = 0$ indicates that no vessels of type i are involved in the operation.

7

Determining an Optimal Fleet Mix and Schedules: Part II — Multiple Sources and Destinations, and the Option of Leasing Transshipment Depots

Hanif D. Sherali and Salem M. Al-Yakoob

CONTENTS

7.1   Introduction

Scheduling the shipment of products to destinations is a challenging problem faced by many organizations that are interested in improving the efficiency of their transportation systems due to the high cost associated with operational and penalty costs. Efficient transportation systems have the potential for enormous savings. For example, in the oil industry, a typical vessel in a fleet of oil tankers usually costs millions of U.S. dollars, and the daily operational costs of an oil tanker amounts to tens of thousands of U.S. dollars. Furthermore, considerable penalties are levied for either shortages in fulfilling customer demands, or requiring customers to carry higher than desired levels of inventory during certain periods. Hence, a formal modeling approach that compromises among these different cost components in an effective manner becomes imperative.

7.1.1   Problem Statement

Without loss of generality, we assume that the product to be transported is crude oil and that the transporting vessels are oil tankers. Accordingly, the terms "crude oil" and "product" will be used interchangeably. Hence, suppose that an organization is concerned with transporting a product from different source points to customers located at various destinations. The source points might represent refineries or storage facilities either owned or leased by the organization. There also exist certain specified delivery time-windows for the product from the sources to the destinations that are governed by a number of factors. These factors include the availability of the product at the sources, storage capacities at the destination facilities, rates of consumption of the product at the destinations, penalties imposed by customers due to unacceptable storage levels, routes available for transportation, vessels available for transportation, transshipment depots available for leasing, and contracted constraints agreed upon between the organization and the respective customers. The fleet of vessels utilized by the organization may be composed of self-owned vessels, chartered vessels, or a mix of both.

The leasing of transshipment depots is a strategic decision that is open to the organization in seeking an overall efficient and cost effective solution. The organization aims to satisfy customer demands for the product based on agreed-upon contracts at a minimum total cost that is comprised of the operational costs of the vessels, chartering expenses, penalties associated with violating certain storage levels, and the cost of leasing storage facilities.

7.1.2   Related Research

This work is an extension to the work of the authors in [6] and [7]. In [6] the authors investigated a problem concerned with the scheduling of oil tankers to ship various products from one source to different destinations. The demand structures that were investigated were determined by the total demand for each product at each destination, along with the respective subdemands that have to be satisfied within specified time intervals. It was assumed that some vessels can carry more than one product. In the previous chapter, the authors considered a vessel scheduling problem to transport a product from a single source to a single destination in which the demand for the product is governed by the consumption rates at the destination, with an allowance for over- or under-shipments, subject to appropriate penalties. Two mixed-integer programming models were developed for this problem: a full-scale representation that incorporates all the detailed features of the problem, and an aggregate model that suppresses the individual identities of the vessels of each type, while incorporating this aspect later when prescribing the actual consignment decisions.

The problem addressed here is similar to that investigated in [7], with the exception that we now accommodate multiple sources and destinations and moreover, we incorporate the consideration that the organization can lease temporary transshipment storage facilities having known locations. Note that by leasing suitable transshipment depots, a cost savings could be realized by stocking the shipments closer to the destinations during slack periods for use during periods of higher demand, while using fewer vessels, and in particular, avoiding the high costs associated with chartering vessels. Incorporating multiple sources and destinations, along with the option of leasing transshipment storage facilities in the problem, necessitates novel modeling and algorithmic considerations in order to derive good quality strategic decisions with a reasonable computational effort.

In general, vessel routing and scheduling problems can be partitioned into four categories (see, for example, Al-Yakoob [1] and Christiansen [2]): a) liner, b) tramp, c) industrial, and d) other related models. The bulk of the water transportation routing and scheduling models deal with the transport and delivery of cargo. Significantly fewer models have been developed to tackle vessel routing and scheduling problems in the context of passengers. This is a consequence of the fact that most vessels transport cargo around the world, whereas passengers mainly travel by air or land. Similar to the vessel scheduling problems that are investigated in [6] and [7], the problem that is considered here falls into the fourth category mentioned above. For further details on vessel scheduling and related models, the reader may refer to Al-Yakoob [1], Christiansen [2], Perakis [3], and Ronen [4, 5], as well as to the more detailed literature review given in Part I of this study [7].

The remainder of this chapter is organized as follows. The next section provides a description of the problem, and Section 7.3 presents a mixed-integer programming model for the problem. Due to the immense size of this model for most practical problem instances, Section 7.4 proposes an aggregate reformulation of the problem. Rolling horizon heuristics along with related computational results based on a set of test instances representing various operational scenarios are presented in Section 7.5. Section 7.6 provides results pertaining to potential enhancements that can be attained via the leasing of transshipment depots, and Section 7.7 concludes this chapter with a summary, future research, and recommendations pertaining to implementation.

7.2 Problem Description

As mentioned earlier, the problem considered here is an extension to the work of the authors in [7]. Accordingly, the demand structure, consumption rates, and penalty representation considered in this part are similar to those used in [7]. These are briefly described below for the sake of completeness and ease in presentation. Pertinent notation related to the present work is also introduced in this section. For the sake of reader's convenience, a glossary of notation (sequenced in the order of appearance) is provided in appendix A.

7.2.1 Demand Structure

The demand structure for the problem is influenced by the following factors: a) the storage capacities at client destinations; b) initial levels of storage at destinations; c) the rates of consumption at client storage facilities; and d) other customer-specific requirements. The rate of consumption of the product might not be fixed throughout the entire time horizon. The storage level at any client facility is desired to lie between some minimum and maximum allowable limits on any given day of the time horizon. Some customers impose penalties when the storage level on a given day falls below the minimum desired level or exceeds the maximum specified level. The latter maximum level might be the storage capacity of the facility, and exceeding this capacity would require resorting to extra temporary storage facilities, thereby resulting in additional cost penalties. More specific details on the characterization of the demand structure are prescribed below.

7.2.2 Problem Notation

Contracts between the transporting company and clients are signed sequentially over time and accordingly, scheduling decisions are updated every time a new contract is signed. Note that each contract specifies a time horizon along with required storage levels to fulfill demand. A time horizon associated with a contract is also referred to as a "contract horizon". Let $CT^{aa}_{hh,HH}$ denote a contract that is signed on day aa having a time horizon given by $\{hh, \dots, HH\}$. Suppose that scheduling decisions need to be made at time aa to fulfill demand requirements for this contract as well as a set of previously signed contracts, say γ of them, that have not been fulfilled by time aa. Let $CT_{hh_1,HH_1}, \dots, CT_{hh_\gamma,HH_\gamma}$ denote these previously signed contracts having time horizons given by $\{hh_1, \dots, HH_1\}, \dots, \{hh_\gamma, \dots, HH_\gamma\}$, respectively.

Note that for $i = 1, \ldots, \gamma$, $HH_i > aa$. Let $AA = \max\{HH_1, HH_2, \ldots, HH_\gamma, HH\}$. Then the collective time horizon for the contracts $CT_{hh_1, HH_1}, \ldots, CT_{hh_\gamma, HH_\gamma}$, and $CT_{hh, HH}^{aa}$ is given by $\{aa, \ldots, AA\}$. For the sake of simplicity, we let $h = 1, \ldots, H$ index the days of the collective time horizon, which will henceforth be referred to as the "time horizon."

Let $t = 1, \ldots, T$ denote the types of vessels in the company's fleet, where Ω_t represents the capacity of a vessel of Type t. For each $t = 1, \ldots, T$, let $n = 1, \ldots, M_t$ index the vessels of Type t, and let O_t and $CH_t = M_t - O_t$ respectively denote the number of company-owned vessels and the number of available vessels of this type that can be possibly chartered. The company-owned vessels are indexed by $n = 1, \ldots, O_t$ while the chartered vessels are indexed by $n = O_t + 1, \ldots, O_t + CH_t \equiv M_t$. The actual numbers of vessels of each type that are chartered are effectively determined by the schedules prescribed by the model. Note that two vessels of the same type might be chartered at different prices. Accordingly, let $\$_{t,n}$ be the cost (in U.S. dollars) of chartering a vessel n of type t, for $n = O_t + 1, \ldots, O_t + CH_t$, and for each $t = 1, \ldots, T$. A vessel may be chartered for the entire duration of the time horizon or for only a specified subset of it, depending on its availability. The chartering cost given by $\$_{t,n}$ is incurred as a fixed-cost whenever the vessel is chartered during the time horizon, regardless of its usage during this time interval. This stems from the fact that the chartered vessels will be made available to the (leased-to) company (in our case, KPC) essentially over the horizon, and the (leased-to) company is free to make any related dispatching decisions during this period. Let $UT_{t,n}$ be the maximum number of days vessel n of type t can be used during the time horizon. This time restriction is typically needed for maintenance purposes.

Let $f = 1, \ldots, F$ denote the storage facilities that are available for leasing over the duration of the contract horizon. Each storage facility is characterized by its location, capacity, and leasing expenses. For $f = 1, \ldots, F$, let C_f denote the capacity of storage facility f and let $\$\$_f$ be the cost (in U.S. dollars) of leasing a storage facility f for the duration of the contract horizon. A storage facility can effectively be either a source or a destination for a given vessel trip, based on whether the vessel happens to be respectively loading or unloading at the storage facility. Let $s = 1, \ldots, S + F$ denote the entire collection of sources comprised of the company's sources and the storage facilities that are available for leasing. The company's sources are indexed by $s = 1, \ldots, S$, while $s = S + 1, \ldots, S + F$ respectively correspond to the storage facilities $f = 1, \ldots, F$. Let Q_s be the maximum daily permitted amount that can be shipped from source s to any destination. This might represent a production or a storage capacity restriction. Likewise, let $d = 1, \ldots, D + F$ denote the destinations (client storage facilities or storage facilities available for leasing). The client destinations are indexed by $d = 1, \ldots, D$, while $d = D + 1, \ldots, D + F$ respectively correspond to the storage facilities $f = 1, \ldots, F$.

Let LEG_{h,t,n,s_1,d,s_2} represent a leg for vessel n of type t leaving source s_1 toward destination d on day h, and then returning to source s_2. Let T_{t,s_1,d,s_2} represent the time required to complete LEG_{h,t,n,s_1,d,s_2}, which includes the time required to load the vessel of type t at source s_1, the time this vessel takes to travel to destination d, the unloading time at destination d, and then the time needed to travel from destination d to source s_2. We assume that there is a unique prescribed route from source s_1 to destination d, and likewise, from destination d to source s_2. Let $T_{t,s_1,d,s_2} = T_{1,t,s_1,d,s_2} + T_{2,t,s_1,d,s_2}$, where T_{1,t,s_1,d,s_2} is the time required to load a vessel of type t at source s_1 plus the travel time to destination d, and T_{2,t,s_1,d,s_2} is the time required to unload at destination d plus the travel time from destination d to source s_2. Vessels of the same type are assumed to have equal values of T_{t,s_1,d,s_2} (and their splits), and this duration is also assumed to be independent of h; i.e., T_{t,s_1,d,s_2} is independent of the day the leg starts (weather effects are neglected). Let $DC_{t,n}$ denote the daily operational cost of vessel n of type t. Let $C_{t,n,s_1,d,s_2} = T_{t,s_1,d,s_2}\,(DC_{t,n})$, which is the cost associated with $\text{LEG}_{h,t,n,s_1,d,s_2}$. Note that this cost is independent of the day the trip starts.

Next, let us define the following sets to identify permissible shipment routing itineraries:

$$I_1 = \{(s_1, d, s_2): \quad \text{where} \quad s_1 \in \{1, \ldots, S\}, d \in \{1, \ldots, D\}, \quad \text{and}$$
$$s_2 \in \{1, \ldots, S + F\}\},$$

$$I_2 = \{(s_1, d, s_2): \quad \text{where} \quad s_1 \in \{1, \ldots, S\}, d \in \{D + 1, \ldots, D + F\}, \quad \text{and}$$
$$s_2 \in \{1, \ldots, S\}\}, \quad \text{and}$$

$$I_3 = \{(s_1, d, s_2): \quad \text{where} \quad s_1 \in \{S + 1, \ldots, S + F\}, d \in \{1, \ldots, D\}, \quad \text{and}$$
$$s_2 \in \{1, \ldots, S + F\}\}$$

Note that I_1 represents trips initiated from any of the company sources $s_1 \in \{1, \ldots, S\}$, headed to a client storage facility $d \in \{1, \ldots, D\}$, and then returning to a company source or to a leased storage facility $s_2 \in \{1, \ldots, S + F\}$. I_2 represents trips initiated from any of the company sources $s_1 \in \{1, \ldots, S\}$, headed to one of the leased storage facilities $d \in \{D + 1, \ldots, D + F\}$, and then returning to a company source $s_2 \in \{1, \ldots, S\}$. I_3 represents trips initiated from any of the leased storage facilities $s_1 \in \{S + 1, \ldots, S + F\}$, headed to a client storage facility $d \in \{1, \ldots, D\}$, and then returning to a company source or to one of the leased storage facilities $s_2 \in \{1, \ldots, S + F\}$. Observe that successive trips between one leased facility and another is disallowed. Hence, $I = I_1 \cup I_2 \cup I_3$ represents all types of permitted trips in the operation.

For $d \in \{1, \ldots, D + F\}$, let w_d denote the storage level at either destination d if $d \in \{1, \ldots, D\}$, or transshipment depot $(d - D)$ if $d \in \{D + 1, \ldots, D + F\}$, at the beginning of the time horizon, i.e., on the first day of the time horizon. Let $SL_{1,d}$ and $SL_{2,d}$ denote the minimum and maximum desired levels, respectively, at the storage facility of destination d. Some customers may allow the storage level to go below or to exceed $SL_{1,d}$ and $SL_{2,d}$, respectively; however, with a penalty based on the corresponding shortage or excess quantities. Let π_d denote the daily penalty for each shortage or excess unit at destination d. This penalizes shortages and excess quantities at destination d with respect to the desired levels $SL_{1,d}$ and $SL_{2,d}$, to the extent given by $A_{1,d}$ and $A_{2,d}$, respectively. Let $b_{1,d} = SL_{1,d} - A_{1,d}$ and $b_{2,d} = SL_{2,d} + A_{2,d}$. Storage levels falling below $b_{1,d}$ or in excess of $b_{2,d}$ (up to a sufficiently large upper bound SUB_d), while permitted, are highly undesirable, and incur a significantly greater penalty $\lambda_d > \pi_d$ per unit, at each destination d.

Let $R_{j,d}$ denote the expected consumption at destination d on day j, for $j \in \{1, \ldots, H\}$. Usually, at most two distinct daily consumption rates exist at a given destination. The different daily consumptions arise from the possible seasonal changes during the horizon, as well as some client-specific consider-ations. Let $TC_{h,d}$ be the total consumption at destination d during the interval $\{1, \ldots, h\}$, which is given by $TC_{h,d} = \sum_{j=1}^{h} R_{j,d}$.

7.2.3 Penalty Representations

Penalties are computed on a daily basis and depend on the storage levels at the customers' facilities. These penalties are similar to those introduced in part I, and are extended to address multiple destinations. Let $S_{h,d}$ be the storage level at destination d on day h. Define Type I and Type II penalty functions as follows.

Type I penalty:

$$P_I(S_{h,d}) = \pi_d \text{ maximum } \{0, (SL_{1,d} - S_{h,d}), (S_{h,d} - SL_{2,d})\} \quad \text{if} \quad S_{h,d} \in [b_{1,d}, b_{2,d}],$$

and

Type II penalty:

$$P_{II}(S_{h,d}) = \begin{cases} \pi_d A_{1,d} + \lambda_d (b_{1,d} - S_{h,d}) & \text{if} \quad S_{h,d} \in (0, b_{1,d}), \\ \pi_d A_{2,d} + \lambda_d (S_{h,d} - b_{2,d}) & \text{if} \quad S_{h,d} \in (b_{2,d}, SUB_d), \end{cases}$$

where π_d and $\lambda_d > \pi_d$ are as defined above. Note that if $S_{h,d} \in [SL_{1,d}, SL_{2,d}]$, then the storage level at destination d lies within the desired bounds and no penalty is induced. If $S_{h,d} \in [b_{1,d}, SL_{1,d}) \cup (SL_{2,d}, b_{2,d}]$, then a penalty is in-curred based on the respective shortage or excess quantity at destination d. On the other hand, if $S_{h,d} \in [0, b_{1,d}) \cup (b_{2,d}, SUB_d]$, then a sufficiently large ad-ditional penalty rate is imposed continuously beyond that of $P_I(.)$ to indicate the undesirabity of such a storage level at destination d on any given day of the time horizon.

PROPOSITION 7.1

Let

$$S_{h,d} = S_{1,h,d} - S_{2,h,d} - S_{3,h,d} + S_{4,h,d} + S_{5,h,d}, \tag{7.1}$$

where

$$SL_{1,d} \le S_{1,h,d} \le SL_{2,d}, \quad 0 \le S_{2,h,d} \le A_{1,d}, \quad 0 \le S_{3,h,d} \le b_{1,d},$$
$$0 \le S_{4,h,d} \le A_{2,d}, \quad \text{and} \quad 0 \le S_{5,h} \le (SUB_d - b_{2,d}). \tag{7.2}$$

Define $P(S_{h,d}) : [0, SUB_d] \to [0, \infty)$ as the linear penalty function.

$P(S_{h,d}) = \pi_d(S_{2,h,d} + S_{4,h,d}) + \lambda_d(S_{3,h,d} + S_{5,h,d})$. *Then any minimization objective formulation that incorporates the term* $P(S_{h,d})$ *defined above along with (7.1) and (7.2) will automatically enforce the sum of the Type I and Type II penalties* $P_I(S_{h,d}) + P_{II}(S_{h,d})$.

PROOF The proof is similar to that of Proposition 7.1 in part I for the case of a single source-destination operation. ∎

7.3   Model Formulation

7.3.1   A Mixed-Integer Programming Model

In order to formulate the problem described in the foregoing section, we define the following sets of binary decision variables. Let

$$X_{h,t,n,s_1,d,s_2} = \begin{cases} 1 & \text{if } LEG_{h,t,n,s_1,d,s_2} \text{ is selected,} \\ 0 & \text{otherwise.} \end{cases}$$

Since a vessel cannot be dispatched from source s on day h unless it is available on this day, another set of binary variables is defined as follows:

$$Y_{h,t,n,s} = \begin{cases} 1 & \text{if vessel } n \text{ of type } t \text{ is available at source } s \text{ on day } h, \\ 0 & \text{otherwise.} \end{cases}$$

Finally, in order to represent the chartering decisions and the storage leasing decisions, let

$$Z_{t,n} = \begin{cases} 1 & \text{if vessel } n \in \{O_t + 1, \ldots, O_t + CH_t\} \text{ of type } t \text{ is selected for} \\ & \text{chartering during the time horizon,} \\ 0 & \text{otherwise,} \end{cases}$$

and

$$W_f = \begin{cases} 1 & \text{if storage facility } f \text{ is leased for the duration of the time horizon,} \\ 0 & \text{otherwise.} \end{cases}$$

Note that several of the above defined binary decision variables are *a priori* known to be inadmissible, or effectively, zero. Naturally, whenever $Y_{h,t,n,s_1} = 0$, this implies that the corresponding variables $X_{h,t,n,s_1,d,s_2} = 0 \,\forall\, (d, s_2)$. Examples of such zero variables are specified below.

(A) If $(s_1, d, s_2) \notin I$ then this implies that LEG_{h,t,n,s_1,d,s_2} is not permitted in the operation $\forall\, h, t, n$. Accordingly, $X_{h,t,n,s_1,d,s_2} \equiv 0 \,\forall\, h, t, n, s_1, d, s_2$ such that $(s_1, d, s_2) \notin I$.

(B) $X_{h,t,n,s_1,d,s_2} \equiv 0 \; \forall h, n$ if source s_1, or destination d, or source s_2, does not admit vessels of type t. This occurs if a source or a destination cannot admit a vessel due to its size or specifications, as in the case of super-tankers that cannot be admitted into certain source or destination ports.

(C) A vessel n of type t may not be available for any new consignments during some subset $[h_1, h_2]$ of the time horizon. Such a situation occurs if the company needs to perform a scheduled maintenance on a vessel during some specified days of the time horizon provided the vessel is in port any time during that duration. This information can be incorporated into the model by letting $Y_{h,t,n,s} \equiv 0 \; \forall h \in \{h_1, \ldots, h_2\}$ and $\forall s$. Note that if a vessel n of type t is required to be inactive during $[h_1, h_2]$, we can set $X_{h,t,n,s_1,d,s_2} = 0 \; \forall \; (s_1, d, s_2) \in I$ and $\forall h$ such that $[h, h + T_{t,s_1,d,s_2}] \cap [h_1, h_2] \neq \emptyset$.

(D) The present demand contract might have been signed *during* a previous contract horizon. In this case, certain decisions for the previous contract might have committed some self-owned vessels over periods concurrent with the present contract horizon. Therefore, proper sets of the X and Y variables should be defined to be zero to indicate the unavailability of vessels during the present contract horizon. Moreover, whenever a vessel n of type t becomes available for the first time on day h in the horizon at source s, we fix $Y_{h,t,n,s} = 1$, and let $Y_{h,t,n,s_1} = 0 \; \forall \; s_1 \neq s$, and $Y_{h_1,t,n,s_1} = 0 \; \forall h_1 \in \{1, \ldots, (h-1)\}$ and for all s_1.

We will let ϕ_X and ϕ_Y denote the sets of X and Y variables, respectively, that are restricted to be fixed at specified binary values by virtue of such considerations. The prescribed vessel scheduling problem can now be stated as follows. (All indices are assumed to take on only their respective relevant values.)

VSP:

$$\text{Minimize} \sum_h \sum_t \sum_n \sum_{s_1} \sum_d \sum_{s_2} C_{t,n,s_1,d,s_2} X_{h,t,n,s_1,d,s_2}$$

$$+ \sum_h \sum_d \pi_d [S_{2,h,d} + S_{4,h,d}] + \sum_h \sum_d \lambda [S_{3,h,d} + S_{5,h,d}]$$

$$+ \sum_t \sum_{n=O_t+1}^{M_t} \$_{t,n} Z_{t,n} + \sum_f \$\$_f W_f,$$

subject to

(C_1) $\quad S_{\overline{h},\overline{d}} = w_{\overline{d}} + \sum_t \sum_n \sum_{s_1} \sum_{s_2} \sum_{\substack{h: \\ h+T_{1,t,s_1,\overline{d},s_2} \in \{1, \ldots, h\}}} \Omega_t \, X_{h,t,n,s_1,\overline{d},s_2} - TC_{\overline{h},\overline{d}},$

$$\forall \, \overline{h} \text{ and } \overline{d} \in \{1, \ldots, D\},$$

(C$_2$) $S_{\overline{h},\overline{d}} = S_{1,\overline{h},\overline{d}} - S_{2,\overline{h},\overline{d}} - S_{3,\overline{h},\overline{d}} + S_{4,\overline{h},\overline{d}} + S_{5,\overline{h},\overline{d}},$

$\quad\quad \forall \overline{h} \text{ and } \overline{d} \in \{1, \dots, D\},$

(C$_3$) $L_{\overline{h},\overline{f}} = w_{D+\overline{f}} + \displaystyle\sum_t \sum_n \sum_{s_1=1}^{S} \sum_{s_2=1}^{S} \sum_{\substack{h: \\ (h+T_{1,s_1,D+\overline{f},s_2})\in\{1,\dots,\overline{h}\}}} \Omega_t \, X_{h,t,n,s_1,D+\overline{f},s_2}$

$\quad\quad -\displaystyle\sum_t \sum_n \sum_{d=1}^{D} \sum_{s_2} \sum_{\substack{h: \\ h\in\{1,\dots,\overline{h}\}}} \Omega_t \, X_{h,t,n,S+\overline{f},d,s_2}, \quad \forall \overline{h}, \overline{f},$

(C$_4$) $Y_{\overline{h},\overline{t},\overline{n},\overline{s}} = Y_{\overline{h}-1,\overline{t},\overline{n},\overline{s}} - \displaystyle\sum_d \sum_{s_2} X_{\overline{h}-1,\overline{t},\overline{n},\overline{s},d,s_2} + \sum_{s_1} \sum_d \sum_{\substack{h: \\ h+T_{\overline{t},s_1,d,\overline{s}}=\overline{h}}} X_{h,\overline{t},\overline{n},s_1,d,\overline{s}},$

$\quad\quad \forall \overline{h} \geq 2, \overline{t}, \overline{n}, \overline{s},$

(C$_5$) $\displaystyle\sum_d \sum_{s_2} X_{\overline{h},\overline{t},\overline{n},\overline{s},d,s_2} \leq Y_{\overline{h},\overline{t},\overline{n},\overline{s}}, \quad \forall \overline{h}, \overline{t}, \overline{n}, \overline{s},$

(C$_6$) $\displaystyle\sum_s Y_{\overline{h},\overline{t},\overline{n},s} \leq Z_{\overline{t},\overline{n}}, \quad \forall \overline{h}, \overline{t}, \overline{n} \in \{O_t + 1, \dots, O_t + CH_t\},$

(C$_7$) $\displaystyle\sum_t \sum_n \sum_d \sum_{s_1} \Omega_t X_{\overline{h},t,n,\overline{s},d,s_1} \leq Q_{\overline{s}}, \quad \forall \overline{h} \quad \text{and} \quad \overline{s} \in \{1, \dots, S\},$

(C$_{8.1}$) $Y_{\overline{h},\overline{t},\overline{n},S+\overline{f}} \leq W_{\overline{f}}, \quad \forall \overline{h}, \overline{t}, \overline{n}, \overline{f},$

(C$_{8.2}$) $\displaystyle\sum_{s_1} \sum_{d=1}^{D} X_{\overline{h},\overline{t},\overline{n},s_1,d,D+\overline{f}} \leq W_{\overline{f}}, \quad \forall \overline{h}, \overline{t}, \overline{n}, \overline{f},$

(C$_9$) $\displaystyle\sum_h \sum_{s_1} \sum_d \sum_{s_2} X_{h,\overline{t},\overline{n},s_1,d,s_2} T_{\overline{t},s_1,d,s_2} \leq UT_{\overline{t},\overline{n}}, \quad \forall \overline{t}, \overline{n},$

(C$_{10}$) $X_{h,t,n,s_1,d,s_2} \in \{0, 1\}, \quad \forall h, t, n, s_1, d, s_2, \text{ if } X_{h,t,n,s_1,d,s_2} \notin \phi_X,$

and fixed at zero or one otherwise,

$\quad\quad Y_{h,t,n,s} \in [0, 1], \quad \forall h, t, n, s, \quad \text{if } Y_{h,t,n,s} \notin \phi_Y,$

and fixed at zero or one otherwise,

$\quad\quad Z_{t,n} \in [0, 1], \quad \forall t, n = O_t + 1, \dots, O_t + CH_t,$

$\quad\quad W_f \in [0, 1], \quad \forall f,$

$\quad\quad S_{h,d} \geq 0, \quad SL_{1,d} \leq S_{1,h,d} \leq SL_{2,d}, \quad 0 \leq S_{2,h,d} \leq A_{1,d},$

$\quad\quad 0 \leq S_{3,h,d} \leq b_{1,d}, \quad 0 \leq S_{4,h,d} \leq A_{2,d},$

and

$$0 \le S_{5,h,d} \le (SUB_d - b_{2,d}), \quad \forall \quad h \quad \text{and} \quad d \in \{1, \ldots, D\},$$
$$0 \le L_{h,f} \le C_f, \quad \forall h, f$$

7.3.2 Objective Function

The objective function of the problem seeks to minimize the total cost comprised of the operational costs of selected legs (both for company-owned and chartered vessels), the penalty costs resulting from shortage or excess levels at the customer storage facilities (both of Type I and Type II), the chartering expenses, and the costs for leasing storage facilities.

7.3.3 Constraints

Revisits to destination $d = 1, \ldots, D$ are determined by the initial storage level (given by w_d), the storage capacity (including minimum and maximum desired levels, given respectively by $SL_{1,d}$ and $SL_{2,d}$), and the daily rates of consumption $R_{j,d}$. Constraint (C_1) computes the storage level at destination d on day h, whereas Constraint (C_2) represents $S_{h,d}$ in terms of $S_{1,h,d}$, $S_{2,h,d}$, $S_{3,h,d}$, $S_{4,h,d}$, and $S_{5,h,d}$ as in Proposition 7.1. The Type I and II penalties are incurred in the objective function based on this representation as stated in Proposition 7.1. In Constraint (C_3), the variable $L_{h,f}$ represents the storage level at facility f on day h, being equal to the difference between the total quantity delivered to storage facility f by day h and the total quantity shipped from this storage facility f by day h. Note that $S_{h,d}$ and $L_{h,f}$ are nonnegative continuous variables, and moreover, $L_{h,f}$ is bounded above by the storage capacity given by C_f, as noted in Constraint (C_{10}).

A vessel n of type t can be dispatched from source s on day h only if it is available at s on that day. This vessel is available at source s on day h if either the vessel was available there on the previous day and it was not dispatched, or this vessel was not available there during the previous day but it arrived on the current day. On the other hand, this vessel is unavailable on day h at source s if it was available there on the previous day and it was dispatched on that day, or it was unavailable there on the previous day and it did not arrive on the current day. Constraint (C_4) examines the availability of vessel n of type t at source s on day h by incorporating these cases (assuming that any trip is at least two days long), and then Constraint (C_5) permits the dispatchment of vessels conditioned on this availability. For the chartered vessels $n \in \{O_t +1, \ldots, O_t + CH_t\}$ of type t, Constraint (C_6) enforces that this vessel is selected for chartering $(Z_{t,n} = 1)$ if it is required to be made available for consignment. Note that once a chartered vessel is made available at some source, then (C_4) accounts for its proper feasible use. Constraint (C_7) ensures that the total amount of the product that can be shipped from a given source s on a given day h to any destination d does not exceed the daily permitted amount being specified by Q_s. Constraint $(C_{8.1})$ examines if storage facility f

is leased for the duration of the time horizon in order to permit the availability of vessels thereat and Constraint $(C_{8.2})$ rules out legs that terminate in non-leased storage facilities. Constraint (C_9) enforces that any vessel n of type t can be used for at most $UT_{t,n}$ days during the time horizon. Finally, (C_{10}) enforces the various bounding and logical constraints.

REMARK 7.1
Assume that the Y-variables corresponding to $h = 1$ are fixed at some binary values. Then Constraint (C_4) ensures the integrality of all the other Y-variables once the integrality of the X-variables is enforced. This follows from the recursive relation in (C_4) that defines $Y_{h,t,n,s}$ for $h \geq 2$ in terms of $Y_{h-1,t,n,s}$ and a subset of the X-variables. Moreover, the integrality conditions on the Z- and W-variables can also be relaxed. This follows because constraints (C_6), $(C_{8.1})$, and $(C_{8.2})$ along with the fourth and fifth terms of the objective function will automatically enforce the integrality of these variables.

7.4 An Aggregated Reformulation for Model VSP and Related Issues

In this section, we derive an aggregated version of Model VSP, denoted by AVSP, which retains the essential operational features of the original problem, while being far more computationally tractable. The aggregate reformulation basically disregards the individual vessel identities, and instead, attempts to decide on the number of vessels of each type to be dispatched on a given day to traverse a prescribed trip. This aggregated reformulation is motivated by the overwhelming number of binary variables resulting from the initial formulation for a typical operational scenario as illustrated in Table 7.1 and Table 7.2 of Section 7.5.1.

7.4.1 Formulation of Model AVSP

Let LEG_{h,t,s_1,d,s_2} represent a leg for a vessel of type t leaving source s_1 to destination d on day h, and then returning to source s_2. The average cost associated with this leg is denoted by c_{h,t,s_1,d,s_2}, and the time required to complete this leg is given by T_{t,s_1,d,s_2}.

Define x_{h,t,s_1,d,s_2} as an integer variable that represents the number of vessels of type t that traverse LEG_{h,t,s_1,d,s_2}. Define $y_{h,t,s}$ to be an integer decision variable that represents the maximum number of vessels of type t that are available for consignment from source s on day h. As mentioned in the foregoing section, vessels might become available for use at different days of the time horizon due to, for example, their involvement in trips from previously signed contracts that will terminate sometime during the current time horizon. Hence, we let $O_{h,t,s}$ be the number of self-owned vessels of type t that will become available for use for the first time at source s on day h of the time horizon, and we let $CH_{h,t,s}$ be the number of vessels of type t that will become available for chartering for the first time at source s on day h of the time horizon. Hence, let $\alpha_{h,t,s} = O_{h,t,s} + CH_{h,t,s}$. Accordingly, we let $z_{h,t,s}$ be the integer variable that represents the number of vessels of type t that are actually selected for chartering on day h of the time horizon at source s. Let $\$_{h,t}$ denote the average chartering cost of a vessel of type t that will become available for use for the first time on day h of the time horizon, regardless of the source at which it first became available.

Hence, $y_{1,t,s} = O_{1,t,s} + z_{1,t,s}$. Note that as before, the y-variables will be used to indicate the supply distribution of vessels during the time horizon, and the actual use of the vessels will be examined via the x-variables.

Let $A_{h,t,s}$ be a subset of vessels of type t (both self-owned and vessels available for chartering) that will become available for use at source s for the first time on day h of the time horizon. Hence, we let $U_{h,t} = \frac{\Sigma_s \Sigma_{n \in A_{h,t,s}} UT_{t,n}}{\Sigma_s \alpha_{h,t,s}}$, which gives the average usage allowance for a vessel of type t that will become available for use for the first time on day h of the time horizon. Accordingly, $U_t = \Sigma_h U_{h,t}$ gives the average usage allowance for a vessel of type t. Also, let $O_t \equiv \Sigma_h \Sigma_s O_{h,t,s}$.

Similar to the index sets ϕ_X and ϕ_Y, we let ϕ_x and ϕ_y denote the sets of x- and y-variables, respectively, that are *a priori* restricted to be zero, or fixed at some known positive integer values.

The formulation AVSP adopts an aggregated viewpoint as exemplified by the integer decision variables that represent the number of vessels assigned to traverse from the source points to the various destinations over the time horizon (without any vessel identities), in lieu of using the previous binary variables. Accordingly, as we shall see, we can represent the Model VSP more compactly using this aggregated representation, at the expense of having to relax the individual vessel's total usage and downtime restrictions (C_9) to the revised representation (AC_9), as we no longer specifically account for each individual vessel's activity. These relaxed constraints need to be dealt with separately while implementing the model-based decision. Because of the relatively soft nature of these constraints, this is an acceptable compromise between model size and representability.

The aggregated model is now stated as follows, where the W, S, and L variables are defined as in Model VSP.

AVSP:

$$\text{Minimize} \sum_h \sum_t \sum_{s_1} \sum_d \sum_{s_2} c_{t,s_1,d,s_2} x_{h,t,s_1,d,s_2}$$

$$+ \sum_h \sum_d \pi_d [S_{2,h,d} + S_{4,h,d}] + \sum_h \sum_d \lambda_d [S_{3,h,d} + S_{5,h,d}]$$

$$+ \sum_h \sum_t \sum_s \$_{h,t} z_{h,t,s} + \sum_f \$\$_f W_f,$$

subject to

$(\mathbf{AC_1})$ $S_{\overline{h},\overline{d}} = w_{\overline{d}} + \sum_t \sum_{s_1} \sum_{s_2} \sum_{\substack{h: \\ h+T_{1,t,s_1,d,s_2} \in \{1,\dots,\overline{h}\}}} \Omega_t \, x_{h,t,s_1,\overline{d},s_2} - TC_{\overline{h},\overline{d}}, \quad \forall \overline{h}$

and $\overline{d} \in \{1,\dots,D\}$,

$(\mathbf{AC_2})$ $S_{\overline{h},\overline{d}} = S_{1,\overline{h},\overline{d}} - S_{2,\overline{h},\overline{d}} - S_{3,\overline{h},\overline{d}} + S_{4,\overline{h},\overline{d}} + S_{5,\overline{h},\overline{d}}, \quad \forall \overline{h}$

and $\overline{d} \in \{1,\dots,D\}$,

$(\mathbf{AC_3})$ $L_{\overline{h},\overline{f}} = w_{D+\overline{f}} + \sum_t \sum_{s_1=1}^{S} \sum_{s_2=1}^{S} \sum_{\substack{h: \\ (h+T_{1,s_1,D+\overline{f},s_2}) \in \{1,\dots,\overline{h}\}}} \Omega_t \, x_{h,t,s_1,D+\overline{f},s_2}$

$- \sum_t \sum_{d=1}^{D} \sum_{s_2} \sum_{\substack{h: \\ h \in \{1,\dots,\overline{h}\}}} \Omega_t \, x_{h,t,S+\overline{f},d,s_2}, \quad \forall \overline{h}, \overline{f},$

$(\mathbf{AC_4})$ $y_{\overline{h},\overline{t},\overline{s}} = y_{\overline{h}-1,\overline{t},\overline{s}} - \sum_d \sum_{s_2} x_{\overline{h}-1,\overline{t},\overline{s},d,s_2} + \sum_{s_1} \sum_d$

$\sum_{\substack{h: \\ h+T_{\overline{t},s_1,d,\overline{s}} = \overline{h}}} x_{h,\overline{t},s_1,d,\overline{s}} + O_{\overline{h},\overline{t},\overline{s}} + z_{\overline{h},\overline{t},\overline{s}}, \quad \forall \overline{h} \geq 2, \overline{t}, \overline{s}$

$(\mathbf{AC_5})$ $\sum_d \sum_{s_2} x_{\overline{h},\overline{t},\overline{s},d,s_2} \leq y_{\overline{h},\overline{t},\overline{s}}, \quad \forall \overline{h}, \overline{t}, \overline{s},$

$(\mathbf{AC_6})$ $y_{1,t,s} = O_{1,t,s} + z_{1,t,s}, \quad \forall \, t, s,$

$(\mathbf{AC_7})$ $\sum_t \sum_d \sum_{s_1} \Omega_t x_{\overline{h},,t,\overline{s},d,s_2} \leq Q_{\overline{s}}, \quad \forall \overline{h} \quad \text{and} \quad \overline{s} \in \{1,\dots,S\},$

$(\mathbf{AC_{8.1}})$ $y_{1,\overline{t},S+\overline{f}} \leq (O_{1,\overline{t},S+\overline{f}} + CH_{1,\overline{t},S+\overline{f}}) W_{\overline{f}}, \quad \forall \overline{t}, \overline{f},$

$(\mathbf{AC_{8.2}})$ $y_{\overline{h},\overline{t},S+\overline{f}} \leq M_{\overline{t}} W_{\overline{f}}, \quad \forall \overline{h} \geq 2, \overline{t}, \overline{f},$

$(\mathbf{AC_{8.3}})$ $\sum_{s_1} \sum_{d=1}^{D} x_{\overline{h},\overline{t},s_1,d,D+\overline{f}} \leq M_{\overline{t}} W_{\overline{f}}, \quad \forall \overline{h}, \overline{t}, \overline{f},$

$(\mathbf{AC_9})$ $\sum_h \sum_{s_1} \sum_d \sum_{s_2} T_{\overline{t},s_1,d,s_2} x_{h,\overline{t},s_1,d,s_2} \leq U_{\overline{t}}(O_{\overline{t}} + \sum_h \sum_s z_{h,\overline{t},s}), \quad \forall \overline{t},$

$(\mathbf{AC_{10}})$ $x_{h,t,s_1,d,s_2} \in \{0,1,\dots,M_t\}, \forall \, h, t, s_1, d, s_2, \text{ if } x_{h,t,s_1,d,s_2} \notin \phi_x,$

and fixed at some positive integer values otherwise,

$0 \leq y_{h,t,s} \leq M_t, \quad \forall h \geq 2, t, s, \quad \text{if} \quad y_{h,t,s} \notin \phi_y,$

and fixed at some positive integer values otherwise,

$$0 \le z_{h,t,s} \le CH_{h,t,s}, \text{ and integer valued } \forall \ h, t, s,$$

$$W_f \in \{0, 1\}, \quad \forall \ f,$$

$$S_{h,d} \ge 0, \quad SL_{1,d} \le S_{1,h,d} \le SL_{2,d}, \quad 0 \le S_{2,h,d} \le A_{1,d},$$
$$0 \le S_{3,h,d} \le b_{1,d}, \quad 0 \le S_{4,h,d} \le A_{2,d},$$

and

$$0 \le S_{5,h,d} \le (SUB_d - b_{2,d}), \forall \ h \quad \text{and} \quad d \in \{1, \ldots, D\},$$
$$0 \le L_{h,f} \le C_f, \quad \forall \ h, f.$$

7.4.2 Objective Function and Constraints

As before, the objective function of Model AVSP represents the total operational costs for both the company owned vessels and the chartered vessels, the penalties resulting from shortage or excess levels at the customer storage facilities, and the cost of leasing storage facilities. Similar to constraints (C_1), (C_2), and (C_3) of Model VSP, the constraints (AC_1), (AC_2), and (AC_3) respectively represent storage levels at customers' facilities, decomposed storage levels for penalty representations, and storage levels at the leased facilities. Constraints (AC_4), (AC_5), (AC_7), and (AC_9) are used as representations of constraints (C_4), (C_5), (AC_7), and (C_9) respectively, in an aggregated sense. Note that (AC_9) is a relaxed version of (C_9) where the right-hand-side of (AC_9) gives the total allowable utilization of all the self-owned and chartered vessels. Constraint (AC_6) in concert with constraints (AC_4) and (AC_5) account for the number of chartered vessels. Restrictions related to the leasing facilities are enforced via the W-variables in constraints $(AC_{8.1})$ to $(AC_{8.3})$. The right-hand-sides of $(AC_{8.1})$ to $(AC_{8.3})$ yield the maximum number of vessels of type t that can be possibly available at a given leasing storage facility during the pertinent days of the time horizon. This number equals $(O_{1,\bar{t},S+\bar{f}} + CH_{1,\bar{t},S+\bar{f}})$ when $h = 1$, and equals M_t when $h \ge 2$, if the particular leasing facility is actually leased during the time horizon ($W_{\bar{f}} = 1$), and equals zero otherwise. Note that Constraint $(AC_{8.3})$ assures that if a storage facility is not selected, then this facility is not the termination point of any vessel, because otherwise, such a vessel would end up idle at this storage facility until the end of the time horizon. Observe that the integrality conditions on the y-variables are relaxed in Model AVSP for similar reasoning as discussed in Remark 7.1 above.

Having solved Model AVSP, the variable $z_t \equiv \Sigma_h \Sigma_s z_{h,t,s}$ at optimality yields the number of vessels of type t to charter, thereby determining the fleet of vessels being composed of company-owned and chartered vessels. The leasing of storage facility is determined at optimality by the binary variables W_f for all f. Now, based on the x-variables, we can begin dispatching different vessels of Type t from different sources on each day h, scheduled for a specified trip as determined by the corresponding values of $(s_1, d, s_2) \in I$. In this process, the downtime unavailabilities of various vessels, and the balancing of days for which the different vessels are put into service could be incorporated while consigning specific vessels of each type. It is worth mentioning that, in practice, there is some flexibility in scheduling maintenance and in the vessel usage constraints. Hence, this facilitates the conversion of the model solution to one that is implemented, without significantly perturbing the solution and its associated cost.

7.5   Solution Algorithms and Computational Results

In this section, we present computational results related to solving Model AVSP directly using the CPLEX package (version 7.5) and via a specialized rolling horizon algorithm based on eighteen test problems that represent various operational scenarios. These test problems are referred to as $I_1, \ldots, I_{10}$ and $NI_1, \ldots, NI_8$. Relevant statistics about the test problems are provided in Appendix B. The reason for considering only Model AVSP in the computational results stems from the fact that we were unable to obtain meaningful solutions for Model VSP in [7] even for the single source-destination test cases, without the consideration of transshipment depots.

Note that test problems $I_1, \ldots, I_{10}$ are constructed by adding three transshipment storage depots to the test cases in [7]. The introduction of the transshipment depots has magnified the number of the integer variables, which basically makes this model computationally intractable. Table 7.1 compares the number of the x-variables in Model AVSP with and without transshipment depots.

Notationally, we will let $\overline{P}$ denote the linear relaxation of any model P. The optimal objective function value of model P will be denoted by $v(P)$. The best upper bound and lower bound found for model P will be respectively denoted by $v_{UB}(P)$ and $v_{LB}(P)$. Note that all runs below are made on a Pentium 4, CPU 1.70 GHz computer having 512 MB of RAM using CPLEX-7.5, with coding in Java.

7.5.1   A Rolling Horizon Algorithm for Model AVSP

Because of the overwhelming number of integer variables in Model AVSP as exhibited in Table 7.1, we were unable to solve this model for any of the test problems directly using CPLEX due to out-of-memory difficulties. Hence, in order to facilitate the derivation of good quality solutions with reasonable effort for Model AVSP, we developed a specialized rolling horizon heuristic, similar to those proposed in [6,7]. This heuristic is based on a sequential fixing of integer variables and is presented below along with related computational results.

In Model AVSP, let the vector x be partitioned as $(x_1, x_2, \ldots, x_H)$, where x_h denotes the vector of x–variables associated with the hth day of the time horizon. Let H_1 be the length of the horizon interval for which the corresponding x variables are restricted to be integer valued, and the remaining variables are declared to be continuous. Accordingly, in a rolling-horizon framework, let H_2 be the duration of the initial subset of this interval for which the determined decisions are permanently fixed. Let $KK = \lceil ((H - H_1)/H_2) + 1 \rceil$ and let AVSP $(H_1, H_2, \text{opt-gap}, k)$, for $k = 1, \ldots, KK$, denote Model AVSP having the following characteristics:

(a) x_h is enforced to be integer valued for $h \leq H_1 + (k - 1)H_2$, and is relaxed to be continuous otherwise, $\forall\, h = 1, \ldots, H$.

(b) x_h for $h \leq (k - 1)H_2$ is fixed at the values found from the solution to Model AVSP$(H_1, H_2, \text{opt-gap}, (k - 1))$.

(c) The optimality gap tolerance for fathoming is set at opt-gap.

The rolling-horizon heuristic RHA then proceeds as follows.

Initialization: Select integers $H_1 \leq H$, $H_2 \leq (H_1/2)$, and initialize $k = 1$. Let opt-gap be some selected optimality gap criterion. Solve Model AVSP $(H_1, H_2, \text{opt-gap}, 1)$.

Main Step: If $k = KK$; then terminate the algorithm; the proposed solution is that obtained from solving Model AVSP $(H_1, H_2, \text{opt-gap}, KK)$. Otherwise, increment k by one and solve the Model AVSP $(H_1, H_2, \text{opt-gap}, k)$. Repeat the Main Step.

Note that the solution value obtained from the first iteration of the algorithm RHA provides a lower bound for problem AVSP that is at least as high as $v(\overline{\text{AVSP}})$. Accordingly, we let v_{LB} equal the objective value obtained from the first iteration of the algorithm RHA. Table 7.2 reports results obtained

TABLE 7.1

Number of the x-Variables for Model AVSP with and without Transshipment Depots

Test Problem	No Transshipment Depots	3 Transshipment Depots
	Total Number of x-Variables	
I_1	30	570
I_2	60	1,140
I_3	120	2,280
I_4	180	3,420
I_5	240	4,560
I_6	480	9,120
I_7	450	8,550
I_8	540	10,260
I_9	840	15,960
I_{10}	960	18,240

TABLE 7.2

Statistics Related to Solving AVSP Using RHA

I_i	$v(\overline{AVSP})$ ($)	CPU Time (seconds)	$H_1 = 6, H_2 = 3$, (days)		perct_opt (v_{RHA}) (Percentage Optimality for RHA)
			v_{RHA} ($)	CPU Time (seconds)	
I_1	350,106	10.23	438,000	15.02	82.93
I_2	600,500	12.78	670,560	123.08	92.09
I_3	598,000	11.89	676,100	62.10	91.00
I_4	1,189,000	20.12	1,582,000	155.76	79.75
I_5	3,930,023	22.61	4,560,000	161.01	89.01
I_6	8,580,232	30.73	11,087,000	256.56	80.11
I_7	5,900,000	31.90	7,780,000	367.08	79.16
I_8	5,130,000	34.83	7,200,000	530.00	72.31
I_9	7,166,000	45.09	8,690,000	741.09	85.50
I_{10}	11,002,166	55.16	14,541,000	1,050.98	78.29
NI_1	614,007	143.98	854,001	1,419.09	74.71
NI_2	1,008,005	201.09	1,218,100	1,684.31	84.50
NI_3	2,145,908	299.54	2,745,908	1,860.90	78.95
NI_4	10,143,000	342.90	12,543,012	2,056.00	81.23
NI_5	10,047,411	486.54	14,547,411	2,952.11	72.01
NI_6	10,054,747	698.98	13,901,000	3,324.00	74.49
NI_7	12,598,000	898.82	16,987,090	3,685.49	75.16
NI_8	19,990,000	1,001.82	27,890,787	5,842.23	72.67

by using RHA for some fixed judicious values of H_1 and H_2 as determined via some computational experimentation, where opt-gap is set to the CPLEX default optimality tolerance. Here, v_{RHA} gives the solution value obtained by the rolling-horizon RHA, and $\text{perct\_opt}(v_{RHA}) \equiv 100 \left(1 - \frac{v_{RHA} - v_{LB}}{v_{RHA}}\right)$ gives the percentage of optimality of v_{RHA} with respect to the lower bound v_{LB}.

TABLE 7.3

Cost Improvement via the Use of Transshipment Depots

Test Problem	v_{RHA} ($)	v_{min} ($)	perct_imp(v_{RHA}, v_{min}) (Improvement of Cost Using the Modeling Approach over RHA)
I_1	420,000	438,000	−4.28
I_2	660,000	670,560	−0.01
I_3	660,000	676,100	−2.43
I_4	1,500,000	1,582,000	−5.46
I_5	4,680,000	4,560,000	2.56
I_6	11,400,000	11,087,000	2.74
I_7	8,000,000	7,780,000	2.75
I_8	7,400,000	7,200,000	2.70
I_9	9,360,000	8,690,000	7.15
I_{10}	16,450,000	14,541,000	11.60

7.5.2 Usefulness of the Proposed Modeling Approach

In [7], we compared the costs of schedules generated via the proposed approach with the cost of schedules generated via an *ad-hoc* procedure in a case study concerning Kuwait Petroleum Corporation (KPC). It was emphasized that significant savings can be achieved via the proposed approach in [7]. In this section, we examine the potential cost improvement via accommodating the consideration of leasing transshipment depots, whereby fewer chartered vessels might need to be acquired. We also emphasize the cost effectiveness of simultaneously examining multiple sources and destinations instead of associating specific vessels with designated source-destination combinations.

The following table compares costs of schedules obtained from [7], where no transshipment depots are utilized, with those obtained here, where three transshipment depots are available for usage. Note that v_{min} gives the solution value obtained from [7], and hence, $perct\_imp(v_{RHA}, v_{min}) = 100\left(\frac{v_{min} - v_{RHA}}{v_{min}}\right)$ yields the percentage of improvement in total cost when allowing the usage of the transshipment depots.

The use of the transshipment depots improved the total cost for test problems $I_5 - I_{10}$, with a percentage reduction in total cost ranging from 2.56 to 11.60. It is worth mentioning that for test problems $I_1 - I_4$, we were able to solve Model AVSP without transshipment depots directly using the CPLEX Package with the default setting in [7], while we utilized the heuristic RHA to solve this model when incorporating the three transshipment depots. Consequently, the advantage of incorporating the three transshipment depots was not discernable in test problems $I_1 - I_4$.

As depicted in Appendix B, each test problem NI_k for $k = 1, \ldots, 8$, combines two single source-destination cases to form a new instance that is composed of two sources and two destinations. In particular, a test problem NI_k that correspond to $I_{i,j}$ as defined in Appendix B indicates that NI_k combines test problems I_i and I_j. Let $v_{RHA}(TP)$ denote the objective value obtained for Model AVSP via algorithm RHA based on test problem TP and define $v_{RHA}^{i,j} = v_{RHA}(I_i) + v_{RHA}(I_j)$. Also, let $perct\_imp(v_{RHA}, v_{RHA}^{i,j}) = 100\left(\frac{v_{RHA}^{i,j} - v_{RHA}}{v_{RHA}^{i,j}}\right)$.

Observe that v_{RHA} is lower than $v_{RHA}^{i,j}$ for all the test problems NI_k for $k = 1, \ldots, 8$, with a percentage reduction in total cost ranging from 2.20 to 19.83.

7.6 Summary, Conclusions, and Future Research

This research is a continuation of the work of the authors in [6] and [7] concerning the scheduling of oil tankers. In particular, this work is an extension of the work in the previous chapter in that we consider multiple sources and destinations, and the leasing of transshipment depots, which can enhance

TABLE 7.4

Cost Improvement via the Use of Multiple Sources and Destinations

Test Problem NI_k	$I_{i,j}$	v_{RHA} ($)	$v_{RHA}^{i,j}$ ($)	perct_imp ($v_{RHA}, v_{RHA}^{i,j}$) (Percentage Improvement)
NI_1	$I_{1,1}$	854,001	876,000	2.50
NI_2	$I_{2,3}$	1,218,100	1,346,660	9.54
NI_3	$I_{4,4}$	2,745,908	2,924,000	6.09
NI_4	$I_{5,6}$	12,543,012	15,647,000	19.83
NI_5	$I_{7,7}$	14,547,411	15,560,000	6.50
NI_6	$I_{8,8}$	13,901,000	14,400,000	3.46
NI_7	$I_{9,9}$	16,987,090	17,380,000	2.20
NI_8	$I_{10,10}$	27,890,787	29,082,000	4.09

the overall efficiency and cost effectiveness of the transportation system. Incorporating these issues requires special modeling considerations and novel solution methods. In this chapter, we have developed mixed-integer programming models for the described scheduling transportation problem that take into account different vessel sizes, multiple sources and destinations, varying consumption rates, while also considering the leasing of transshipment depots. In particular, two models were developed — a full-scale representation VSP that incorporates all the detailed features of the problem, and an aggregated version AVSP that suppresses the individual identities of the vessels of each type, while incorporating this aspect subsequently when prescribing the actual consignment decisions. A specialized rolling-horizon heuristic was developed in concert with Model AVSP to derive good approximate solutions. The results indicate that the use of transshipment depots can improve the total cost. For the set of more challenging test problems, this resulted in a percentage reduction in total cost ranging from 2.56% to 11.60%. Also, combining two single source-destination test problems into a single joint decision instance was demonstrated to yield a percentage reduction in total cost ranging from 2.20% to 19.83%. One extension of the present work is to determine optimal transshipment depot locations, instead of assuming fixed locations, a scenario that might need to be examined, for example, in the context of oil-export vessel scheduling operations. This will further complicate the problem; however, better storage locations might lead to more cost effective schedules. Another extension is to deal with random demand structures instead of deterministic demand structures. Continued research and applications will address these issues.

Acknowledgments

This research work was supported by Kuwait University under Research Grant No. SM-06/02 and the National Science Foundation under Research Grant No. DMI-0094462.

References

1. Al-Yakoob, S.M., Mixed-integer mathematical programming optimization models and algorithms for an oil tanker routing and scheduling problem, PhD. dissertation, Department of Mathematics, Virginia Polytechnic Institute and State University, Blacksburg, 1997.
2. Christiansen, M., Fagerholt, K., and Ronen, D., Ship routing and scheduling: status and prospective, *Transportation Science*, 38(1), 1, 2004.
3. Perakis, A.N., Fleet operations optimization and fleet deployment, in *The Hand Book of Maritime Economics and Business*, Grammenos, C.T., Ed., Lloyds of London Publications, London, 580, 2002.
4. Ronen, D., Cargo ships routing and scheduling: survey of models and problems, *European Journal of Operational Research*, 12, 119, 1983.
5. Ronen, D., Ship scheduling: the last decade, *European Journal of Operational Research*, 71, 325, 1993.
6. Sherali, H.D., Al-Yakoob, S.M., and Merza, H., Fleet management models and algorithms for an oil-tanker routing and scheduling problem. *IIE Transactions*, 31, 395, 1999.
7. Sherali, H.D. and Al-Yakoob, S.M. Determining an optimal fleet mix and schedules: Part I—single source and destination. Department of Mathematics and Computer Science, Kuwait University, 2004. (Part II of this sequence is the present chapter.)

Appendix A: Glossary of Notation

- $CT^{aa}_{hh,HH}$: A contract that is singed on day aa having a time horizon given by $\{hh, \ldots , HH\}$.

- $CT_{hh_1,HH_1}, \ldots , CT_{hh_y,HH_y}$: Contracts that are signed prior to Contract $CT^{aa}_{hh,HH}$ having time horizons given by $\{hh_1, \ldots , HH_1\}, \ldots , \{hh_y, \ldots , HH_y\}$, respectively, where $HH_i > aa$ for $i = 1, \ldots , y$.

- $AA = \max\{HH_1, HH_2, \ldots , HH_y, HH\}$.

- $\{aa, \ldots , AA\}$: Collective time horizon for the contracts $CT_{hh_1,HH_1}, \ldots , CT_{hh_y,HH_y}$, and $CT^{aa}_{hh,HH}$.

- H : Number of days in the time horizon.

- $h = 1, \ldots , H$: Days of the time horizon.

- $t = 1, \ldots , T$: Types of vessels.

- Ω_t : Capacity (in barrels) of a vessel of type t.

- M_t : Total number of vessels of type t.

- $M = \sum\limits_{t=1}^{T} M_t$.

- O_t : Total number of self-owned vessels of type t.

- $O = \sum_{t=1}^{T} O_t$.

- CH_t : Total number of vessels of type t that are available for chartering.

- $CH_t = M_t - O_t$.

- $CH = \sum_{t=1}^{T} CH_t$.

- $n = 1, \ldots, M_t$: Vessels of type t.

- $n = 1, \ldots, O_t$: Self-owned vessels.

- $n = O_t + 1, \ldots, O_t + CH_t \equiv M_t$: Vessels of type t that are available for chartering.

- $\$_{t,n}$: Cost (in U.S. dollars) of chartering a vessel n of type t, for $n = O_t + 1, \ldots, O_t + CH_t \equiv M_t$ and for each $t = 1, \ldots, T$.

- $UT_{t,n}$: Maximum number of days that vessel n of type t can be used during the time horizon.

- $f = 1, \ldots, F$: Storage facilities that are available for leasing over the duration of the contract horizon.

- C_f : Capacity (in barrels) of storage facility f.

- $\$\$_f$: Cost (in U.S. dollars) of leasing a storage facility f for the duration of the contract horizon.

- $s = 1, \ldots, S+F$: Entire collection of sources comprised of the company's sources and the storage facilities that are available for leasing.

- $s = 1, \ldots, S$: Indices for the company sources.

- $s = S + 1, \ldots, S + F$: Source indices for the storage facilities $f = 1, \ldots, F$.

- $d = 1, \ldots, D + F$: Entire collection of destinations (client storage facilities or storage facilities available for leasing).

- $d = 1, \ldots, D$: Indices for the client destinations.

- $d = D + 1, \ldots, D + F$: Destination indices corresponding to the storage facilities $f = 1, \ldots, F$.

- LEG_{h,t,n,s_1,d,s_2} : A leg for vessel n of type t leaving source s_1 to destination d on day h, and then returning to source s_2.

- T_{t,s_1,d,s_2} : Time (in days) required to complete LEG_{h,t,n,s_1,d,s_2}.

- $T_{t,s_1,d,s_2} = T_1, t, s_1, d, s_2 + T_2, t, s_1, d, s_2$, where T_{1,t,s_1,d,s_2} is the time required to load a vessel of type t at source s_1 plus the travel time to destination d, and T_{2,t,s_1,d,s_2} is the time required to unload at destination d plus the travel time from destination d to source s_2.

- $DC_{t,n}$: Daily operational cost (in U.S. dollars) of vessel n of type t.

- $C_{t,n,s_1,d,s_2} = T_{t,s_1,d,s_2} (DC_{t,n})$: Cost (in U.S. dollars) associated with LEG_{h,t,n,s_1,d,s_2}.

- $I_1 = \{(s_1, d, s_2) : \text{where } s_1 \in \{1, \ldots, S\}, d \in \{1, \ldots, D\}, \text{ and } s_2 \in \{1, \ldots, S + F\}\}$.
- $I_2 = \{(s_1, d, s_2) : \text{where } s_1 \in \{1, \ldots, S\}, d \in \{D + 1, \ldots, D + F\}, \text{ and } s_2 \in \{1, \ldots, S\}\}$.
- $I_3 = \{(s_1, d, s_2) : \text{where } s_1 \in \{S + 1, \ldots, S + F\}, d \in \{1, \ldots, D\}, \text{ and } s_2 \in \{1, \ldots, S + F\}\}$.
- $I = I_1 \cup I_2 \cup I_3$.
- w_d : Storage level (in barrels) at destination d at the beginning of the time horizon.
- $SL_{1,d}$ and $SL_{2,d}$: Minimum and maximum desired levels (in barrels), respectively, at the storage facility of destination d.
- π_d : Type I daily penalty (in U.S. dollars) for each shortage or excess unit at destination d.
- $A_{1,d}$ and $A_{2,d}$: Permitted shortage and excess quantities (in barrels) at destination d with respect to the desired levels $SL_{1,d}$ and $SL_{2,d}$, respectively.
- $b_{1,d} = SL_{1,d} - A_{1,d}$.
- $b_{2,d} = SL_{2,d} + A_{2,d}$.
- λ_d : Type II penalty (in U.S. dollars) associated with destination d.
- SUB_d : A sufficiently large upper bound (in barrels) on the storage level at destination d.
- $TC_{h,d} = \sum_{j=1}^{h} R_{j,d}$.
- Q_s : Production capacity (in barrels) at source $s \in \{1, \ldots, S\}$.
- $S_{h,d}$: A continuous variable that represents the storage level at destination d on day h.
- $P_I(S_{h,d}) = \pi_d \text{ maximum } \{0, (SL_{1,d} - S_{h,d}), (S_{h,d} - SL_{2,d})\}$ if $S_{h,d} \in [b_{1,d}, b_{2,d}]$.
- $P_{II}(S_{h,d}) = \begin{cases} \pi_d A_{1,d} + \lambda_d (b_{1,d} - S_{h,d}) & \text{if } S_{h,d} \in (0, b_{1,d}), \\ \pi_d A_{2,d} + \lambda_d (S_{h,d} - b_{2,d}) & \text{if } S_{h,d} \in (b_{2,d}, SUB_d), \end{cases}$
- $S_{h,d} = S_{1,h,d} - S_{2,h,d} - S_{3,h,d} + S_{4,h,d} + S_{5,h,d}$, where
  $$SL_{1,d} \le S_{1,h,d} \le SL_{2,d}, \quad 0 \le S_{2,h,d} \le A_{1,d},$$
  $$0 \le S_{3,h,d} \le b_{1,d}, \quad 0 \le S_{4,h,d} \le A_{2,d}, \quad 0 \le S_{5,h} \le (SUB_d - b_{2,d}).$$
- $P(S_{h,d}) = \pi_d(S_{2,h,d} + S_{4,h,d}) + \lambda_d(S_{3,h,d} + S_{5,h,d})$.
- $X_{h,t,n,s_1,d,s_2} = \begin{cases} 1 & \text{if } LEG_{h,t,n,s_1,d,s_2} \text{ is selected,} \\ 0 & \text{otherwise.} \end{cases}$
- $Y_{h,t,n,s} = \begin{cases} 1 & \text{if vessel } n \text{ of type } t \text{ is available at source } s \text{ on day } h, \\ 0 & \text{otherwise.} \end{cases}$

- $Z_{t,n} = \begin{cases} 1 & \text{if vessel } n \in \{O_t + 1, \ldots, O_t + CH_t\} \text{ of type } t \text{ is selected} \\ & \text{for chartering during the time horizon,} \\ 0 & \text{otherwise.} \end{cases}$

- $W_f = \begin{cases} 1 & \text{if storage facility } f \text{ is leased for the duration of the} \\ & \text{time horizon,} \\ 0 & \text{otherwise.} \end{cases}$

- ϕ_X and ϕ_Y : Sets of X and Y variables, respectively, that are restricted to be fixed at specified binary values.

- LEG_{h,t,s_1,d,s_2} : A leg for a vessel of type t leaving source s_1 to destination d on day h, and then returning to source s_2.

- c_{t,s_1,d,s_2} : Average cost (in U.S. dollars) associated with LEG_{h,t,s_1,d,s_2}.

- x_{h,t,s_1,d,s_2} : An integer variable that represents the number of vessels of type t that traverse LEG_{h,t,s_1,d,s_2}.

- $y_{h,t,s}$: An integer decision variable that represents the maximum number of vessels of type t that are available for consignment from source s on day h.

- $O_{h,t,s}$: Number of self-owned vessels of type t that will become available for use for the first time at source s on day h of the time horizon.

- $\alpha_{h,t,s}$: Number of vessels of type t that are available (self-owned or those that would be possibly chartered) for use at source s on day h of the time horizon.

- $CH_{h,t,s}$: Number of vessels of type t that will become available for chartering for the first time at source s on day h of the time horizon.

- $\alpha_{h,t,s} = O_{h,t,s} + CH_{h,t,s}$.

- $z_{h,t,s}$: An integer variable that represents the number of vessels of type t that are actually selected for chartering on day h of the time horizon at source s.

- $\$_{h,t}$: Average chartering cost (in U.S. dollars) of a vessel of type t that will become available for use for the first time on day h of the time horizon, regardless of the source at which it first became available.

- $y_{1,t,s} = O_{1,t,s} + z_{1,t,s}$

- $A_{h,t,s}$: A subset of vessels of type t (both self-owned and vessels available for chartering) that will become available for use at source s for the first time on day h of the time horizon.

- $U_{h,t} = \dfrac{\sum_s \sum_{n \in A_{h,t,s}}^{UT_{t,n}}}{\sum_s \alpha_{h,t,s}}$,

- $UT_t = \sum_h UT_{h,t}$.

- $O_t = \sum_h \sum_s O_{h,t,s}$.

- ϕ_x and ϕ_y : Sets of x- and y-variables, respectively, that are *a priori* restricted to be zero, or fixed at some known positive integer values.

- $\overline{P}$: Linear relaxation of any model P.
- $v(P)$: Optimal objective function value of model P.
- $v_{UB}(P)$: Best upper bound (solution) found for model P.
- $v_{LB}(P)$: Best lower bound found for model P.
- I_i : Test problem i, for $i = 1, \dots, 10$.
- NI_k : Test problem k, for $i = 1, \dots, 8$.
- RHA : The rolling horizon algorithm.
- v_{RHA} : Solution value obtained by RHA.
- perct_opt$(v_{RHA}) \equiv 100(1 - \frac{v_{RHA} - v_{LB}}{v_{RHA}})$: Percentage of optimality of v_{RHA}.
- perct_imp$(v_{RHA}, v_{min}) = 100(\frac{v_{min} - v_{RHA}}{v_{min}})$: Percentage of improvement in total cost when allowing the usage of the transshipment depots.
- $v_{RHA}(TP)$: Objective value obtained for Model AVSP via algorithm RHA based on test problem TP.
- $v_{RHA}^{i,j} = v_{RHA}(I_i) + v_{RHA}(I_j)$.
- perct_imp$(v_{RHA}, v_{RHA}^{i,j}) = 100(\frac{v_{RHA}^{i,j} - v_{RHA}}{v_{RHA}^{i,j}})$.

Appendix B: Test Problems

This appendix presents statistics related to eighteen test problems.

Assume that there are four vessel types ($T = 4$) and three transshipment depots ($F = 3$). The daily consumption rate for all destinations is assumed to be the same for the duration of the given time horizon, and given by $R = 150,000$. We also assume that $SL_{1,d} = 2,250,000$, $SL_{2,d} = 10,000,000$, and $A_{1,d} = A_{2,d} = 1,000,000$ for all d. The type I and type II penalties are fixed at $\pi_d = 3$ and $\lambda_d = 500$ for all d. Moreover, we assume that the daily operational costs of vessels of a given type are the same, and hence, $DC_{t,n} = DC_t$ for $n = 1, \dots, M_t$. Specific details pertaining to the vessel types and transshipment depots are presented in the remainder of this appendix.

TABLE B.1

Vessel-Types

t	Ω_t (barrels)	$\$_t$ ($\$$)	DC_t ($\$$)
1	400,000	\$3,000,000	\$6,000
2	600,000	\$4,000,000	\$8,000
3	800,000	\$5,500,000	\$10,000
4	1,200,000	\$7,000,000	\$13,500

TABLE B.2

Transshipment Depots

f	C_f (barrels)	$\$\$_f$ ($)
1	3,000,000	$1,000,000
2	4,000,000	$1,500,000
3	5,000,000	$2,000,000

Table B.3 presents ten single source-destination test problems, denoted by $I_1, \ldots, I_{10}$. The number of self-owned vessels and the number of vessels available for chartering of type t that are considered in test problem I_i are respectively denoted by O_t^i and CH_t^i. Hence, we let $O_*^i = (O_1^i, O_2^i, O_3^i, O_4^i)$ and $CH_*^i = (CH_1^i, CH_2^i, CH_3^i, CH_4^i)$. Let T_{1,s_1,d,s_2}^i, T_{2,s_1,d,s_2}^i, and T_{s_1,d,s_2}^i respectively denote the parameters corresponding to T_{1,s_1,d,s_2}, T_{2,s_1,d,s_2} and T_{s_1,d,s_2} in test problem I_i. Assume that $T_{1,t,1,1,1}^i = T_{2,t,1,1,1}^i = (0.5)T_{t,1,1,1}^i$, $T_{1,t,1,1+f,1}^i = T_{2,t,1,1+f,1}^i = \lceil (0.5)T_{1,t,1,1,1}^i \rceil + f$, $T_{1,t,1+f,1,1+f}^i = T_{2,t,1+f,1,1+f}^i = (0.5)T_{1,t,1,1,1}^i + 2f$, where $(1 + f)$ is the index for the fth transshipment depot. Hence, all the travel times for a particular test problem I_i are defined in terms of $T_{t,1,1,1}^i$. Let $T_*^i = (T_{1,1,1,1}^i, T_{2,1,1,1}^i, T_{3,1,1,1}^i, T_{4,1,1,1}^i)$ and let w^i represent the initial storage level associated with test problem I_i. Note that the time required to travel directly from a source to a destination is always less than the time required to travel from that source to a transshipment depot plus the time required to travel from the transshipment depot to the destination. Also, $T_{t,1,1,1}^i = 0$ indicates that no vessels of type t are considered in test problem I_i.

Note that the above test problems are those considered in part I, in addition to incorporating three transshipment storage depots. Next, we define a set of eight test problems, each involving two sources and two destinations. Let (s^k, d^k) denote the source-destination pair associated with test problem I_k, for $k = 1, \ldots, 10$. Let $I_{i,j}$ for $i, j \in \{1, \ldots, 10\}$ denote the test problem that has the sources s^i and s^j, and the destinations d^i and d^j. For the sake of consistency in notation, we let $s = 1, 2$ respectively index the sources s^i and s^j, and we let $d = 1, 2$ respectively index the destinations d^i and d^j.

TABLE B.3

Description of the Test Problems Having a Single Source and Destination

Test Problem I_i	H (days)	T	O_*^i	CH_*^i	T_*^i (days)	w^i (barrels)
I_1	30	1	(4,0,0,0)	(5,0,0,0)	(10,0,0,0)	4,000,000
I_2	60	1	(2,0,0,0)	(2,0,0,0)	(10,0,0,0)	7,000,000
I_3	60	2	(2,3,0,0)	(2,7,0,0)	(10,20,0,0)	7,000,000
I_4	90	2	(2,2,0,0)	(2,3,0,0)	(10,20,0,0)	7,000,000
I_5	120	2	(3,4,0,0)	(8,7,0,0)	(20,30,0,0)	7,000,000
I_6	120	4	(1,1,1,1)	(3,4,2,6)	(20,30,40,50)	7,000,000
I_7	150	3	(2,5,2,0)	(1,5,2,0)	(20,30,40,0)	4,000,000
I_8	180	3	(5,1,6,0)	(2,7,9,0)	(20,30,40,0)	7,000,000
I_9	210	4	(4,2,1,4)	(3,2,1,4)	(20,30,40,50)	7,000,000
I_{10}	240	4	(5,2,2,2)	(1,2,2,6)	(30,40,50,60)	7,000,000

TABLE B.4

Description of the Test Problems Having Multiple Sources and Destinations

Test Problem	H (days)	T	O*	CH*
NI_1	$I_{1,1}$	30	1	(8,0,0,0)	(10,0,0,0)
NI_2	$I_{2,3}$	60	2	(4,3,0,0)	(4,7,0,0)
NI_3	$I_{4,4}$	90	2	(4,4,0,0)	(4,6,0,0)
NI_4	$I_{5,6}$	120	4	(4,5,1,1)	(11,11,2,6)
NI_5	$I_{7,7}$	150	3	(4,10,4,0)	(2,10,4,0)
NI_6	$I_{8,8}$	180	3	(15,3,18,0)	(6,21,27,0)
NI_7	$I_{9,9}$	210	4	(12,6,3,12)	(9,6,3,12)
NI_8	$I_{10,10}$	240	4	(15,6,6,6)	(3,6,6,18)

The direct source-destination and destination-source trips that test problem $I_{i,j}$ incorporates are those described by test problems I_i and I_j, in addition to four more trips, two that traverse from s^i to d^j, and two others that traverse from s^j to d^i. Let $T^{i,j}_{t,s_1,d,s_2}$ denote the travel time associated with LEG_{h,t,s_1,d,s_2} in test problem $I_{i,j}$.

Note that for the sake of simplicity and ease in comparison between costs associated with test problem $I_{i,j}$, and the summation of costs associated with its component test problems I_i and I_j, we make the following assumptions:

a) We only combine test problems that have the same time horizon.

b) If (s_1, d, s_2) is a feasible leg in test problem I_i, then $T^{i,j}_{t,s_1,d,s_2} = T^i_{t,s_1,d,s_2}$, $T^{i,j}_{1,t,s_1,d,s_2} = T^i_{1,t,s_1,d,s_2}$, and $T^{i,j}_{2,t,s_1,d,s_2} = T^i_{2,t,s_1,d,s_2}$. In a similar fashion, if (s_1, d, s_2) is a feasible leg in test problem I_j, then $T^{i,j}_{t,s_1,d,s_2} = T^j_{t,s_1,d,s_2}$, $T^{i,j}_{1,t,s_1,d,s_2} = T^j_{1,t,s_1,d,s_2}$, and $T^{i,j}_{2,t,s_1,d,s_2} = T^j_{2,t,s_1,d,s_2}$.

c) The travel times for trips that do not belong to either test problem I_i or I_j (i.e., the trips that involve s_1 and d_2, or s_2 and d_1), are given as follows: $T^{i,j}_{1,t,1,2,1} = T^{i,j}_{2,t,1,2,1} = T^{i,j}_{1,t,1,1,1} + 2$ and $T^{i,j}_{1,t,2,1,2} = T^{i,j}_{2,t,2,1,2} = T^{i,j}_{1,t,2,2,2} + 3$.

d) The total number of self owned and chartered vessels of type t that are considered in test problem $I_{i,j}$ are respectively given by $O^{i,j}_t = O^i_t + O^j_t$ and $CH^{i,j}_t = CH^i_t + CH^j_t$, for $t \in \{1, \dots, 4\}$. Accordingly, we denote $O^{i,j}_* = (O^i_1 + O^j_1, O^i_2 + O^j_2, O^i_3 + O^j_3, O^i_4 + O^j_4)$ and $CH^{i,j}_* = (CH^i_1 + CH^j_1, CH^i_2 + CH^j_2, CH^i_3 + CH^j_3, CH^i_4 + CH^j_4)$.

8

An Integer Programming Model for the Optimization of Data Cycle Maps

David Panton, Maria John, and Andrew Mason

CONTENTS

8.1   Introduction

In this chapter we discuss an application of integer programming which involves the capturing, storage, and transmission of large quantities of data collected during a variety of possible testing scenarios which might involve military ground vehicles, cars, medical applications, large equipment, missiles, or aircraft. The particular application on which the work in this chapter is based involved the flight testing of military aircraft, however the process and procedures discussed might well be applied to any of the other scenarios mentioned above.

The flight testing of military aircraft is a time consuming and expensive process. In some cases, for example with fighter aircraft such as an F16, the aircraft's weapon systems must be removed and replaced by test instruments. In any event, data acquisition equipment must be installed in addition to the standard operational equipment. These instruments are used to collect a large amount of information relating to things such as speed, altitude, mechanical stress, pressure, radar, video, or perhaps internal communications data in planned test missions. Typically, several hundred or possibly thousands of *parameters* will be continuously sampled during the flight, with a subset of these being telemetered (i.e., transmitted) to a ground receiving station as they are measured. In planning for a test flight, the parameters to be transmitted need to be multiplexed into a data structure called a *data cycle map* (*DCM*), a sequence of digital words, each represented by a number of bits. The individual parameters may vary by the number of words required to store them and the length of each word. In addition, sampling rates may also vary from one parameter to another. Data Cycle Maps are sometimes referred to as Telemetry Frames or PCM Formats. A major reason for using DCMs is that their synchronous nature provides an efficiency not attainable in asynchronous packetized telemetry.

The way in which the data cycle map can be constructed is subject to certain standards prescribed by the Inter Range Instrument Group (IRIG) [1]. The essential requirements of the IRIG Class I regulations which are relevant to this discussion can be summarized as follows. The basic building block of a data cycle map is an array called a *major frame* whose rows are called *minor frames*. Each row is a sequence of words, where all words in a map are of the same size (typically 16 bits). Minor frames can be no longer than 512 16-bit words in length, while major frames can consist of no more than 256 minor frames. Each minor and major frame must start with frame synchronization words and contain a frame ID. A data cycle map is a repetition of major frames transmitted through the duration of the test. In building a data cycle map, parameters must be placed on the map in accordance with their sample rates and the number of data words they require. Perhaps the most constraining feature of the DCM construction process is that each parameter must appear periodically within the map. That is, there must be a constant interval between successive occurrences of the same parameter in a data cycle map, even when crossing minor and major frame boundaries.

This chapter discusses the process of data cycle map construction from the data input phase to the generation of optimal telemetry frames. In Section 8.2 we discuss the calculation of key DCM design factors based on the nature of the input data and the constraints imposed by the IRIG regulations. This step typically gives rise to a number of different candidate solutions from which we seek the most efficient. These candidate solutions are constructed to satisfy many of the periodicity requirements, but their associated design factors do not fully specify a solution and so feasibility cannot be guaranteed at this stage. In Section 8.3 we use the DCM design factors calculated in Section 8.2 to formulate a set packing model that will either construct a feasible data cycle map, or show that no feasible data cycle map can be generated for the given design factors. Section 8.4 describes how the set packing model is strengthened in order to speed up the solution process, while results and conclusions are discussed in Section 8.5.

8.2   The Data Cycle Map Structure

The construction of a data cycle map is a complex process. The complexity lies not only in the number of parameters that need to be placed on the map, but more particularly in the fact that, individually, they must be equally spaced. Manual methods for constructing maps are faced with avoiding coincident placements. This avoidance becomes more and more difficult as the map construction process proceeds. As we will see, a coincident placement can always be resolved by changing to some 'next most efficient' DCM design. Manually constructed maps typically make this change many times, and consequently contain a very high percentage of 'empty' space.

Table 8.1 shows an example of a major frame in a data cycle map used to transmit $P = 5$ parameters, A through E. This major frame consists of $n_{\text{minor}} = 4$ minor frames (rows), numbered 1 through 4, each of which contains $S = 20$ positions, labeled as 'slot 1,' 'slot 2,' ... , 'slot 20'. Each slot can contain one 16-bit digital word of data. (For IRIG Class I the largest legal word size is 16 bits, and this is the size we will adopt in all our data cycle maps.) During the transmission of a minor frame, the contents of the slots are transmitted in the sequence slot 1, slot 2, ... , slot S. The transmission of the 4 minor frames in sequence constitutes a complete major frame transmission. The data cycle map consists of repeated transmissions of these major frames. We note that the major frame is the smallest repeating block in the data cycle map, and so the major frame uniquely determines the full data cycle map.

If we examine this example, we see that the first 2 slots in each minor frame are occupied by header data, while the third slot contains a minor frame ID. (This ID can, in fact, be placed anywhere in the minor frame.) The rest of each frame consists of either empty unused slots, or slots containing one of the

TABLE 8.1

A Possible Major Frame Consisting of $n_{\text{minor}} = 4$ Minor Frames, Each with $S = 20$ 16-Bit Words, That Is Used for Transmitting $P = 5$ Parameters. Each Minor Frame Takes 0.125 Seconds to Transmit, Giving a Transmission Rate of $r_{\text{minor}} = 1/0.125 = 8$ Minor Frames per Second

$S = 20$ slots

	1	2	3	4	5	6	7	8	9	10	11	12	13	14	15	16	17	18	19	20
1:	h_1	h_2	ID	E	D	C	C	E		A	A	E			D	E				E
2:	h_1	h_2	ID	E	D	C	C	E		B		E			D	E				E
3:	h_1	h_2	ID	E	D	C	C	E				E			D	E				E
4:	h_1	h_2	ID	E	D	C	C	E		B		E			D	E				E

$n_{\text{minor}} = 4$

Transmission time = 0.125 seconds per minor frame

TABLE 8.2

Summary Statistics for Each Parameter in the Data Cycle Map
Defined by Table 8.1 Assuming a Transmission Rate of $r_{minor} = 8$
Minor Frames per Second

Parameter	A	B	C	D	E
Period (slots)	80	40	20	10	4
Occurrences per minor frame	0.25	0.5	1	2	5
Occurrences per major frame	1	2	4	8	20
Resultant sampling rate (samples/second)	2	4	8	16	40

5 parameters being transmitted. We note that each sampling of parameters B, D, and E generates data that occupies just a single slot; we say their *word requirement* is 1 slot. In this example, however, parameters A and C have a word requirement of 2 slots, and so we see that each occurrence of these parameters occupies 2 adjacent slots in the map. Perhaps the most important observation is that each parameter appears with a regular period within the major frame. For example, parameter D appears in every 10th slot in the major frame, while parameter A appears just once in the major frame, and so appears in the data cycle map with a period of 80 slots.

Let us assume that for this example, we have a transmission rate of $r_{minor} = 1/0.125 = 8$ minor frames per second, and so 2 complete major frames are transmitted every second. By counting the occurrences of each parameter in the major frame, we can compute the resultant sampling rates for each parameter; this data is given in Table 8.2. In practice, the desired sampling rates will be given as part of the input data, and we then have to design a data cycle map that efficiently realizes (or exceeds) these desired rates for each parameter.

We note that parameters A and B do not appear on every minor frame. Parameters such as these are said to be *sub-commutated*. On the other hand, parameters C, D, and E, which occur at least once on each minor frame, are said to be *super-commutated*. The use of these terms relates to the fact that the original flight test instruments were electro-mechanical devices which used commutators to log the data.

In considering the application of optimization methods to data cycle map construction we need to take into account our ability to manage the task in terms of memory and computational requirements. For example, the construction of an entire major frame, with its many repetitions of minor frames, will almost certainly require a model which is too unwieldy and unmanageable. For this reason it will be necessary to decompose the problem into smaller tasks whose solutions can be used as building blocks for solving the larger problem. Our approach therefore will be to construct minor frames whose replication can be used to define the entire major frame. In doing this it will be essential that parameter sample rates are preserved and that parameter periodicity is also achieved, along with other IRIG standards to be discussed next.

8.3   Designing a Minor Frame

The design of a suitable minor frame is dependent on several factors. Apart from periodicity and minimum required sample rates, IRIG Class I conditions also require that transmission bit rates fall within a minimum and maximum level, each minor frame has two header synchronization words and a minor frame ID word, there is a common word length for each parameter, minor frames contain no more than 512 16-bit words, and that there are no more than 256 minor frames in a major frame. Although we will not be concerned with many of these operational issues it is important that they are taken into account when designing a suitable minor frame. We will assume that the two header words are placed at the beginning of each minor frame, but the frame ID may be placed anywhere in the frame. Within these constraints the task is to place the measured parameters onto a frame according to their required sample rates and periodicity. As we will see the periodicity will depend on the size of the frame. There are normally several options available on the size of the frame, however it is our task to select the most efficient frame size. The *efficiency* E of a major frame is the ideal number of parameter words (excluding header words) transmitted per second divided by the actual total transmission rate:

$$E = \frac{\text{ideal parameter words transmitted per second}}{\text{total words transmitted per second}}.$$

For any data cycle map, the number of words (including unused words corresponding to empty slots) that are transmitted per second is given by the product of the minor frame length S and the minor frame rate r_{minor}. The number of words that actually need to be transmitted per second for some parameter p is given by $r_p \times w_p$, where r_p is the *required sample rate* (samples/second) for parameter p, and w_p is the *word requirement* (number of words that need to be transmitted per sample) for parameter p. Thus, the efficiency E of a solution with minor frame length S and minor frame rate r_{minor} can be written

$$E(S, r_{\text{minor}}) = \frac{\sum_{p=1}^{P} (r_p \times w_p)}{S \times r_{\text{minor}}}, \tag{8.1}$$

where P is the number of parameters. We note that the required number of words (in the numerator) does not include the header or frame ID words, even though these are needed in the data cycle map, and so, under this definition, 100% efficiency can never be achieved. Alternative definitions are possible, but they do not significantly change the nature of the optimization problem.

   For any given set of parameters, the efficiency E depends on the minor frame length S and the minor frame transmission rate r_{minor}. Thus we can split our problem into a sequence of steps, being (1) to find a range of values for S and r_{minor} that define a number of candidate solutions, (2) choosing the most efficient solution from among these candidates, and (3) then attempting to find a placement of data parameters within the slots that satisfies the periodicity requirements. Step (2) and Step (3) of this process may need to be repeated if a proposed best candidate is shown to be infeasible. Note that to keep the solution times manageable, we will restrict ourselves to candidate solutions that possess a specific intuitive structure. Although we believe this structure is likely to be possessed by an optimal solution, we cannot guarantee that this is so, and so we cannot offer any formal guarantee of optimality.

We will show that once we have determined the minor frame rate r_{minor}, it is reasonably straightforward to calculate a set of values for the minor frame length S that give rise to a family of candidate solutions. These candidate solutions can be ordered from most efficient (but least likely to give a feasible placement) to least efficient (but guaranteed to be feasible). We also show how to calculate the number of minor frames n_{minor} required per major frame for these solutions.

To illustrate our process, we will demonstrate the construction of a data cycle map for the following example.

Example 8.1

Consider $P = 5$ parameters with sample rates r_p and word requirements w_p as shown in Table 8.3.

Our focus is on building minor frames, and hence we require that successive minor frames be essentially the same in content. To achieve this, the transmission rate of each parameter needs to be an integer multiple (or fraction) of the minor frame transmission rate. This suggests that the minor frame transmission rate should be some integer multiple of at least one of the required parameter transmission rates. Transmitting a minor frame incurs the overhead of transmitting header and ID words, and so lower minor rates are likely to be better. Therefore, we choose to explore a set of candidate solutions in which the minor frame transmission rate matches the transmission rate of at least one of the parameters.

In the process we will describe, we assume that the r_p values are in increasing order, as is the case in Table 8.3. We now choose the minor frame rate r_{minor} to be equal to the transmission rate r_q of some parameter q, $1 \le q \le P$, giving $r_{minor} = r_q$. (For ease of notation, we assume that parameter q is the first such parameter with sampling rate r_q, i.e., we break ties in transmission rates by choosing the smallest q.) Note that some choices of q will be infeasible if they

TABLE 8.3

A 5-Parameter Data Set for Example 8.1

Parameter		A	B	C	D	E
Required sample rate (samples/second)	r_p	2	5	12	25	33
Number of words required per sample	w_p	1	1	1	1	2

give a minor frame rate outside the minimum and maximum allowable bit transmission rates; these values should not be considered.

Given this choice of the minor frame rate r_{minor}, all parameters p with $p < q$ and hence $r_p < r_{minor}$ will be sub-commutated, appearing no more than once per minor frame, while all parameters $p \geq q$ implying $r_p \geq r_{minor}$ will be super-commutated. If we were allowed to place each parameter some fractional number of times on a minor frame, then parameter p would occur r_p/r_{minor} times per minor frame, giving (possibly fractional) per-minor-frame counts for the parameters of

$$\left\{ \frac{r_1}{r_{minor}}, \frac{r_2}{r_{minor}}, \dots, \frac{r_p}{r_{minor}} \right\},$$

where the qth element $r_q/r_q = 1$, all elements to the left of the qth are less than 1 and all elements to the right of the qth are greater than (or equal to) 1. To illustrate this, suppose we choose $q = 3$ (i.e., q is our third parameter, parameter C), giving $r_{minor} = r_q = 12$, and hence defining the vector

$$\left\{ \frac{2}{12}, \frac{5}{12}, \frac{12}{12}, \frac{25}{12}, \frac{33}{12} \right\} = \left\{ \frac{1}{6}, \frac{1}{2.4}, 1, 2.08\dot{3}, 2.75 \right\}.$$

To resolve the fractional per-minor-frame counts, we now need to perform an integerization step that determines new integer counts for both the sub-commutated and super-commutated parameters. These counts will be chosen to ensure each parameter p is sampled at a rate no lower than its desired rate r_p. To ensure each super-commutated parameter $p \geq q$ occurs at least r_p/r_{minor} times per minor frame, we specify a count of $n_p = \lceil r_p/r_{minor} \rceil$ per minor frame for these parameters. Each sub-commutated parameter $p < q$ would, if fractional counts were possible, occur once every r_{minor}/r_p minor frames. To ensure we achieve a sampling rate no less than this, we require that parameter $p < q$ occurs once every $g_p = \lfloor r_{minor}/r_p \rfloor$ minor frames. We can combine these to characterize our parameter counts using the shorthand vector

$$\mathbf{n} = \left\{ n_1 = \frac{1}{g_1}, n_2 = \frac{1}{g_2}, \dots, n_{q-1} = \frac{1}{g_{q-1}}, \frac{1}{1}, n_{q+1}, n_{q+2}, \dots, n_p \right\}.$$

In our example, the integerizing of the per-minor-frame counts $\left\{ \frac{1}{6}, \frac{1}{2.4}, 1, 2.08\dot{3}, 2.75 \right\}$ gives

$$\mathbf{n} = \left\{ n_1 = \frac{1}{6}, n_2 = \frac{1}{2}, n_3 = 1, n_4 = 3, n_5 = 3 \right\},$$

with $g_1 = 6$ and $g_2 = 2$. Thus parameter C will occur once in each minor frame and parameter D three times, for example. Since the minor frame is repeated at a rate of $r_{minor} = 12$ times per second, parameter D will appear $n_4 \times r_{minor} = 36$ times per second, which is acceptable as it is no less than the required sample rate of 25 per second. The value of $n_1 = 1/6$ for parameter A means that this parameter will occur only once in every $g_1 = 6$ minor frames, giving a realized sampling rate of $n_1 \times r_{minor} = 2$ samples per second, exactly matching the required rate.

The next step is to calculate the number of minor frames n_{minor} that will define a complete major frame. Recall that n_{minor} is simply the number of minor frames we have to observe before we see a repetition in the data being sent. If sub-commutated parameter p, $p < q$ occurs once every g_p minor frames, then the number of minor frames required is in general given by

$$n_{minor} = \text{LCM}(g_1, g_2, \ldots, g_q),$$

where the LCM() function computes the least common multiple of its arguments. This value will guarantee that the sub-commutated parameters can be periodically placed on the minor frames. In our example $g_1 = 6$, $g_2 = 2$, and hence $n_{minor} = \text{LCM}(6, 2) = 6$. Thus, parameter A must be placed in any one of the 6 duplications of the minor frame which will make up the major frame. Parameter B must be placed in the same slot in every second minor frame, and so will appear 3 times in the major frame.

We next need to calculate the number of slots S required in the minor frame to ensure we can feasibly place all the parameters onto the data cycle map. We said earlier that there is a family of possible values of S, where smaller values of S from this family are more efficient (i.e., more of the minor frame contains parameter data), but are less likely to permit a feasible parameter placement (i.e., fewer slots are available to avoid coincident placements between parameters). Our approach is to initially try a small value for S, and then, if this does not allow a feasible parameter placement, keep increasing S until feasibility is achieved. This process will terminate by either finding a minor frame length that is feasible, or by generating a bit transmission rate that exceeds the IRIG standards.

In our search for the best S, we need some sensible lower bound from which we can start. A simple lower bound on S can be calculated by ignoring the periodicity requirements, and simply counting up the number of data words that need to be transmitted per minor frame. Each super-commutated parameter $p \geq q$ has to occur n_p times per minor frame, thereby occupying a total of $n_p w_p$ slots. Thus, the space required in each minor frame by the super-commutated data is given by

$$S^{super} = \sum_{p=q}^{P} n_p w_p.$$

For our example, we have $S^{super} = (1 \times 1) + (3 \times 1) + (3 \times 2) = 10$. In addition to this, we need to reserve space for the sub-commutated parameters to ensure there are no clashes between them and the other parameters. We note that different sub-commutated parameters can share the same slot position in the minor frames as long as we never need to transmit both parameters in the same minor frame. Thus, in general, we need to solve some form of set packing problem to determine how best to allocate the sub-commutated parameters to slots and minor frames. However, a simple lower bound based on slot usage used can be easily calculated as

$$\hat{S}_1^{\text{sub}} = \left\lceil \sum_{p=1}^{q-1} n_p w_p \right\rceil.$$

Summing these values, and adding another 3 words for the header and ID words, gives a lower bound on the minor frame length S:

$$\hat{S}_1 = \hat{S}_1^{\text{sub}} + 3 + S^{\text{super}}.$$

For our example (Table 8.3), we have $\hat{S}_1^{\text{sub}} = \lceil \frac{1}{6} \times 1 + \frac{1}{2} \times 1 \rceil = 1$, $S^{\text{super}} = 10$, and so $\hat{S}_1 = 10 + 3 + 1 = 14$.

We can strengthen this bound by considering the periodicity requirements. Each super-commutated parameter $p \geq q$ must appear periodically in the data cycle map, and so the minor frame length must be some multiple of each super-commutated parameter's per-minor-frame count n_p. Thus, we have

$$S = k\text{LCM}(n_p, p = q, q + 1, \ldots, P), \tag{8.2}$$

for some positive integer k. Thus, the lower bound $\hat{S}_1$ can be improved by increasing it until it is a multiple of $\text{LCM}(n_p, p = q, q + 1, \ldots, P)$, giving a new bound of

$$\hat{S}_2 = \hat{k}_2\text{LCM}(n_p, p = q, q + 1, \ldots, P),$$

i.e., we have put $k = \hat{k}_2$ in (8.2), where $\hat{k}_2$ is the smallest integer value giving $\hat{S}_2 \geq \hat{S}_1$. For our example, $\text{LCM}(n_3, n_4, n_5) = \text{LCM}(1, 3, 3) = 3$, and so we increase our bound from $\hat{S}_1 = 14$ to $\hat{S}_2 = 15$. We note, as an aside, that increasing n_p for some parameter $p \geq q$ could actually allow us to decrease the minor frame length S if it decreased this LCM value. This is not something we have considered, but is left as a subject for future research.

We start our search by putting $k = \hat{k}_2$, $S = \hat{S}_2$, and then try to place all the parameters into a major frame consisting of n_{minor} minor frames of length S. We use integer programming to solve this problem; further details are provided in the next section. If the integer program determines that no feasible solution exists, then we increase k by 1 in Equation (8.2), thereby increasing the minor frame length S to the next multiple of the LCM. This process is repeated until a feasible solution is obtained.

In Table 8.4 we show that it is possible to place all five parameters into an $(S \times n_{\text{minor}}) = (15 \times 6)$ major frame constructed from minor frames defined by Table 8.5, and thus we have a feasible solution to our problem. The key parameter statistics for this solution, including the parameter periods $m_p = S/n_p$, are shown in Table 8.6. To summarize, we have 6 minor frames, each containing 15 words. Each word contains 16 bits. Parameter A will be placed in only one of these minor frames (any one will do), parameter B will be placed three times, once in every second minor frame, while parameters

TABLE 8.4

The Full Major Frame for the Example 1 Problem Defined in Table 8.3.
In this Data Cycle Map, We Have $n_{minor} = 6$ Minor Frames per Major Frame,
a Transmission Rate of $r_{minor} = 12$ Minor Frames per Second, and $S = 15$
Slots per Minor Frame

	1	2	3	4	5	6	7	8	9	10	11	12	13	14	15
1:	h_1	h_2	D	E	E	ID	C	D	E	E	B		D	E	E
2:	h_1	h_2	D	E	E	ID	C	D	E	E	A		D	E	E
3:	h_1	h_2	D	E	E	ID	C	D	E	E	B		D	E	E
4:	h_1	h_2	D	E	E	ID	C	D	E	E			D	E	E
5:	h_1	h_2	D	E	E	ID	C	D	E	E	B		D	E	E
6:	h_1	h_2	D	E	E	ID	C	D	E	E			D	E	E

C, D, and E will be placed 1, 3, and 3 times respectively in each minor frame at
appropriate positions that ensure periodicity. In this solution, minor frames
are transmitted at the rate of $r_{minor} = 12$ times per second. Using Equation
(8.1), the efficiency $E(S, r_{minor})$ of this solution is given by $E(15, 12) = 110/180 = 61.1\%$.

The above process assumed we started by calculating a lower bound of
$\hat{S}_1 = \hat{S}_1^{sub} + 3 + S^{super}$. In practice, we do not attempt to optimally allocate the
sub-commutated parameters, and so we do not use $\hat{S}_1^{sub}$. Instead, we replace
the lower bound $\hat{S}_1^{sub}$ by the number of slots that must be reserved in each
minor frame for a heuristically generated parameter placement, which we
construct as follows. Let D^{sub} be the number of *distinct* sub-commutated parameters, where we say two such parameters are distinct if they have different
g_p or w_p values. Let $\{(g'_d, w'_d), d = 1, 2, \ldots, D^{sub}\}$ be the set of distinct (g_p, w_p)
pairs found in the sub-commutated parameters, and let $n'_d, d = 1, 2, \ldots, D^{sub}$
be the number of sub-commutated parameters having $(g_p, w_p) = (g'_d, w'_d)$.
A feasible allocation of the sub-commutated parameters is to pack the distinct
parameters into their own slots within the minor frame. Therefore, the number of minor-frame slots that we reserve for sub-commutated parameters is
given by

$$\sum_{d=1}^{D^{sub}} \left\lceil \frac{n'_d}{g'_d} \right\rceil w'_d.$$

In practice, we replace $\hat{S}_1^{sub}$ by this value.

TABLE 8.5

One Possible Minor Frame Configuration for the Example 8.1 Problem
Defined in Table 8.3. Note That Slot 11 Is Reserved for Sub-Commutated
Parameters A and B Who Share this Slot Position in the Major Frame

1	2	3	4	5	6	7	8	9	10	11	12	13	14	15
h_1	h_2	D	E	E	ID	C	D	E	E	AB		D	E	E

TABLE 8.6

Sampling Rates for Each Parameter in The Data Cycle Map Defined by Table 8.5 Assuming a Transmission Rate of $r_{minor} = 12$ Minor Frames Per Second

Parameter		A	B	C	D	E
Period	m_p	90	30	15	5	5
Desired sampling rate	r_p	2	5	12	25	33
Occurrences per minor frame	n_p	1/6	1/2	1	3	3
Resultant sampling rate	$n_p r_{minor}$	2	6	12	36	36

The process described above for Example 8.1 can be carried out for all possible values of r_q and an efficiency value E calculated for each. A number of factors may influence the value of E. These include the degree of rounding up required to determine the n_p values leading to over-sampling, the gap between the minimal slot usage requirement $\hat{S}_1$, and the minor frame length $\hat{S}_2$ required to satisfy periodicity, and any increases in k in Equation (8.2) required to permit a feasible parameter placement.

For very large data sets all or many of the values for S may violate the maximum minor frame length of 8192 bits. In this situation a simple device for restoring legality is to introduce header splits into the minor frame. In this situation the header can be thought of as another parameter with a sample rate determined by the required number of splits. For example, suppose that the most efficient frame design gives a frame length of 842 16-bit words, including the original header of two words and a frame ID of one word. We introduce another two word header and treat it as a parameter with a sample rate of 2. The additional 2 words can be added to the original nominal minor frame length which will in most cases still give a frame length less than or equal to 842. If necessary the revised minor frame length S may need to be adjusted up by adding another LCM factor of the super-commutated parameters. When the new S value is computed (assume for the sake of this argument this is still 842), our original minor frame of length 842 can now be replaced by two minor frames each of length 421. Periodic placement of the header will guarantee that it is placed at the start of each of these two new frames. The header splitting strategy will work as long as the total number of minor frames does not exceed the limit of 256. If this limit is exceeded, the data set is too large to conform to IRIG class I conditions for this r_{minor}.

Preprocessing of the data will examine all options for r_{minor} and their relative efficiencies. Under normal circumstances the option which has the highest efficiency will be chosen, and we will then be able to create a feasible minor frame for this configuration using the integer programming model discussed next. Sometimes, however, the integer program will fail for the desired option, and so k in (8.2) will need to be increased by 1, changing that option to its next most efficient configuration. The new most efficient option

TABLE 8.7

A Data Set for Example 8.2 with 4 Parameters

Parameter	A	B	C	D
Required sample rate r_i	1	3	3	5
Number of words per sample w_p	1	1	1	1

will then be solved using integer programming, with this process being repeated until a feasible solution is found. Example 8.2 shows the results of this process.

Example 8.2

Consider the data given in Table 8.7.

Table 8.8 shows all possible rates, minor frame lengths, and efficiencies. Of the three possible frame rates the most efficient is a single frame with efficiency 80%.

The natural choice in this case is a frame rate of 1 with no sub-commutation. The periods $m_p = S/n_p$ for each parameter p are given by {15, 5, 5, 3} with a space of 2 words reserved at the beginning for the header and an additional space for the frame ID. It is easy to show by construction that placement of the parameters on this frame is impossible without avoiding coincidence. In this case we are forced to try a less efficient frame construction with $S = 2 \times \mathrm{LCM}(1, 3, 3, 5) = 30$. This $S = 15$ example fails because of the relative prime relationship between the periods for parameters B and D (i.e., because these periods have no factors in common other than 1). In fact it can be shown [2] that a sufficient condition for the coincidence of parameters i and j in the frame is that periods m_i and m_j are relatively prime. Panton et al. [2] explore the fact that other more complex relationships between the periods also lead to coincidence. An interesting parallel with juggling is also discussed.

In the discussion of the optimization models used to generate feasible placements which follows, we will assume that k in equation (8.2) has been increased to give an S for which no pair of parameters have periods that are relatively prime.

TABLE 8.8

Alternative Frame Rates and Efficiencies for Data from Table 8.7

Option	q	r_{minor}	S	E
1:	1	1	15	80%
2:	2	3	8	50%
3:	4	5	7	34%

8.4   Optimization Models

We now turn to the problem of finding a feasible placement of the parameters within a minor frame of length S with desired parameter periods m_p. We assume that the data set has been modified by removing all sub-commutated parameters, and inserting in their place a set of new 'aggregated' parameters, each with per-minor-frame count $n_p = 1$, that represent the packed sub-commutated parameter solution generated by our heuristic.

Three models were considered for the generation of feasible placements. The first considered was an Integer Programming formulation in which parameters were directly assigned to a minor frame with all desired properties. In this model we let x_{ps} be a binary decision variable whose value is 1 if parameter p *starts* at position s on the minor frame and is 0 otherwise. This method proved to be too slow to be operationally useful. It is well known [4], that although this modeling approach is economic in terms of its proliferation of variables, the structure of the polytope is not conducive to finding solutions efficiently. The second approach considered to generate feasible parameter placements involved the use of Genetic Algorithms. A complete description of this model can be found in [3].

8.4.1   Generalized Set Packing Model

The third approach was to employ a set packing model in which the columns in the model represent all feasible placements for each individual parameter. Having determined the length of the minor frame S we can now enumerate all possible placements for a given parameter on this frame. Each placement pattern will be referred to as a *tour*. For example, consider the 4th parameter, parameter D in Table 8.6, which we require to appear $n_4 = 3$ times in a minor frame of length $S = 15$ with a period of $m_4 = 5$. This parameter will have the following tours as shown in Table 8.9.

Our problem is to select one tour for each parameter so that each slot contains no more than one parameter. Generation of the parameter tours automatically ensures that they are periodic. Consider the following model.

Define the binary variable $x_{p,j}$ to be 1 if column j is selected for parameter p, and 0 otherwise. Our objective is then to:

$$[\text{DCM}] \text{ Maximize } \sum_{p,j} x_{p,j}$$

$$\text{subject to } \sum_{p,j} a_{s,(p,j)} x_{p,j} \leq 1 \quad \forall\, s,$$

$$\sum_{j} x_{p,j} = 1 \quad \forall\, p,$$

$$x_{p,j} \text{ binary } \quad \forall\, p,\, j.$$

TABLE 8.9

Enumeration of All Tours for a Parameter
with Period $m_p = 5$ in a Frame of Length
$S = 15$

	Tour				
	1	2	3	4	5
slot 1	1	.	.	.	.
slot 2	.	1	.	.	.
slot 3	.	.	1	.	.
slot 4	.	.	.	1	.
slot 5	.	.	.	.	1
slot 6	1	.	.	.	.
slot 7	.	1	.	.	.
slot 8	.	.	1	.	.
slot 9	.	.	.	1	.
slot 10	.	.	.	.	1
slot 11	1	.	.	.	.
slot 12	.	1	.	.	.
slot 13	.	.	1	.	.
slot 14	.	.	.	1	.
slot 15	.	.	.	.	1

where $a_{s,(p,j)} = 1$ if the jth column for parameter p covers slot s and 0 otherwise. Each row in the first block of constraints represents a slot in the minor
frame. The second block of constraints ensure that only one column is selected
from each parameter set. This is a feasibility problem, and so the objective
function is essentially irrelevant.

Savings in the number of variables can be made by noting that in the generation of the parameter tours there is considerable duplication since parameters
with the same minor frame sample rate and word requirements will have identical tour sets. Parameters can be grouped into parameter classes and tours
generated for each class. In this case the constraints which ensure that only
one parameter is selected from each set are modified so that the appropriate number are selected from each class. Once the optimal solution is found,
members of each class are assigned arbitrarily to each tour and the minor
frame map reconstructed.

An interesting feature of this model is that we know *a priori* what the
optimal solution objective value is, since we know that all P parameters must
be allocated. This enables us to set the tightest possible bound on the optimal solution, thereby greatly reducing the size of the branch and bound tree.
Two other observations are important. First, when these models are solved
using CPLEX^{TM} 7.0, [5], we have the option of seeking feasible rather than
optimal solutions since in this case any feasible solution is automatically optimal. Second, Dual Simplex is used as the default linear program (LP) solver,
however when the number of columns in the model becomes significantly
greater than the number of rows we switch to the Primal LP solver. Both of
these strategies contribute significantly to reductions in execution times.

8.5 Strengthening the Set Packing Integer Programming Model

In addition to the issues discussed at the end of the last section, the nature of this problem lends itself to the addition of sets of clique inequalities which will considerably tighten the LP bounds and hence accelerate the solution process. To summarize the requirements, we require that the pth parameter, $p = 1, 2, \ldots, P$, be regularly spaced in the minor frame, and appear n_p times. We assume that n_p, $p = 1, 2, \ldots, P$, are all positive integers, and the n_p values are relatively prime, meaning that their greatest common divisor is 1. (This assumption is consistent with the n_p values defining relative, not absolute, transmission rates.) We note that to ensure periodicity, we have a minor frame length $S = k\mathrm{LCM}(n_p, p = 1, 2, \ldots, P)$ where $k \in \{1, 2, 3, \ldots\}$. This means that parameter p will occur every $m_p = S/n_p$ slots in the minor frame, where m_p is the period of parameter p. We show in Appendix A, Theorem 8.1 that

$$\mathrm{LCM}(m_p, p = 1, 2, \ldots, P) = S.$$

The set packing integer programming model [DCM] for this problem has S 'slot' constraints ensuring no more than 1 parameter occupies any slot, followed by P generalized upper bound (GUB) constraints that ensure that, for each parameter, just one possible placement option from those that are legal is chosen. We will assume, until Section 8.5.2, that all $w_p = 1$. Model [DCM] can be written in the form

$$\begin{aligned}
[\text{DCM}] \quad \text{Maximize} \quad & c^T x \\
\text{subject to} \quad & Ax \le 1 \\
& Gx = 1 \\
& x \in \{0, 1\}.
\end{aligned}$$

For ease of notation in this section, we number the slots, and the corresponding slot constraints, from 0, giving slot constraints $0, 1, 2, \ldots, S-1$. A parameter p with period m_p has m_p columns, and so we can consider the x vector to be partitioned by parameters in the form $x = (x_{p,0}, x_{p,1}, \ldots, x_{p,m_p-1})$, $(p = 1, 2, \ldots, P)$. Variable $x_{p,j} = 1$ if and only if parameter p occupies slots $j, j + m_p, j + 2m_p, \ldots, j + (n_p - 1)m_p$. In $Ax \le 1$, we have $A = (a_{s,(p,j)})$ with rows $s = 0, 1, 2, \ldots, S - 1$ (one per slot) and columns indexed by (p, j), $p = 1, 2, \ldots, P$, $j = 0, 1, 2, \ldots, m_p - 1$ where

$$a_{s,(p,j)} = \begin{cases} 1 & s \in \{j, j + m_p, j + 2m_p, \ldots, j + (n_p - 1)m_p\} \\ 0 & \text{otherwise.} \end{cases}$$

That is, column (p, j) has 1's in every m_p'th row starting with row j, indicating that if column (p, j) is selected, parameter p will occupy every m_p'th row starting with row j.

The slot constraints are of the form $\sum_{p=1}^{P} \sum_{j=0}^{m_p-1} a_{s,(p,j)} x_{p,j} \leq 1$, $s = 0, 1,$ $2, \ldots, S-1$. From the definition of $a_{s,(p,j)}$, we see that for any parameter p and any slot s, there is one and only one possible value of j which results in $a_{s,(p,j)} = 1$, i.e., in slot s containing parameter p. This occurs when $s \bmod m_p = j$. Hence, we see that the slot constraints $Ax \leq 1$ can be written in the form

$$\sum_{p=1}^{P} x_{p,s \bmod m_p} \leq 1, \quad s = 0, 1, \ldots, S-1. \tag{8.3}$$

The GUB constraints $Gx = 1$ specify that exactly one placement is chosen for each parameter, where $g_{i,(p,j)} = 1$ for $i = p$, $j = 1, 2, \ldots, m_{p-1}$ and $g_{i,(p,j)} = 0$ otherwise. The $Gx = 1$ constraints can also be expressed as

$$\sum_{j=0}^{m_p-1} x_{p,j} = 1, \quad p = 1, \ldots, P.$$

8.5.1 Pairwise Slot Constraints

In this section, we show how we can derive new pairwise slot constraints from (8.3). These are examples of clique inequalities, and as such could be discovered using normal conflict graphs, for example. However, we present an explicit representation of the constraints, thereby allowing the discovery process to be bypassed. We also present a physical interpretation of the inequalities.

Let us consider two parameters, parameter $p \in \{1, 2, \ldots, P\}$ and parameter $q \in \{1, 2, \ldots, P\}$, $p \neq q$. Dropping the other parameters from (8.3) gives the relaxed constraint:

$$x_{p,s \bmod m_p} + x_{q,s \bmod m_q} \leq 1, \tag{8.4}$$

for any row $s \in \{0, 1, 2, \ldots, S-1\}$.

It is useful, at this stage, to define $\text{GCF}_{pq} = \text{GCF}(m_p, m_q)$ to be the greatest common factor (i.e., greatest integer divisor) for m_p and m_q. We can then define $m'_p = m_p/\text{GCF}_{pq}$ and $m'_q = m_q/\text{GCF}_{pq}$, and we note that m'_p and m'_q are relatively prime (i.e., their greatest common divisor is 1). The ratio $m'_p : m'_q$ defines the interaction between parameters p and q in their slot usage. We also note that the least common multiple of m_p and m_q is given by $\text{LCM}(m_p, m_q) = \text{GCF}_{pq} m'_p m'_q$.

Recall that $S = \text{LCM}(m_j, j = 1, 2, \ldots, P) \geq \text{LCM}(m_p, m_q)$. For all values of $s = 0, 1, \ldots, S-1$ where $s \geq \text{LCM}(m_p, m_q)$, we have both $s \bmod m_p = s'$ $\bmod m_p$ and $s \bmod m_q = s' \bmod m_q$ where $s' = s \bmod \text{LCM}(m_p, m_q)$. Therefore, there are only $\text{LCM}(m_p, m_q)$ unique equations defined by (8.4), being those given by $s = 0, 1, \ldots, \text{LCM}(m_p, m_q) - 1$. Consider every m_p'th of these constraints, starting with some constraint row (slot) r, $0 \leq r < m_p - 1$.

There are $\text{LCM}(m_p, m_q)/m_p = (\text{GCF}_{pq} m'_p m'_q)/(m'_p \text{GCF}_{pq}) = m'_q$ of these constraints. We can sum these m'_q constraints to give:

$$\sum_{s=r,r+m_p,r+2m_p,\ldots}^{r+(m'_q-1)m_p} (x_{p,s \bmod m_p} + x_{q,s \bmod m_q}) \leq m'_q, \quad r = 0, 1, \ldots, m_p - 1,$$

$$\Rightarrow \sum_{s=r,r+m_p,r+2m_p,\ldots}^{r+(m'_q-1)m_p} (x_{p,r} + x_{q,s \bmod m_q}) \leq m'_q, \quad r = 0, 1, \ldots, m_p - 1,$$

$$\Rightarrow m'_q x_{p,r} + \sum_{s=r,r+m_p,r+2m_p,\ldots}^{r+(m'_q-1)m_p} x_{q,s \bmod m_q} \leq m'_q, \quad r = 0, 1, \ldots, m_p - 1,$$

$$\Rightarrow m'_q x_{p,r} + \sum_{j=0}^{m'_q-1} x_{q,(r+jm_p) \bmod m_q} \leq m'_q, \quad r = 0, 1, \ldots, m_p - 1,$$

$$\Rightarrow m'_q x_{p,r} + \sum_{j=0}^{m'_q-1} x_{q,r \bmod \text{GCF}_{pq} + j\text{GCF}_{pq}} \leq m'_q, \quad r = 0, 1, \ldots, m_p - 1.$$

(8.5)

where the last step follows from Theorem 8.3 in Appendix B.

This equation has an intuitive physical interpretation that follows from adopting parameter p's view of the minor frame. This parameter's slot usage repeats every m_p slots, and so from this parameter's viewpoint there are only m_p different 'positions of interest'. Each of these m_p positions is realized as n_p equally spaced slots in the minor frame. (Because of repetition, only m'_q of these have a unique interaction with parameter q. However, the equation is perhaps more intuitive if thought of as summing over repeated sets of m'_q slots, giving n_p slots in total.) By summing our equations, we are counting how many of the slots used to realize this position are occupied by parameters p and q, for each of these m_p positions of interest. The equation above specifies that the number of slots occupied in total by the parameters p and q cannot exceed the number of slots available.

Equation (8.5) can be strengthened by noting that if parameter q occupies one of p's positions of interest even just once in the minor frame, then parameter p cannot occupy that position of interest without a clash occurring somewhere in the minor frame. This motivates the following lifting of the above equations.

Recalling that all the variables are binary, and only one $x_{q,j}$ variable for parameter q can be at value 1, we see that the constraints in (8.5) can be strengthened by increasing the coefficients on the $x_{q,r \bmod \text{GCF}_{pq} + j\text{GCF}_{pq}}$ variables from 1 to m'_q, and then dividing through by m'_q to give

$$x_{p,r} + \sum_{j=0}^{m'_q-1} x_{q,r \bmod \text{GCF}_{pq} + j\text{GCF}_{pq}} \leq 1, \quad r = 0, 1, \ldots, m_p - 1. \quad (8.6)$$

This expression gives m_p equations. We can form a set of new constraints from (8.6) through appropriate summation of every GCF_{pq}'th of these equations. For each value of e, $e = 0, 1, \ldots, \text{GCF}_{pq} - 1$, we can sum the equations in (8.6) defined for $r = e + j\text{GCF}_{pq}$, $j = 0, 1, 2, \ldots, m'_p - 1$, where each summation includes $m_p/\text{GCF}_{pq} = m'_p$ equations. This gives:

$$\sum_{r=e,e+\text{GCF}_{pq},e+2\text{GCF}_{pq},\ldots}^{e+(m'_p-1)\text{GCF}_{pq}} \left(x_{p,r} + \sum_{j=0}^{m'_q-1} x_{q,r \bmod \text{GCF}_{pq}+j\text{GCF}_{pq}} \right) \leq m'_p,$$
$$e = 0, 1, \ldots, \text{GCF}_{pq} - 1$$

$$\Rightarrow \sum_{k=0}^{m'_p-1} \left(x_{p,e+k\text{GCF}_{pq}} + \sum_{j=0}^{m'_q-1} x_{q,(e+k\text{GCF}_{pq}) \bmod \text{GCF}_{pq}+j\text{GCF}_{pq}} \right) \leq m'_p,$$
$$e = 0, 1, \ldots, \text{GCF}_{pq} - 1$$

$$\Rightarrow \sum_{k=0}^{m'_p-1} \left(x_{p,e+k\text{GCF}_{pq}} + \sum_{j=0}^{m'_q-1} x_{q,e \bmod \text{GCF}_{pq}+j\text{GCF}_{pq}} \right) \leq m'_p,$$
$$e = 0, 1, \ldots, \text{GCF}_{pq} - 1$$

$$\Rightarrow \sum_{k=0}^{m'_p-1} x_{p,e+k\text{GCF}_{pq}} + m'_p \sum_{j=0}^{m'_q-1} x_{q,e+j\text{GCF}_{pq}} \leq m'_p,$$
$$e = 0, 1, \ldots, \text{GCF}_{pq} - 1.$$

As before, the GUB constraints allow us to lift these constraints by increasing the coefficients on $x_{p,e+k\text{GCF}_{pq}}$ from 1 to m'_p, and then dividing through by m'_p to give

$$\sum_{j=0}^{m'_p-1} x_{p,e+j\text{GCF}_{pq}} + \sum_{j=0}^{m'_q-1} x_{q,e+j\text{GCF}_{pq}} \leq 1, e = 0, 1, \ldots, \text{GCF}_{pq} - 1,$$

$$\Rightarrow \sum_{j=0}^{\alpha} x_{p,e+j\text{GCF}(m_p,m_q)} + \sum_{j=0}^{\beta} x_{q,e+j\text{GCF}(m_p,m_q)} \leq 1,$$
$$e = 0, 1, \ldots, \text{GCF}(m_p, m_q) - 1. \qquad (8.7)$$

where $\alpha = m_p/\text{GCF}(m_p, m_q) - 1$, and $\beta = m_q/\text{GCF}(m_p, m_q) - 1$. Equation (8.7) gives a set of valid inequalities for each pair of parameters. These can be used to strengthen the relaxed linear programming form of [DCM], thus making it easier to solve to integrality.

8.5.2 Multi-Word Parameters

We now consider a model where the number of bits required to transmit a parameter exceeds those of a slot, and so the parameter occupies more than one consecutive slot in the data cycle map.

Let us assume that parameter p requires w_p consecutive slots (words) each time it is transmitted. Clearly, we require the number of words to be less than the parameter's period m_p, i.e., $w_p < m_p$. The jth column $x_{p,j}$, $j = 0, 1, \ldots, m_p - 1$, for parameter p now defines the case where parameter p occupies slots $h, h+1, \ldots, h+w_p - 1$ for all $h = j, j+m_p, j+2m_p, \ldots, j + (n_p - 1)m_p$.

To keep the model form consistent with the $w_p = 1$ case, we include columns in the model in which a parameter occupies 'consecutive' slots that wrap around from the end of the frame back to the beginning. For example, some parameter p with $w_p = 3$ might occupy the slots $S - 1, 0$, and 1. Any solution like this can be interpreted as a valid solution by renumbering the slots so that slot 0 occurs at the start, not the middle, of the parameter's slot usage. Alternatively, if reducing symmetry in the formulation is important, then variables such as these can be forced to zero using additional constraints that leave the underlying model form intact.

In the $w_p = 1$ model, slot s was used by parameter p if parameter p was placed in position $j = s \bmod m_p$, i.e., if slot j was the first slot occupied by parameter p. In the new model, the same slot s will be used if parameter p is placed in position j, or any of the $w_p - 1$ preceding positions. Thus, the constraint for slot s, Equation (8.3), now becomes

$$\sum_{p=1}^{P} \sum_{h=0}^{w_p-1} x_{p,(s \bmod m_p)-h} \leq 1, \quad s = 0, 1, \ldots, S-1,$$

where we assume modulo m_p wrap around for $x_{p,j}$ in the sense that $x_{p,-1} = x_{p,m_p-1}$, $x_{p,-2} = x_{p,m_p-2}$, etc.

Proceeding as before, we relax this constraint to consider only the two parameters p and q, giving

$$\sum_{h=0}^{w_p-1} x_{p,(s \bmod m_p)-h} + \sum_{h=0}^{w_q-1} x_{q,(s \bmod m_q)-h} \leq 1, \quad s = 0, 1, \ldots, S-1.$$

We can then add the equations as before, noting that each occurrence of $x_{p,j \bmod m_p}$ is replaced by $\sum_{h=0}^{w_p-1} x_{p,(j \bmod m_p)-h}$, and similarly for $x_{q,j \bmod m_q}$. This gives the following general form of Equation (8.5):

$$\sum_{h=0}^{w_p-1} m_q' x_{p,r-h} + \sum_{j=0}^{m_q'-1} \sum_{h=0}^{w_q-1} x_{q,(r \bmod \mathrm{GCF}_{pq} + j\mathrm{GCF}_{pq})-h} \leq m_q', \quad r = 0, 1, \ldots, m_p - 1.$$

$$(8.8)$$

Before we lift this equation, we note that if $w_q < \mathrm{GCF}_{pq}$, then each $x_{q,k}$ variable that appears in the summation appears exactly once. If $w_q = \mathrm{GCF}_{pq}$, then, by considering the $r = 0$ case, we see that the second summation includes variables $x_{q,k}$ for all $k \in \{j\mathrm{GCF}_{pq} - h : j = 0, 1, \ldots, m_q' - 1; h = 0, 1, \ldots, \mathrm{GCF}_{pq} - 1\}$. Given $m_q' \mathrm{GCF}_{pq} = m_q$, we see that this includes all of the m_q possible variables defined for parameter q. In this case, the GUB constraint on the $x_{q,k}$ variables implies $\sum_{j=0}^{m_q'-1}\sum_{h=0}^{w_q-1} x_{q,(r \bmod \mathrm{GCF}_{pq}+j\mathrm{GCF}_{pq})-h} = 1$ for all r, and so $x_{p,k} = 0 \ \forall \ k$ is the only solution to this equation. Thus, there is no feasible solution to the problem if $w_q = \mathrm{GCF}_{pq}$, or indeed if w_q is larger than this, implying that physically parameter q requires more consecutive slots, and so the problem is more tightly constrained. The same argument applies, by symmetry, to w_p, and so we see that a feasible solution can only exist if $w_p < \mathrm{GCF}_{pq}$ and $w_q < \mathrm{GCF}_{pq}$. From now on, we will restrict ourselves to this case.

As before, we proceed to lift Equation (8.8), which gives:

$$\sum_{h=0}^{w_p-1} x_{p,r-h} + \sum_{j=0}^{m_q'-1}\sum_{h=0}^{w_q-1} x_{q,r \bmod \mathrm{GCF}_{pq}+j\mathrm{GCF}_{pq}-h} \leq 1, \quad r = 0, 1, \ldots, m_p-1. \quad (8.9)$$

This expression gives m_p equations. As before, we can form a set of new constraints from (8.9) through appropriate summing of every GCF_{pq}'th of these equations. For each value of c, $c = 0, 1, \ldots, \mathrm{GCF}_{pq} - 1$, we can sum the equations in (8.9) defined for $r = e + j\mathrm{GCF}_{pq}$, $j = 0, 1, 2, \ldots, m_p' - 1$. Each summation includes $m_p/\mathrm{GCF}_{pq} = m_p'$ equations. This gives:

$$\sum_{r=e,e+\mathrm{GCF}_{pq},e+2\mathrm{GCF}_{pq},\ldots}^{e+(m_p'-1)\mathrm{GCF}_{pq}} \left(\sum_{h=0}^{w_p-1} x_{p,r-h} + \sum_{j=0}^{m_q'-1}\sum_{h=0}^{w_q-1} x_{q,(r \bmod \mathrm{GCF}_{pq}+j\mathrm{GCF}_{pq})-h} \right) \leq m_p',$$
$$e = 0, 1, \ldots, \mathrm{GCF}_{pq} - 1$$

$$\Rightarrow \sum_{k=0}^{m_p'-1} \left(\sum_{h=0}^{w_p-1} x_{p,e+k\mathrm{GCF}_{pq}-h} + \sum_{j=0}^{m_q'-1}\sum_{h=0}^{w_q-1} x_{q,(e+k\mathrm{GCF}_{pq}) \bmod \mathrm{GCF}_{pq}+j\mathrm{GCF}_{pq}-h} \right) \leq m_p',$$
$$e = 0, 1, \ldots, \mathrm{GCF}_{pq} - 1$$

$$\Rightarrow \sum_{k=0}^{m_p'-1} \left(\sum_{h=0}^{w_p-1} x_{p,e+k\mathrm{GCF}_{pq}-h} + \sum_{j=0}^{m_q'-1}\sum_{h=0}^{w_q-1} x_{q,e \bmod \mathrm{GCF}_{pq}+j\mathrm{GCF}_{pq}-h} \right) \leq m_p',$$
$$e = 0, 1, \ldots, \mathrm{GCF}_{pq} - 1$$

$$\Rightarrow \sum_{k=0}^{m_p'-1}\sum_{h=0}^{w_p-1} x_{p,e+k\mathrm{GCF}_{pq}-h} + m_p' \sum_{j=0}^{m_q'-1}\sum_{h=0}^{w_q-1} x_{q,e+j\mathrm{GCF}_{pq}-h} \leq m_p',$$
$$e = 0, 1, \ldots, \mathrm{GCF}_{pq} - 1.$$

Using the same arguments as before, we can lift the coefficients to give:

$$\sum_{k=0}^{m_p'-1}\sum_{h=0}^{w_p-1} x_{p,e+k\mathrm{GCF}_{pq}-h} + \sum_{j=0}^{m_q'-1}\sum_{h=0}^{w_q-1} x_{q,e+j\mathrm{GCF}_{pq}-h} \leq 1, \quad e = 0, 1, \ldots, \mathrm{GCF}_{pq} - 1.$$

Thus, if we have $w_p < \mathrm{GCF}_{pq}$ and $w_q < \mathrm{GCF}_{pq}$ then we can strengthen our model using the above valid inequalities. If the restrictions on w_p and w_q are not satisfied, then no feasible solution exists to the problem.

8.6 Results and Conclusions

A total of 1870 data sets representing bit requirements ranging from 640 to 1119968 bits per second were selected for analysis from data sets provided by the United States Air Force.

All results were generated on a PC with clock speed 733 MHz. and run under Red Hat Linux v6.2.

In order to benchmark the results obtained using the set packing approach we have provided two sets of results which compare both the relative efficiency (E) of the major frames generated, and the speed of execution, with commercial software AutoTelem^{TM}, [6]. AutoTelem uses a local search procedure to produce near optimal data cycle maps. The relative efficiencies of our set packing approach and AutoTelem^{TM} are displayed in Figure 8.1. These graphs represent the average efficiency as a function of the size of the input file required bits. All results were sorted in order of increasing required bits and the efficiency values averaged over batches of size 100 (except for the last batch which is of size 70). The median batch size is shown on the figure for each batch. It can be observed that for batches in the mid to high range of required bits the set packing approach generally produces solutions with a higher efficiency than for AutoTelem^{TM}, whereas for data sets in the low to mid range of required bits the reverse is true.

A comparison of execution times is shown in Figure 8.2. The results in this case are also sorted in order of increasing required bits and times are averaged over batches of size 100. It is clear that the set packing approach executes faster than AutoTelem^{TM} across the range of data set batches. Both algorithms have low execution times in absolute terms compared with current frame generation practice.

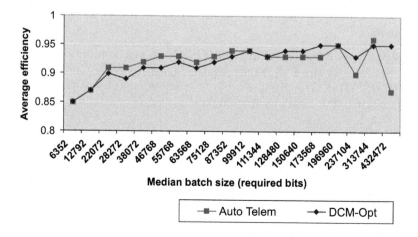

FIGURE 8.1

Average efficiency for each algorithm, over batches of 100 data sets which have been ordered in size of increasing bit requirements.

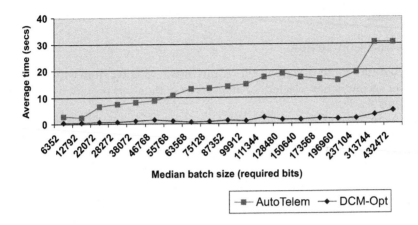

FIGURE 8.2
Average execution time in seconds for AutoTelem^{TM} and the set packing approach, over batches of 100 data sets which have been ordered in size of increasing bit requirements.

We have described a fast and flexible method for the generation of Telemetry Frames or Data Cycle maps which has provided high efficiencies for large operationally realistic data sets. The method is flexible in the sense that changes in frame construction rules can easily be implemented using this modeling approach. Execution times are sufficiently small to suggest that even larger data sets can be accommodated in the future, and are several orders of magnitude lower than times taken at present to create Telemetry Frames using computer assisted methods. The set packing approach has also been shown to compare very favorably with the commercial software AutoTelem^{TM} both in terms of speed and efficiency.

Acknowledgments

This work was partly sponsored by an internal ITEE Divisional Grant from the University of South Australia and also by United States Air Force Contract number F04700-00-P-1181. The authors wish to acknowledge the valuable improvements made to this paper by the referees.

References

1. Telemetry Group, Range Commanders Council, IRIG Standard 106-96, Telemetry Standards, Secretariate, Range Commanders Council, U.S. Army White Sands Missile Range, NM, May 1996.
2. Panton, D., John, M., Lucas, S., and Mason, A., Flight test data cycle map optimisation, *EJOR*, 146, 486, 2003.

3. John, M., Optimisation models for the generation of data cycle maps and regional surveillance, Ph.D. thesis, University of South Australia, Mawson Lakes, Australia, 2003.

4. Johnson, E. L., Nemhauser, G. L., and Savelsbergh, M. W. P., Progress in linear programming-based algorithms for integer programming: An exposition, *INFORMS Journal on Computing*, 12, 1, 2000.

5. ILOG CPLEX 7.0, ILOG INC., Mountain View CA.

6. AutoTelem, QUEST Integrated Inc.

Appendix A: LCM Proof

In this appendix, we show that $\text{LCM}(m_p, p = 1, 2, \ldots, P) = S$ where $S = k\,\text{LCM}(n_p, p = 1, 2, \ldots, P)$, $k \in \{1, 2, 3, \ldots\}$, and $m_p = S/n_p$.

LEMMA 8.1

Consider the calculation of $\text{LCM}(n_p, p = 1, 2, \ldots, P)$. *Let* $n_p = \Pi_{z \in \Omega} z^{r_z^p}$, $p = 1, 2, \ldots, P$, *where* $\Omega = \{2, 3, 5, \ldots\}$ *is the set of prime numbers, and* $r_z^p \in \{1, 2, 3, \ldots\}$ *is the integer power associated with* $z \in \Omega$ *in the prime factorization of* n_p. *(For example, if we have* $n_1 = 60 = 2^2 3^1 5^1$, *then* $r_1^2 = 2, r_1^3 = 1, r_1^5 = 1, r_1^7 = 0, \ldots$ *) By definition, we have* $\text{LCM}(n_p, p = 1, 2, \ldots, P) \equiv \Pi_{z \in \Omega} z^{\max_{p=1,2,\ldots,P} r_z^p}$.

LEMMA 8.2

$\text{LCM}(k \times n_p, p = 1, 2, \ldots, P) = k \times \text{LCM}(n_p, p = 1, 2, \ldots, P)$ *for* $k \in \{1, 2, 3, \ldots\}$.

PROOF   This follows easily from Lemma 8.1 by considering the prime factorization of k and then rearranging. ∎

THEOREM 8.1

$\text{LCM}(m_p, p = 1, 2, \ldots, P) = S.$

PROOF   From the definition $m_p = S/n_p$, we need to show that $\text{LCM}(S/n_p, p = 1, 2, \ldots, P) = S$, where $S = k\,\text{LCM}(n_p, p = 1, 2, \ldots, P)$ for any $k \in \{1, 2, 3, 4, \ldots\}$. Firstly, we note that all $S/n_p, p = 1, 2, \ldots, P$ are strictly positive integer values, and so the LCM is well defined. As in Lemma 8.1, we construct prime factorizations giving $n_p = \Pi_{z \in \Omega} z^{r_z^p}$, $p = 1, 2, \ldots, P$, where Ω is the set of prime numbers, and $r_z^p \in \{0, 1, 2, \ldots\}$ is the power associated with $z \in \Omega$ in the factorization of n_p. Similarly, let $S = \Pi_{z \in \Omega} z^{r_z}$, where $\{r_z = \max_{p=1,2,\ldots,P} r_z^p, z \in \Omega\}$ defines the prime factorization of S. We note that because the n_p values, $p = 1, 2, \ldots, P$ are relatively prime (i.e., their only common divisor is 1) we have the key observation that

$$\min_{p \in \{1, 2, \ldots, P\}} r_z^p = 0 \;\; \forall z \in \Omega. \qquad (8.10)$$

Using Lemma 8.1, we have

$$
\mathrm{LCM}\left(\frac{S}{n_p}, p = 1, 2, \ldots, P\right) = \mathrm{LCM}\left(\frac{\prod_{z\in\Omega} z^{r_z}}{\prod_{z\in\Omega_p} z^{r_z^p}}, p = 1, 2, \ldots, P\right),
$$

$$
= \mathrm{LCM}\left(\prod_{z\in\Omega} z^{r_z - r_z^p}, p = 1, 2, \ldots, P\right),
$$

$$
= \prod_{z\in\Omega} z^{\max_{p=1,2,\ldots,P}(r_z - r_z^p)},
$$

$$
= \prod_{z\in\Omega} z^{r_z - \min_{p=1,2,\ldots,P} r_z^p},
$$

$$
= \prod_{z\in\Omega} z^{r_z},
$$

where the last simplification follows from (8.10). As $S = \prod_{z\in\Omega} z^{r_z}$, the result is proven. ∎

Appendix B: Equivalency Proofs

In this section, we prove several equivalencies between sets of values.

THEOREM 8.2
The set $\{(jm'_p)\mathrm{mod}\, m'_q, j = 0, 1, \ldots, m'_q - 1\} = \{0, 1, 2, \ldots, m'_q - 1\}$ when m'_p and m'_q are relatively prime.

PROOF We note that this set has m'_q values, each between 0 and $m'_q - 1$ inclusive. We will show, by contradiction, that these values are all unique, and hence the result will follow.

Assume that two of these values are equal, i.e., $(jm'_p) \bmod m'_q = ((j + k)m'_p) \bmod m'_q = ([(jm'_p) \bmod m'_q] + [(km'_p)\bmod m'_q]) \bmod m'_q$, where integers j and k satisfy $0 \leq j < j + k < m'_q$. For this to be true, we require $(km'_p) \bmod m'_q = 0$, implying km'_p is an integer multiple of m'_q, i.e., $km'_p = k'm'_q$ for some positive integer k'. But, because m'_p and m'_q are relatively prime, the smallest value of k for which this is true is $k = m'_q$ (from $\mathrm{LCM}(m'_p, m'_q) = m'_p m'_q$ when m'_p and m'_q are relatively prime). This value of k is outside the range permitted, and thus we have a contradiction. Therefore, all the values, between 0 and $m'_q - 1$ must be distinct, and so the result follows. ∎

THEOREM 8.3
The set of indices $R = \{(r + jm_p)\bmod m_q, j = 0, 1, \ldots, m'_q - 1\} = \{r_1, r_2, \ldots, r_{m'_q}\}$ with $r_{i+1} = r_i + \mathrm{GCF}_{pq}$, and $r_1 = r\bmod\mathrm{GCF}_{pq}$, where $\mathrm{GCF}_{pq} = \mathrm{GCF}(m_p, m_q)$, the greatest integer divisor common to m_p and m_q, and we define $m'_p = m_p/\mathrm{GCF}_{pq}$ and $m'_q = m_q/\mathrm{GCF}_{pq}$.

PROOF We observe that $(r + jm_p) \bmod m_q = (r + [(jm_p) \bmod m_q]) \bmod m_q$. We also note that $(jm_p) \bmod m_q = (j\mathrm{GCF}_{pq} m'_p) \bmod (\mathrm{GCF}_{pq} m'_q) = \mathrm{GCF}_{pq}$ $[(jm'_p) \bmod m'_q]$. This allows us to write

$$R = \{(r + \mathrm{GCF}_{pq}[(jm'_p) \bmod m'_q]) \bmod m_q, \quad j = 0, 1, \ldots, m'_q - 1\}.$$

Using Theorem 8.2, we can rewrite this as

$$R = \{(r + j * \mathrm{GCF}_{pq}) \bmod m_q, \quad j = 0, 1, \ldots, m'_q - 1\}. \tag{8.11}$$

Noting that $m_q = \mathrm{GCF}_{pq} m'_q$, we see that $j\mathrm{GCF}_{pq} < m_q \; \forall j = 0, 1, \ldots, m'_q - 1$, and so the successive values in set R must differ by GCF_{pq}, allowing us to write $R = \{r_1, r_2, \ldots, r_{m'_q}\}$ where $r_{i+1} = r_i + \mathrm{GCF}_{pq}, 1 \le i < m'_q - 1$. We note that the set 'wraps' in the sense that $(r_{m'_q} + \mathrm{GCF}_{pq}) \bmod m_q = r_1$. We also observe that $r_1 = r \bmod \mathrm{GCF}_{pq}$. This r_1 value follows from the observation that the smallest value in R must lie between 0 and $\mathrm{GCF}_{pq} - 1$ inclusive. The wrap property of R means that we can remove the bounds on j in (8.11) without changing the contents of R. By putting $j = -\lfloor r/\mathrm{GCF}_{pq} \rfloor$ we have $r + j\mathrm{GCF}_{pq} = r - \lfloor r/\mathrm{GCF}_{pq} \rfloor \mathrm{GCF}_{pq} = r \bmod \mathrm{GCF}_{pq}$, showing that r_1 as defined above is indeed a member of R. This completes the proof. ∎

9

Application of Column Generation Techniques to Retail Assortment Planning

Govind P. Daruka and Udatta S. Palekar

CONTENTS

9.1 Introduction

The retail industry is one of the largest sectors of commerce worldwide and is the second largest in the U.S. both in the number of establishments and the number of employees. In the U.S. the retail trade presently generates about

3.8 trillion dollars in retail sales annually and accounts for about 12.9% of all business establishments (U.S. Census [42]).

Retailers must estimate the likelihood of sales for each item and also evaluate the trade-off between different items due to limited availability of retail space in the store and other constraints. This task is very complex due to the almost infinite product variety, inherent variability in product demand, and significant errors in forecasting demand. Since space constraints do not allow all products to be carried, the retailer must select a limited number of products. The set of products carried in a store is termed as the *assortment* of products. The products in a store typically form a hierarchy. A line of product can have multiple items in it and each item can have multiple stock keeping units (SKUs). SKUs are at the lowest level of the product hierarchy. For example, if shoes of a particular pattern are an item then the different shoe sizes form the SKUs for the item.

The product assortment decision is one of the key issues faced by retailers. Retailers need to decide the number of different merchandizing categories they would like to keep within a store or department, referred to as product variety or the breadth of the assortment, as well as the number of (SKUs) they should carry within a category, referred to as the depth of the assortment or inventory depth. Retailers also need to decide the brands they would carry for the chosen categories namely store brands, local brands, and national brands, which affects the pricing across categories.

Retailers are driven by consumers who have high expectations for service and product excellence and yet demand lower prices. Intense competition and rapidly changing consumer tastes have resulted in increased turnover rate in most product categories (Chong et al. [9]). In response, retailers are addressing these new demands in different ways. Some retailers are focusing on aggregating their overall needs from a supplier or group of suppliers to establish greater buying leverage but the process, time, and technology constraints introduce significant difficulty in accomplishing and sustaining this task. Some retailers change their assortment frequently according to perceived changes in demographic, economic, or seasonal demand patterns, but the consequent investment in retail space and inventory can be prohibitively expensive. Some retailers carry larger inventories with a higher risk of these inventories becoming obsolete and simultaneously incur higher inventory costs. Meeting marketing and financial planning obligations, forces most retailers to take the easy approach to merchandizing repeating the assortment breadth and depth from previous seasons, creating store assortments based on store volume and ranking items by sales volume alone. This leads to the creation of a few standard product assortments typically based on store size, which are then deployed throughout the chain of stores. However, mismatches of chain wide strategies and local consumer needs can lead to significant lost sales due to out-of-stock conditions or distress sales due to overstock conditions. Pashigan [31] shows that the costs incurred by retailers due to markdowns have been increasing in the U.S. apparel industry.

Retailers must address the customer's needs by aligning the merchandizing processes such as strategic and financial planning, assortment and space planning, inventory deployment, and in-season management. The merchandizing process is a systems approach, aimed at maximizing return on investment, through planning sales and inventory in order to increase profitability, by maximizing sales potential and minimizing losses from markdowns and stock-outs. A comprehensive approach would be to balance the demands of consumers and the constraints of operating a retail business profitably by transforming the merchandizing process by having an intelligent assortment and category-planning process that accommodates the entire merchandizing processes.

Assortment planning involves asking questions such as: Which product? How much of it? What colors? What sizes? Where to place it? Who is the target customer? and so on. Assortment planning is the decision making process that a retailer undertakes to determine the correct mix of items to be carried by each store so as to maximize the opportunity for sale. Decisions are made regarding what to keep (assortment width) and how much to keep (assortment depth) in a store. Assortment planning directly affects the product selection, pricing, timing, and micro-merchandizing. By employing a proper assortment planning solution the retailer can gain immensely since the product assortment matched with market potential and inventory aligned with sales lead to increased sales, fewer markdowns, and improved margins. This detailed plan of action also helps in improving employee morale and the overall efficiency. Thus the retailer can provide better service to customers and simultaneously avoid carrying large inventory and the consequent higher inventory costs, and reduce the risk of inventory obsolescence.

In the previous decade, large retailers have invested heavily in information technology. Point-of-sale scanners coupled with massive databases allow retailers to gather information in real time. However, there are very few tools that make effective use of this data for the purpose of planning assortments to maximize sales. Most commercial software facilitate the assortment planning process by providing database connectivity and user-friendly interfaces. However, very little optimization and decision making capability is provided.

In this paper we present a mathematical model of assortment planning which provides a decision support system that addresses the problem of assortment width as well as the assortment depth. We address the problem of a single period assortment planning problem with stochastic demand under shelf space, inventory turns, and various *returns-on* constraints. We formulate the problem as a Linear Integer Programming problem and use a column generation technique to solve the model. The model was implemented by a large chain of department stores for assortment planning.

9.2 Literature Review

Research on issues related to assortment planning dates back to the early years of Operations Research. Whitin [43] noted that for the retail stores, the inventory control problem for style goods is further complicated by the fact that inventory and sales are not independent of one another. An increase in inventories may bring about increased sales of some items but a decrease in sales of other items. Cairns [7] proposes that the business of a retailer can be regarded as selling the retail space at their disposal to their suppliers. The retail space should, therefore, be allocated to those products that give the maximum profit.

A significant amount of research has been done since then on the importance of inventory and shelf space in assortment decisions. Fox [14] describes the importance of assortment planning process in retail industry. Cox [11] measured the relationship between sales of an individual product brand and its shelf space. The model was tested in six supermarkets where the shelf space inventory of the two test products — powdered coffee cream and salt — over two leading brands in each category were varied each day, while keeping the total shelf space for these two categories constant. Shelf space was shown to affect sales volume while location on shelves was not significant. Frank and Massy [15] take sixty three weeks of store audit data and use multiple regression analysis to estimate the parameters of their two basic models. They conclude that by varying the number of facings of a brand in a shelf display increases sales for high but not low volume SKU's. They also conclude that there is minimal or no effect of the shelf level or height and there is absolutely no effect whether the product is kept at floor or any other shelf row. Curhan [12] proposes a model to explain the space elasticity as a function of several product-specific variables and tests this model under actual operating conditions for nearly five hundred grocery products over five to twelve weeks before and after changing the space allocation. He finds there is a positive relationship between shelf space and unit sales. Similar results were obtained by Wolfe [45] who concludes based on his empirical experiment that the sales of style merchandize, such as women's dresses or sports clothes, are proportional to the amount of inventory displayed. Other studies by Levin et al. [26], Schary and Becker [33], Silver and Peterson [36], Larson and DeMarais [24] show that the retail display inventory has an stimulating effect on the customers which can lead to increased sales. Urban [41] presents a generalization of the inventory-level-dependent demand inventory models that explicitly models the demand rate as a function of the displayed inventory level and makes an explicit distinction between the back room and displayed inventories. The special case of full-shelf merchandizing, in which the product is replenished as soon as the back room inventory is depleted, is also presented. Then they consider the product assortment and shelf-space allocation problems by extending this model into a multi-item, constrained environment.

Marketing models treat a product as a vector of attributes with customers' utility function defined over those attributes. Lancaster [23] provides a summary of this research. These models focus on choosing a product line to maximize the expected utility of the customer by either designing products with desirable attributes or choosing preexisting products which possess such attributes (Green and Krieger [16], Urban [40], Shocker and Srinivasan [35]). Urban [39] develops a mathematical model for product line decisions to identify which products should be included in a firm's product line. Product line decisions are difficult to make because the products in the line are not usually independent. Interdependency is the key consideration in such decision making. A comprehensive discussion of marketing models for product line selection can be found in Lilien et al. [27].

A large body of literature addresses the issue of assortment selection and allocation of limited retail shelf space for established commodity products typically sold in a supermarket. Anderson and Amato [1] formulate a model to determine the most profitable, short run brand mix concomitant with a determination of the optimum allocation of a fixed product display area among available brands. This model when solved gives the composition of the specific brands that should be displayed, and the amount of display area that should be assigned to these brands, in order to maximize profit. Bultez and Naert [6] consider this problem by using marginal analysis of the profit function. They assume individual product demand is known and develop a generalized framework to allocate shelf space for an assortment with interacting demand. Based on this framework, they develop a practical allocation rule called Shelf Allocation for Retailers' Profit (SHARP). They discuss the special conditions under which SHARP reduces to the various rules of thumb commonly employed by retailers. However, in this framework, they do not explicitly consider which products to include in their assortment and assume that substitution between products is symmetric.

Borin et al. [3] consider the joint assortment selection and space allocation problem. It is assumed that demand is generated from the inherent characteristics of a product, the allocated shelf space, the selection of the assortment, and the stock levels of other products. The estimates of the parameters capturing these effects were estimated using regression analysis. Demand is assumed to be known and deterministic through functions involving these parameters. The objective was to maximize return on inventory subject to space constraints. The model was solved based on a heuristic using the simulated annealing technique, to select an assortment of ketchup in the supermarket.

Mahajan and Ryzin [29] developed a theoretical model to determine the number and type of colors for seasonal and replenishable products. They suggest that with each additional item, the total demand for the product line increases, but demand uncertainty across each item also increases, thus increasing overstocking and understocking costs. To determine the number and type of items before this increase in revenue is offset by such costs, the model considers groups of items corresponding to a product defined as a category and assumes that the demand across these items is generated by combining a multinomial logit choice process with a Poisson arrival process for the customer. The resulting Poisson distribution of demand was approximated to a normal distribution and was linked to the retailers' inventory costs by a newsboy model. Based on the assumptions of the demand process, structural results were developed to answer the question regarding how these items should be included in the assortment corresponding to this product. Kahn and Lehmann [20] found preference for an assortment to be positively related to individual item preference, an additional item's uniqueness relative to the existing assortment, and the total number of items in the assortment.

For fashion products, assortment planning issues were addressed more directly by Smith and Agrawal [38] who considered the joint problem of assortment selection and optimal stocking for a group of products at a given level. Demand was modelled as negative binomial random variable. The nonlinear model developed has as its objective function, the cost function of the newsboy problem which represents the total inventory costs of overstocking and understocking products. The model is solved using numerical methods. Mahajan and Ryzin [30] analyze a single-period, stochastic inventory model (newsboy-like model) in which a sequence of heterogeneous customers dynamically substitute among product variants within a retail assortment when inventory is depleted. The customer choice decisions are based on a utility maximization criterion. The retailer must choose initial inventory levels for the assortment to maximize expected profits. The authors propose and analyze a stochastic gradient algorithm for the problem, and prove that it converges to a stationary point of the expected profit function under mild conditions. The authors give numerical examples to show that the substitution effects can have a significant impact on an assortment's profits.

There are several assortment models dealing with issues similar to the one presented in this chapter. Rajaram [32] models the problem of fashion retail assortment planning as a nonlinear integer programming problem with assumption that the retailer places a single order for all the products. The demand is modelled as a random variable and the constraint set includes a budget constraint which is later expanded to include shelf space constraint. To solve the model an upper bound for the problem is developed by relaxing the budget constraint. The solution obtained corresponding to the upper bound may not be feasible due to violation of budgetary constraint and so to achieve feasibility they develop a Lagrangian heuristic and a two-phase heuristic as an alternative approach. Corstjens and Doyle [10] developed a shelf-space allocation model in which they considered the demand rate as a function of shelf space allocated to the product and also applying a polynomial functional form of demand with main- and cross-elasticities of shelf space, they utilized signomial geometric programming to solve the model. Zufryden [48] suggested the use of dynamic programming to solve the shelf-space allocation problem, as it will allow general objective-function specifications and will provide integer solutions.

In the above two articles an assumption is made that all of the inventory is on display which is generally not appropriate for retail stores such as grocery stores. Also these models do not account for the location factors. In view of these limitations Yang and Chen [47] proposed a simplified integer programming model based on the model developed by Corstjens and Doyle [10]. The objective function is linearized by assuming that the profit of any product is linear with respect to a range of facings for which it is displayed by controlling the lower and upper bounds for that product. They also do not consider the product availability constraint assuming that the product availability is no longer an issue in the present market scenario. The model then resembles a multi-knapsack problem which is solved in multiple stages. Yang [46] presents a heuristic algorithm that extends the greedy algorithm used for knapsack problems to solve the above model. The profit of each item per display length on a particular shelf is treated as a weight, and the ranking order of weight is used as a priority index in the process of space allocation. The heuristic procedure can allocate a product with lower profit to a shelf in order to satisfy the lower bounds for the product. To improve the heuristic procedure solution, an adjustment phase consisting of three adjustment methods is proposed. These adjustment methods improve the heuristic solution but are not useful if there is variation in the length of facings of the products. To overcome this shortcoming, Lim et al. [28] propose new neighborhood moves namely multi-shift, multi-exchange, and multi-add-exchange and solve the Yang and Chen [47] model employing a strategy of combining a strong local search with metaheuristics. They also develop a network flow solution approach to solve the model as a minimum cost maximum flow problem. They further incorporate the cross product affinity effect among products such that there will be additional profit by having two particular products on the same shelf and considering a nonlinear base profit function. They obtained near-optimal solutions for small number product-shelf ranges and consistently good results for all other ranges tested.

Hoch et al. [18] present a general mathematical model of variety based on the complete information structure of an assortment, defined by the multi-attribute structure of the objects and their spatial locations. The model is used to develop assortments that vary widely in terms of their information structure. The influence of variety perception and organization on stated satisfaction and store choice was investigated. Information structure has a big impact on variety perceptions, though diminishing returns accompany increases in the number of attributes on which object pairs differ. Customers are more influenced by local information structure (adjacent objects) than nonlocal information structure and the organization of the display can either increase or decrease variety perceptions. Jayaraman et al. [19] address the problem of jointly determining the product variety decision with the ordering decision for the variations of a brand of a certain product that comprise a particular product line by modelling it as a nonlinear optimization model. Assumptions are made that the consumer preferences for products are known and are ordinally scaled for the subset of products in the product line and can be added or deleted according to profitability criterion.

The impact of product assortment on an individual's decisions is examined by Chernev [8] by conducting four experiments using fifty to one hundred undergraduates from Northwestern University as respondents for different experiments. They conclude that the impact of assortment on individual decision processes and choice is moderated by the degree to which individuals have articulated attribute preferences. Individuals with an articulated ideal point are more likely to prefer larger assortments than individuals without articulated preferences. Shelby and Miller [34] consider the shopper's joint decision of item selection and pricing to determine the best assortment to carry in terms of profit maximization. They specifically consider item complementarity and substitution with respect to profit implications for the retailer. Koelemeijer and Oppewal [22] tackle the question of what is the optimal assortment in a very different purchasing situation — those involved with a frequently purchased category like cut flowers. They choose this category because there is no branding, no stockpiling, and no packaging. In this type of category, variety seeking and bundling issues become relevant. Their model provides a tool to allow for optimization of retail assortments by considering substitution, complementarity, and asymmetric dominance effects. Simonson [37] reviews and synthesizes recent empirical evidence indicating that product assortments can play a key role, not only in satisfying customers' wants, but also in influencing what they want. He shows that retailers can use the considered assortment to change the likelihood that a consumer will make a purchase, and to affect the probability that a specific option will be chosen.

Stassen et al. [25] consider assortment decisions across multiple stores. When consumers are shopping among several stores, a retailer needs to consider whether his or her assortment should compete with the competition (thinking of the stores as substitutes) or complement the competition (thinking of stores as complements). In this sense, the assortment decision can be thought of as an element of the marketing mix in attracting consumers into the store. Their empirical results based on market data for a specific market show that the assortment decision reflects a competitive relationship rather than a complementary one. Furthermore, the assortment decision was shown to be as important, if not more important, as the price.

Evaluation of competing assortments has also received attention in literature. Chong et al. [9] present an empirically based modelling framework for managers to assess the revenue and lost sales implication of alternative category assortments. The framework consists of a category-purchase-incidence model and an extended version of the classical brand-share model of Guadagni and Little [17]. The brand-share model predicts which brand the customer chooses if a purchase incidence occurs in the category. Dhruv et al. [13] use Data Envelopment Analysis to account for sales differences in stores for a multi-chain retailer given differences in category assortments and regional preferences. Because regional preferences vary and the assortments carried by each store are different it is not appropriate to compare them directly to assess performance. This method establishes a best practice benchmark so stores within the chain can be fairly evaluated. A number of studies have shown that brands and brand sizes can be reduced without affecting sales or consumers' perceptions of variety (Williard Ltd. and Inc. [44], Broniarcyzk et al. [5]). What constitutes efficient assortments is likely to vary in different categories and in different shopping environments (Kahn and McAlister [21]). Brijs et al. [4] tackle the problem of product assortment analysis by introducing a concrete microeconomic integer programming model for product selection (PROFSET) based on the use of frequent item sets. They demonstrate its effectiveness on real-world sales transaction data obtained from a fully-automated convenience store.

From marketing literature it is known that the optimal product assortment should meet two important criteria. First, the assortment should be qualitatively consistent with the store's image. The retailer's background knowledge with regard to basic products should be easily incorporated in the model by means of additional constraints. A store's image distinguishes the retailer from its competition and is projected through its design, layout, services, and of course its products. Second, because retailing organizations are profit seeking companies, the product assortment should be quantitatively appealing in terms of the profitability it generates for the retailer. Most of the models that address the problem of assortment planning are mathematical models that are either nonlinear or involve integrality constraints which makes these models computationally hard to solve. Accordingly, most of them use heuristics or relaxations of the problem to reach a solution. Most of the models do not consider the inter-dependence between decisions for two or more items except for satisfying them on a more or less ad-hoc basis during postprocessing of solution. Finally, the problem is solved for one store at a time. This leads to individually customized assortments. This proliferation of assortments makes it difficult for chain stores to devise a consistent presentation across all stores. Most of the models also do not adequately treat the availability of back-room space or the retailers desire to maintain high inventory turnover rates. Finally, many of the models are unwieldy in terms of the number of parameters that must be estimated making it difficult to scale up to large practical problem solving.

9.3   Problem Description

In this chapter we focus on the retail assortment planning problem. Given a set of stores and a set of SKUs along with their demand distribution over a planning period, the problem is to find the best assortment of SKUs to keep in the stores so as to maximize the profit of all stores while satisfying financial, spacial, logistical, and marketing related constraints. Generally, such assortment planning decisions are made at an item or a SKU level. For our purposes, we assume that all decisions are made at a SKU level. If the decisions are to be made at an item level, the problem can be solved by aggregating sales for SKUs within the item.

While demand elasticity due to substitution among products is an important consideration, it is generally a very difficult effect to measure. Retailers are more comfortable giving estimates of base demand for a line, i.e., the amount of demand that will substitute to staple items in the product line. This allows us to consider a fraction of an item's demand as substitutable and therefore unaffected by assortment decisions. Items then have to justify inclusion in the assortment based on their fraction of nonsubstitutable demand. This also allows us to assume that demand for items is independent.

Items are assumed to be sold at multiple discrete price levels. For example, there may be a regular retail price, a sales price, and a clearance price. We assume that the retailer has historical information and can provide forecasts for each price level. Further, we make the assumption that the item is sold at the highest price level first, items not sold at that level are then sold at the next level and so on.

We assume the demand distribution at each price level to have a unimodal probability distribution. We use either Normal or Poisson distribution for most of our analysis. The Normal distribution is a good representation for high volume SKUs while the Poisson represents slow moving SKUs well. The mean and variance (if required) are obtained from the forecasted demand information. Demand for items which do not have history is based on history of items with similar characteristics.

9.3.1   Expected Revenue/Profit Functions (ER/PF)

Given the demand distribution at each price level, we define an expected revenue or profit function. Mathematically, the ER/PF can be computed as

$$ERF(q) = \sum_{y_1=1}^{q-1} \left((y_1 * p_1 + \Xi_2(y_1)) * g_y^1 \right) + q * p_1 * \left(\sum_{y_1=q}^{\infty} g_y^1 \right) \tag{9.1}$$

$$\Xi_2(y_1) = \sum_{y_2=1}^{q-y_1-1} \left((y_2 * p_2 + \Xi_3(y_1 + y_2)) * g_y^2 \right) + (q - y_1) * p_2 * \left(\sum_{y_2=q-y_1}^{\infty} g_y^2 \right) \tag{9.2}$$

$$EPF(q) = ERF(q) - q * c - H(q) \tag{9.3}$$

where q is the quantity kept in the store and $q/\rho \in Z^+$, ρ is the package quantity of the SKU, g_y^j is the probability of selling y_j units at level j, p_j is the selling price at level j, and c is the purchase price. The g_y^j's are computed from the demand distribution that is used. $\Xi_j(.)$ are calculated recursively by varying the parameters in each recursion. The depth of recursion is equal to the price levels in the problem.

In the equation for $EPF(q)$, $H(q)$ is the cost of holding a quantity q over the specified period of time. It is calculated as

$$H(q) = \frac{q}{2n} * h \qquad (9.4)$$

where h is the holding cost and n is the number of replenishments. The holding cost for q units as calculated above is an approximation to the expected holding cost since it assumes uniform demand over the planning horizon.

Given the demand distribution, these functions estimate the variation of revenue and profit with increasing SKU quantity in the store. EPF is the single period news-vendor objective function which has been modified to accommodate demand at different price levels. It is assumed that any residual inventory is discarded at the end of the period, incurring a loss equal to the cost of the SKU. ER/PF functions are nonlinear and therefore difficult to use. However, we simplify the problem by using a piecewise linear approximation. For low volume SKUs we linearize the problem by considering break points at each discrete value. For a high volume SKU we divide the ER/PF function curve into several discrete segments on both sides of the maximum point and consider break points at those segments. The accuracy of this method depends on the number of segments used to approximate the EPF.

The revenue function $ERF(q)$ is a nondecreasing function that asymptotically approaches the value $ERF(\infty)$ whereas profit function $EPF(q)$ is unimodal. This implies that as the quantity of a particular SKU increases in a store, the profit initially increases but after reaching a maximum, starts to fall. This is because beyond the point of maximum profit, the probability of selling an extra unit of the SKU is sufficiently small so that the expected realized profit becomes smaller than the expected loss due to not selling that additional unit. This results in an overall decline in the Profit function.

We denote the point of maximum profit on the profit function by $EPF(q^*)$ and the corresponding point on the revenue function by $ERF(q^*)$. The average sale price of a SKU is then obtained by calculating $ERF(q^*)/q^*$.

9.3.2 Constraints on Assortment

Since it is important to ensure that stores present a consistent appearance to customers to create brand recognition, it is important that the number of different assortments that are created is limited. Accordingly, a retailer may limit the number of different assortments that can be used. Such a restriction will obviously reduce the expected profit but it also reduces the cost associated with maintaining a large number of assortments. It is also easier to create a store image that is consistent. One way to permit more customization for stores is to use a modular planogram. In such a planogram, one constructs smaller sub-assortments and the store then selects from the sub-assortments to create its planogram. Since several sub-assortments may carry the same item, it is important that an item is in only one of the sub-assortments selected.

The choice of assortment can also be restricted based on financial criteria. There may be a limit on the total dollar value of inventory that a store can carry. This may limit the use of certain planograms which have multiple facings of the same item and therefore require a larger amount of inventory to appear filled. Inventory may also be constrained in the form of constraints on the number of inventory turns that the store must provide. Thus an assortment must be picked such that the ratio of the sales to the average inventory is above a given threshold. Thus a planogram with multiple facings may require a large amount of inventory and not perform as well as one with fewer facings. An overall budget constraint may be used across the entire line of items and across the entire chain of stores. Such constraints can be used to model what retailers call "open to buy". The budget is the maximum dollar value that the chain will invest in system-wide inventory for a specific product line.

Space considerations limit the number of planograms or sub-assortments that can be stored in a store. Space can either be a display space or back room space. Backroom or storage space is used when the amount of display space is inadequate to accommodate the inventory amount that must be stored in the store. Depending on the maximum number of replenishments that are permissible for a store and the package quantities for the item, the display space may not be adequate to hold the peak amount of inventory. In such cases, it is important to ensure that sufficient back room space is provided. Clearly, back room space is to be used only if the item is part of the assortment and therefore assumed to be on display.

9.4 Mathematical Model

Given a set of SKUs and stores, there are an exponential number of ways to generate strategies which makes it impossible to generate all possible strategies and pick the best among them. Therefore, we formulate the assortment planning problem as a Mixed Integer Programming problem and decompose it into two parts, the master problem and the subproblem, which can be solved using a Column Generation technique (Bazaraa et al. [2]). The subproblem returns improved strategies based on the set of space related dual variables passed to it by the master problem. The method is discussed in greater detail in subsequent sections.

The assortment for a store is constrained by several different requirements. These constraints result from the availability of display and storage space,

from advertising related legal requirements, and from performance requirements. There are also further restrictions caused by the presentation of items
in the store. This last aspect of the problem is called planogramming. A
planogram is a preset collection of related items that are arranged on display
fixtures to gain maximum visibility and impact. While commercial software
to draw the planograms exists, such software does not take a global view of
the problem in terms of system performance. The assortment planning model
must therefore also consider the problem of generating planograms.

Our model uses a two-level approach. This allows us to divide the problem
into what we call the Buyer's problem (technically, the master problem) and
the Plannogrammer's problem (technically, the pricing sub-problem). The
lower level Planogrammer's problem generates planogram proposals or sub-
assortments based on information supplied with placing items in a store. The
Buyer's problem then evaluates the sub-assortments so generated and assigns
them to appropriate stores such that all restrictions are met and the objective
function is maximized. This is an iterative process in which the two problems
are alternately solved until the optimal solution is obtained.

The objective of the model is to select an assignment of sub-assortments
to stores so as to maximize the expected profit while ensuring that the assignment uses only a prespecified number of assortments, satisfies space
availability at the store, and achieves on the aggregate level a prespecified
number of inventory turns.

Details of the mathematical model and explanation of the various constraints in the two problems are given below.

9.4.1 Problem Decomposition

The Buyer's Problem is to assign existing sub-assortments to stores so as
to maximize the overall revenue or profit of the system. Any assignment
can use at most a prespecified number of sub-assortments. The assignment
of sub-assortments to stores must meet various restrictions including those
on inventory values and turns and various space constraints. These various
restrictions and constraints in the model are discussed in the following subsection.

The Planogrammer's Problem (PP) is to use the solution of the Buyer's
Problem and generate a suitable new sub-assortment or plan which will increase the revenue/profit in the Buyer's Problem.

When the model for the Buyer's Problem is solved, along with the assignment of sub-assortments to store, we obtain information that can be used to
find the value of having a new sub-assortment for each store. The Planogrammer's Problem is to come-up with a new sub-assortment that fits in the display
space available for the store and has the maximum possible value. Solving
this model yields a new sub-assortment which can be added to the Buyer's
Problem. This process can be repeated until it is not possible to find a new sub-
assortment for any store. At this point, the final assignment of sub-problem
to stores can be made.

We use the following notation in the mathematical description of the model.

Parameters:

I : number of SKUs.

J : number of stores.

K : number of planograms.

T : number of linear segments used to approximate the EPF.

i : index of SKUs.

j : index of stores.

k : index of planograms.

t : index of linear segments.

$\hat{s}$: system or retailer wide stock to sales ratio.

$\bar{S}$: maximum system stock in dollars.

$\bar{M}$: maximum number of planograms allowed.

$\bar{Z}$: A large integer value.

p_i : average sale price of SKU i.

l_i : length of SKU i.

K_i: set of planograms containing SKU i.

$\hat{s}_j$: stock to sales ratio for the store j.

b_j : back-room linear space in store j.

l_j^f : fixture length in store j.

$\bar{I}_j$: maximum inventory in dollars of aggregate of SKUs in store j.

$\bar{d}_{ij}$: maximum depth of SKU i allowed in store j.

$\bar{n}_{ij}$: maximum number of replenishments for SKU i in store j per year.

l_k^p : length of planogram k.

W_{ijt} : piecewise linear approximation of q_{ij}.

r_{ijt} : profit (revenue) for W_{ijt} amount in store.

Decision variables:

X_{ij} : amount of SKU i in back-room in store j.

q_{ij} : total quantity of SKU i in store j.

Y_{ij} : number of facings of SKU i in store j.

a_{ik} : number of facings of SKU i in planogram k.

$$Z_k = \begin{cases} 1 & : \quad \text{if planogram } k \text{ is selected by some store} \\ 0 & : \quad \text{otherwise} \end{cases}$$

$$Z_{kj} = \begin{cases} 1 & : \quad \text{if store } j \text{ selects planogram } k \\ 0 & : \quad \text{otherwise} \end{cases}$$

9.4.2   Mathematical Formulation of Buyer's Problem

$$max \sum_t \sum_j \sum_i r_{ijt} W_{ijt}$$

$$\sum_t \sum_j \sum_i W_{ijt} - q_{ij} = 0 \quad \forall \ i, j \tag{9.5}$$

$$q_{ij} - \bar{d}_{ij} Y_{ij} - X_{ij} \leq 0 \quad \forall \ i, j \tag{9.6}$$

$$X_{ij} + Y_{ij} - q_{ij} \leq 0 \quad \forall \ i, j \tag{9.7}$$

$$\sum_i p_i q_{ij} \le \bar{I}_j \quad \forall \; j \tag{9.8}$$

$$\sum_i p_i (0.5 - \hat{s}_j \bar{n}_{ij}) q_{ij} \le 0 \quad \forall \; j \tag{9.9}$$

$$\sum_j \sum_i p_i q_{ij} \le \bar{S} \tag{9.10}$$

$$\sum_j \sum_i p_i (0.5 - \hat{s} \bar{n}_{ij}) q_{ij} \le 0 \tag{9.11}$$

$$\sum_i l_i X_{ij} \le b_j \quad \forall \; j \tag{9.12}$$

$$X_{ij} - \bar{Z} Y_{ij} \le 0 \quad \forall \; i, j \tag{9.13}$$

$$Y_{ij} - \sum_{k \in K_i} a_{ik} Z_{kj} = 0 \quad \forall \; i, j \tag{9.14}$$

$$\sum_k l_k^p Z_{kj} \le l_j^f \quad \forall \; j \tag{9.15}$$

$$\sum_{k \in K_i} Z_{kj} \le 1 \quad \forall \; i, j \tag{9.16}$$

$$\sum_k Z_k \le \bar{M} \tag{9.17}$$

$$Z_{kj} - Z_k \le 0 \quad \forall \; k, j \tag{9.18}$$

In the above model, the objective function is to maximize the expected revenue or profit where we are using the piece-wise linear approximation of the ER/PF. Constraint (9.5) ensures that the linearization of ER/PF function is consistent.

Constraint (9.6) to Constraint (9.9) are inventory related constraints. Constraint (9.6) ensures that the total store inventory for an SKU cannot exceed the amount of space on display shelves plus the amount kept in the back-room. Thus, with Y_{ij} facings the maximum amount of SKU i on display shelves of store j can be at most $\bar{d}_{ij} Y_{ij}$ and then the maximum amount of SKU i in the store j becomes the sum of $\bar{d}_{ij} Y_{ij}$ and the back-room quantity X_{ij}. Constraint (9.7) ensures that the total quantity is at least equal to the back-room quantity and as many items as there are facings in the front room. Constraint (9.8) is made active if the maximum inventory value for the store has been specified. The value of the inventory in the store is calculated as the sum of selling price of each SKU times the total quantity of the SKU in the store. The selling price of the SKU can vary from store to store. This constraint ensures that the aggregate dollar value of inventory of all the SKU's in store j is less than the maximum dollar value of inventory specified for that store. Constraint (9.9) is store level stock-to-sales ratio constraint. This constraint ensures that the average inventory (half the total inventory) must be less than the stock-to-sales ratio specified times the total demand that can be satisfied. If the target number of turns cannot be met then no item will be carried in the store. Note that, there is an inherent assumption in this constraint that the average sales rate is constant over the period of interest. While this may not be entirely accurate, stock-to-sales ratio constraints are applied over the average inventory over longer intervals of times such as months or quarters and hence the constant sales rate assumption is practically reasonable. Further note that the inventory is valued at selling price. It is relatively easy to convert inventory valuation to cost price.

Constraint (9.10) and Constraint (9.11) are budget related constraints. Constraint (9.10) is the retailer or system level investment constraint and it ensures that the total investment by the retailer for all the SKUs in all the stores is less than the specified maximum retailer investment stock in dollar value. Constraint (9.11) is retailer or system level stock-to-sales ratio constraint.

Constraint (9.12) to Constraint (9.16) are space related constraints. Constraint (9.12) is the back-room space constraint and it ensures that the amount of space used by inventory of all the SKUs in the store does not exceed the back-room space available for the store. Constraint (9.13) is the front-back space constraint and it ensures that inventory of a SKU is kept in the back-room only if at least one facing of the SKU is displayed in the front-room or the display space of the store. Constraint (9.14) is the facing space constraint and it calculates the number of facings of each SKU based on the assignment of sub-assortments to the store and the facings in each sub-assortment. Notice, that each planogram has a given number of facings, a_{ik}, for each SKU i. Thus the total number of facings for an SKU i is given by $\Sigma_{k \in K_i} a_{ik} Z_{kj}$. This number affects the total displayed inventory in Constraint (9.7). Since the planograms are designed by the subproblem, the number of facings in a planogram are designed to best utilize the available display space. Constraint (9.15) is the front-room or display space constraint and it ensures that the total display space used by the sub-assortments assigned to the store is no more than the space available. Constraint (9.16) is the only-one SKU constraint and it ensures that two sub-assortments assigned to a store cannot have a SKU in common. This restriction has been included from marketing point of view for aesthetic reasons and it avoids the problem of having the same SKU at two ends of the assortments.

Constraint (9.17) and Constraint (9.18) are planogram related constraints. Constraint (9.17) ensures that the total number of sub-assortments or plans selected for all the stores combined do not exceed the maximum number of plans allowed. Constraint (9.18) ensures that the total number of sub-assortments or plans selected for a store do not exceed the maximum number of plans allowed for a store.

9.4.3 Mathematical Formulation of the Subproblems

For subproblems, we use the following additional notations.

Parameters:
  s : index of shelf.
  $\bar{M}_{is}$: maximum amount of SKU i on a shelf s.

$\bar{L}_p$: maximum length of planogram allowed for a store .
F_{ij} : dual variable for the Constraint (9.14).
D_j : dual variable for the Constraint (9.15).
O_{ij} : dual variable for the Constraint (9.16).

Decision variables:
x_i : multiplicity of SKU i in a planogram.
y_{is} : multiplicity of SKU i on shelf s.

$$z_i = \begin{cases} 1 & : & \text{if SKU } i \text{ is in planogram} \\ 0 & : & \text{otherwise} \end{cases}$$

The model is:

$$v_j = min - \sum_i F_{ij}x_i + D_j \bar{L}_p + \sum_i O_{ij}z_i \qquad (9.19)$$

$$\sum_i l_i y_{is} \le \bar{L}_p \quad \forall\ s \qquad (9.20)$$

$$\sum_s y_{is} = x_i \quad \forall\ i \qquad (9.21)$$

$$y_{is} \le \bar{M}_{is}z_i \quad \forall\ i, s \qquad (9.22)$$

Constraint (9.20) is planogram length constraint and it ensures that the calculated sum of length of each SKU times the number of SKUs on the shelf
does not exceed the maximum allowable length of the planogram for all the
shelves. Constraint (9.21) ensures that the aggregate number of SKU i on all
shelves is equal to the total number of SKU i in the planogram. Constraint
(9.22) ensures that the total number of SKUs i on shelf s does not exceed the
maximum allowable amount of SKU i on shelf s.
    The length $\bar{L}_p$ depends on the physical shelves that are available in a store.
If more than one type of shelving system is available, then the length $\bar{L}_p$ can
be varied to create planograms of different lengths. This allows for creation
of standardized modules that can be used across multiple stores. Thus, even
if a planogram module is created for a single store, it is possible to use it
across all stores. Alternatively, it is possible to restrict the stores in which a
particular planogram is used. This can be achieved by not creating columns
corresponding to the restricted stores in the master problem.

9.4.3.1   Carry/No Carry Rules

Carry/No Carry rules allow the user to force the decision for a specific SKU.
The user may require that certain SKUs are carried in specific stores due to
marketing or advertising requirements. Alternatively, certain SKUs may not
be allowed in certain stores due to local restrictions. The sub-problem can be
modified in this case to set the specific z_i to be either 1 or 0 depending on
whether the SKU must be carried or not. In the event that we set $z_i = 1$ we
also require $x_i \ge 1$. Moreover, the resulting planogram is assigned to only
those stores which match the carry/no carry rule.

9.4.4   Algorithmic Considerations

The column generation scheme outlined above would be adequate if the problem was a linear programming problem. Unfortunately, the master problem has integer variables and thus column generation is not guaranteed to give the optimal solution and may not even give a feasible integer solution. Ideally one would use a branch-and-price methodology, which uses a column generation at each node of the branch and bound tree. However, given the size of the problem we are solving, it is unlikely to be computationally feasible to solve the problem in a reasonable amount of time using branch-and-price. Our experience thus far shows that solving an integer program using the columns generated for the LP relaxation almost always gives a feasible solution and the solution so obtained is generally quite good. However, even in this case, the column generation schema generates a large number of sub-assortments which results in an enormous master problem. Accordingly, we have to restrict the number of sub-assortments that are generated. Since the problem is solved across a large number of stores, it is possible that such a restriction may or may not be able to generate assortments that are suitable for every store. To ensure that many stores are able to contribute to the sub-assortment generation problem, we use one of several strategies:

- Restriction : Each store is allowed to generate no more than a fixed number of sub-assortments. This allows more stores to generate sub-assortments. Moreover, stores with differing amounts of sales can be given priority in generating the sub-assortments to get wider representation in the sub-assortments that are generated.

- Grouping by Sales : The stores can be grouped into a number of groups using the average demand vector for the different items. This will allow stores with similar sales patterns to cluster together. A composite store can be constructed using the average of demands across each store within a cluster and this composite store can be used to generate the sub-assortments.

- Grouping by Size : Stores can alternatively be grouped by the total sales for the product line in the store. This usually works well if there are significant differences in available space at different stores. A composite store can again be created. Notice that this approach looses most of the individual differences between stores of a similar size.

- Grouping by Sales and Size : This approach combines both of the above criteria for clustering stores. This is done by adding the total sales as one more of the attributes of the store.

The use of the grouping strategies requires the problem to be solved first with the composite stores to generate candidate planograms. The planograms so generated can be used to solve the master problem with a fixed number of columns but using the original larger number of stores. In this schema, no new columns are generated when the model is run a second time.

9.5 Computational Results

The algorithm described was implemented in Microsoft Visual C/C++ and executed on a Pentium Xeon 550 Mhz processor using a Windows 2000 Server operating system and 512 MB of RAM. CPLEX version 7.0 is used as the linear programming and integer programming solver.

To test the model we used data provided by a large national retail chain. The model was tested on five different data sets, which are shown in Table 9.1. We refer to these different product line data sets as 1, 2, 3, 4 and 5. Set 3 has data corresponding to 780 stores and 271 items with one SKU per item. All other data sets correspond to 880 stores though the number of items vary. Except for set 5 all product lines have one SKU per item. Set 5 contains 21 items each having a varying number of SKU's such that the aggregate number of SKU's is 669. The data sets are thus arranged in order of increasing number of SKU's. Except for set 4 all lines have moderate to high sales volume for most items in most stores. Set 4 is a low sales volume line.

9.5.1 Algorithm Performance on Base Data

To test how the algorithm performs both in terms of computational times and results with respect to algorithmic options, we tested the following variations in the algorithms:

1. We solved each data set as is without making any changes in the given data. We limited each store to generate no more than two assortments and restricted the maximum number of assortments to 500 across all stores. In this case it is possible that certain stores are never polled for generating assortments.

2. Next, for each data set, we grouped the stores into 5, 10, 20, 40 and 80 groups. We used a composite store for each group as explained earlier and used the composite store to generate planograms. The generated assortments were then used to solve the master problem with each individual store allowed to select from among the generated assortments. To group the stores we used three separate criteria as explained before:

   • For each data set we used the vector of sales as an attribute vector to group the stores. We refer to this strategy as Grouping by Sales.

TABLE 9.1

Base Problem Sets

	1	2	3	4	5
Stores	880	880	780	880	880
Items	29	33	271	368	21
SKUs	29	33	271	368	669

- We summed up the forecasted demand for the entire product to represent the store size. The generated data file was then used to group the stores on the basis of sales volume. We refer to this as Grouping by Size.
- Next, for each data set, we embedded the aggregated forecasted demand as a dummy SKU along with the original demand file for each store to create an attribute vector that represents the size of the store while simultaneously maintaining the effect of individual SKUs in each store. We refer to this as Grouping by Sales and Size.

It is to be noted that we are aggregating the demand information for a large number of stores distributed all over a wide geographical area. Stores in similar geographic and demographic settings usually have similar sales profiles and tend to group together. Because the number of groups is limited, it is possible that a store may get grouped improperly when sales data alone are used. However, the model allows each store to select any assortment generated by any of the groups. Computational results show that by not restricting the stores to choose the same planogram as the group, results in significantly improved profits.

In Table 9.2, we show the relative performance of the algorithms under the different algorithmic options described above. All values are shown as the % difference from the No Grouping profit value for each product line. Table 9.3 shows the amount of computational time taken by each method. Except for product line 4, we see that there is very little difference in the relative performance of the algorithms. However, product line 4 shows a

TABLE 9.2

Comparison of Group Profits

Grouping	Number of Groups	Product Line				
		1	2	3	4	5
None	—	0.00	0.00	0.00	0.00	0.00
	5	−1.54	0.39	−1.46	−5.04	−1.25
	10	−0.10	0.52	−0.30	56.51	−1.09
By sales	20	0.02	0.38	0.00	55.60	−0.99
	40	0.02	0.65	−0.06	18.15	−1.75
	80	−0.01	0.57	0.03	15.35	−0.98
	5	−0.98	0.28	−1.89	12.02	−1.41
	10	−0.66	0.53	−0.60	35.77	−1.49
By size	20	−0.14	0.52	−0.61	68.38	−0.41
	40	−0.07	0.61	−0.45	45.11	0.67
	80	−0.02	0.71	0.10	15.35	0.64
	5	−1.03	0.29	−1.24	−0.16	−1.22
	10	−0.59	0.29	−0.04	17.66	−1.58
By sales and size	20	−0.05	0.54	−0.23	76.71	−0.79
	40	0.00	0.62	−0.01	8.29	0.78
	80	−0.02	0.76	0.06	15.35	−0.21

TABLE 9.3

Total Computational Time (Seconds)

Grouping	Number of Groups	Product Line				
		1	2	3	4	5
None	–	1396	1529	1541	1324	2483
	5	227	478	312	337	899
	10	397	731	461	427	1128
By sales	20	641	775	674	680	1794
	40	633	1087	825	742	1836
	80	749	1529	1098	873	2361
	5	226	310	267	272	667
	10	319	422	323	292	827
By size	20	477	795	457	361	1285
	40	499	869	428	360	1300
	80	754	1346	516	402	1592
	5	366	449	487	483	1321
	10	559	528	667	598	1572
By sales and size	20	866	986	965	892	2236
	40	1074	1431	1142	974	2411
	80	1168	1590	1402	1135	2905

huge improvement in expected profits. This results from the fact that product line 4 is a low sales volume line and as such stores may have very different sales for different items. When there is no grouping, stores which do not get a chance to generate sub-assortments are forced to select from assortments generated for other stores and as such perform rather poorly and drag down overall profits. Grouping similar stores and using a composite store helps generate assortments that are more acceptable to all stores and this leads to a significant improvement in performance. It appears that in such cases creating 20 or 40 store groups gives much better performance. For the other product lines, even when stores do not get to participate in sub-assortment generation when there is no grouping, the stores can use the generated sub-assortments by suitably adjusting the assortment depth or inventory and still manage to perform well. In this case the grouping of stores may occasionally lead to slightly reduced performance because the composite stores do not generate as many sub-assortments and therefore reduce the options available when stores select their assortments.

The benefit of grouping is seen in the reduced run times for the algorithms in Table 9.3. As can be seen the run times increase with the number of groups generated. This is due to the fact that the number of sub-assortments generated increases with the number of groups. Thus with 80 groups we sometimes end up with more time to solve the overall problem since we spend time grouping the stores and still have roughly the same number of sub-assortments to consider.

Grouping by size seems to offer the best results in general because both the solve times and the results are in general very good. The smaller solve times are the result of much smaller times to do the grouping since, grouping is done based on a single attribute.

TABLE 9.4

Effect of Grouping and Fixing Assortments

	Fixed		Not Fixed	
Number of Assortments	Profit Value	Time (sec)	Profit Value	Time (sec)
2	1654327.23	2179	4818450.68	2459
4	1983453.53	2180	4865612.69	2675
6	2563685.88	2181	4881610.37	2665
8	2885437.46	2181	4889297.39	2850
10	3451765.26	2180	4894723.55	2749
12	3762453.51	2181	4898851.89	2894
14	3956443.34	2181	4901553.37	2817
16	4123536.74	2182	4903748.86	2865
18	4125466.34	2182	4905577.19	2843
20	4125466.34	2183	4906899.97	2905

Grouping, usually by size, is routinely used by retailers for assorting purposes. However, in many of these cases retailers force all stores in the group to use the same assortment. We consider the performance of such a policy for product line 5 and for the case where 80 groups are formed. The results when the stores in each group are forced to carry a single assortment are contrasted with the results when each store has free choice. We designate this as Fixed and Not Fixed in Table 9.4. The table gives actual values of the profits and run times for both cases.

The table shows how performance changes when the number of assortments that can be selected is varied from 2 to 20 assortments. The profits for both cases increase with the number of assortments that are permitted. However the difference between having 2 and 20 assortments for the fixed case is significantly larger than that for the not fixed case. Accordingly, when 2 assortments are allowed the profits for the fixed case are only 35% of the profits for the not fixed case. The fixed case is able to improve to about 84% of the profits of the not fixed case when 20 assortments are allowed. The amount of time is a bit smaller for the fixed case. Clearly, using a grouping strategy to force stores into the same assortment within a group may lead to lower profitability. It is also interesting to see how having a larger number of assortments impacts profitability. In the fixed case, there is significant improvement as more assortments are allowed until about 16 assortments. On the other hand, the not fixed case outperforms the best fixed case result with only two assortments. There is an improvement in profitability for the not fixed case as well but the increases are nominal.

9.5.2 Effect of Problem Parameters

We next consider the effect of different problem parameters to study their effect on the solution quality. Clearly, profitability will be affected by the choice of these parameters. We consider the effect of package quantities for each item. Package quantities affect the amount of inventory that must be carried by each store and they also affect the amount of inventory that may have to be sold at lower prices or scrapped at the end of the horizon. Similarly, inventory holding costs will affect the overall profitability. The maximum number of replenishments allowed per store also affects the average inventory carried by the store. In all these cases, the actual effect on profitability depends on the profit margins enjoyed by the retailer. If profit margins are very large the effect of changes in these parameters is relatively small and assortments are not expected to change radically.

9.5.2.1 Effect of Package Quantity

To test the effect of package quantity we consider product line 4 and adjust the costs so that the average profit margin for individual items is 35%. The package quantity is varied to take on values of 1, 2, 4 and 6. The lowest profit was obtained for the case when package quantity was 6 and two assortments were permitted. Table 9.5 shows the fraction deviation of the profits obtained for each package quantity and the number of assortments used from the lowest profit case. From the table it is clear that the profits are lower when the package quantities are larger. Adding extra assortments marginally reduces the negative effects of the package quantity. Thus having a package quantity of 1 gives a 20 fold increase in profit margin in comparison to that obtained for a package quantity of 6 when the number of assortments are 2. The fraction deviation for a package quantity of 1 increases to 26.39 when the number of assortments is increased to 20. There is a similar increase for a package quantity of 6 but the increase is much smaller. There is a more insidious effect of package quantities. Since the inventory costs are higher, many stores simply do not find it profitable to carry many of the products. This reduces the revenue that is generated as well. Thus both the revenue and profit are negatively impacted.

TABLE 9.5

Variation in Profit with Package Quantity

Number of Assortments	Package Quantity			
	1	2	4	6
2	20.05	12.04	1.98	0.00
4	23.90	15.02	2.73	0.39
6	24.97	15.81	3.00	0.70
8	25.66	16.48	3.07	0.84
10	26.03	16.71	3.12	0.95
12	26.13	16.80	3.13	1.00
14	26.22	16.86	3.13	1.01
16	26.29	16.90	3.13	1.01
18	26.35	16.93	3.13	1.01
20	26.39	16.96	3.13	1.01

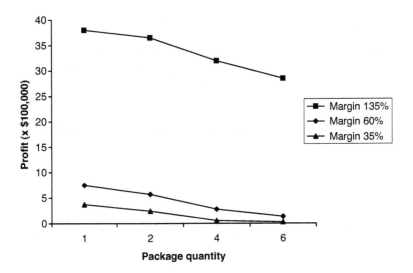

FIGURE 9.1
Effect of package quantity.

Figure 9.1 shows the effect of package quantities as the profit margin is changed. A more significant effect of package quantities occurs when the profit margin is higher because more items get included in the assortment. Many more items are therefore affected by the increased package quantity. Since fewer items are carried by the stores for lower margins the effect is somewhat less pronounced even though the extra inventory holding cost has a more sizeable effect on the profits.

9.5.2.2 Effect of Number of Replenishments

The number of replenishments also affects the inventory holding cost. The more frequently replenishments are done the less the inventory that is carried. Table 9.6 shows fraction deviation from the lowest value obtained for product line 4. In this case we use a package quantity of 1. The number of replenishments allowed is varied from 3 to 52. Notice that the profit increases with the number of replenishments. However, there are diminishing returns as the number of replenishments increases beyond a point. This is important to note since in our model we do not solve for the optimal number of replenishments and so do not consider the fixed replenishment cost. Clearly, beyond a certain point having extra replenishments may not be able to overcome the cost of replenishments.

When package quantities are varied with the number of replenishments, the profits in all cases continue to increase with the number of replenishments. However package quantities seem to have more of an effect when there are fewer replenishments. Figure 9.2 shows a plot of the profits against the number of replenishments for a number of different package quantities.

TABLE 9.6

Variation in Profit with Number of Replenishments

Number of Assortments	Number of Replenishments					
	3	4	7	13	26	52
2	0.00	1.09	2.86	4.39	5.31	5.81
4	0.23	1.50	3.55	5.37	6.47	7.07
6	0.26	1.58	3.70	5.65	6.80	7.44
8	0.26	1.62	3.83	5.82	6.99	7.64
10	0.26	1.64	3.90	5.92	7.13	7.79
12	0.26	1.65	3.92	5.94	7.16	7.83
14	0.26	1.65	3.93	5.97	7.19	7.87
16	0.26	1.66	3.95	5.99	7.22	7.90
18	0.26	1.66	3.95	6.00	7.24	7.92
20	0.26	1.66	3.96	6.01	7.26	7.94

9.5.2.3 Effect of Inventory Holding Cost

To study the effect of holding cost on overall profitability we consider product line 4 again. However, we simultaneously vary the item costs to get different profit margins for the SKUs. We consider a profit margin of 35%, 60% and 135%. Table 9.7 shows the results of our test runs in the form of percent deviations. However, to highlight a number of interesting observations the table shows in columns 2 through 4 the percent deviation when the holding cost is decreased from a base of 30 to 20%. In columns 5 through 7 we show the percent deviation when the holding cost is decreased to 30% from a base of 40%. Clearly, holding costs affect the solution much more when the profit

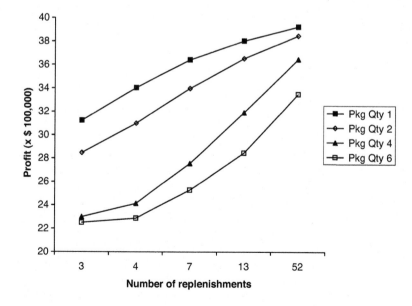

FIGURE 9.2
Effect of replenishments and package quantity.

TABLE 9.7

Percentage Variation in Profit as Holding Cost

Number of Assortments	20% vs. 30%			30% vs. 40%		
	Percentage Profit Margin					
	135	60	35	135	60	35
2	1.52	5.62	17.04	1.54	5.65	17.23
4	1.72	5.89	17.39	1.74	5.92	17.29
6	1.85	6.06	17.49	1.88	6.05	17.71
8	2.48	6.34	17.43	1.52	6.32	17.50
10	1.97	6.48	17.45	1.98	6.40	17.51
12	1.89	6.60	17.49	1.95	6.54	17.42
14	1.98	6.73	17.54	1.99	6.64	17.37
16	2.00	6.80	17.58	2.01	6.62	17.35
18	2.04	6.16	17.59	2.05	6.53	17.39
20	2.10	6.91	17.65	2.11	6.69	17.39

margins are smaller. Thus a 10% increase in holding cost affects the solution much more when the profit margin is 35% (approximately 17% deviation) than when the profit margin is 135% (between 1.52% to 2.11%). This is due to the fact that holding cost has a more significant effect on assortment decisions when margins are lower. It is also interesting to observe that number of assortments plays a more significant part when the margin is higher. When the margins are higher and holding costs are lower, more products qualify to be part of an assortment. Thus allowing more assortments makes it possible to tailor the assortments to individual store requirements giving higher increases in profit. When margins are low, very few products qualify at higher holding costs and permitting more assortments does not increase profitability since stores do not care to carry most of the products.

9.5.2.4 Effect of Number of Assortments

The number of assortments permitted across all stores is a major determinant of the profit. Figure 9.3 shows the percent deviation of the profit with the number of assortments for all five product lines. In each case the profit with 2 assortments is taken as the base line to calculate percent deviations. In all cases increasing the number of assortments leads to increase in the profit but with diminishing returns. Indeed in all cases having more than 10 assortments does not materially affect the profit. A striking feature of Figure 9.3 is the significantly higher increase that product line 4 exhibits in comparison with the other product lines. The reason for this is that product line 4 is a low volume line with a large number of SKUs and there are significant differences between the stores. When very few assortments are permitted many stores experience large losses. As more assortments are permitted stores are able to find assortments that are more suitable to their sales patterns. This allows such stores to pick profitable items and avoid loss making items.

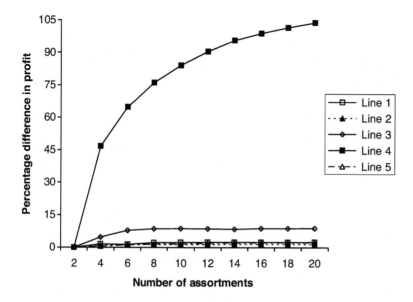

FIGURE 9.3
Effect of number of assortments on profits.

9.5.2.5 Effect of Profit Margin

In much of the discussion thus far, we have mentioned the importance of the average profit margin on assortment decisions. We conclude our computational results by looking at a plot of the profit to the profit margin. Figure 9.4 shows the variation in profit with profit margin. First notice that the y-axis

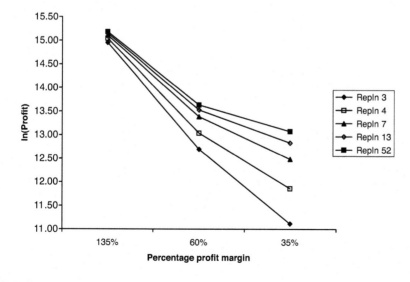

FIGURE 9.4
Effect of profit margin.

has a logarithmic scale. This was done to scale the data so that we could show the relative importance of number of replenishments with changing profit margins. As is to be expected decreasing the profit margin leads to decreased profits. However, it is interesting to note that there is a much smaller effect of the number of replenishments when the profit margin is larger. This stems from the fact that the extra holding cost is a smaller fraction of the overall cost and, thus, does not make products so unprofitable that they are dropped from the assortment.

9.6   Conclusions

In this chapter we present the assortment planning problem. We develop a model and solution algorithm for the problem when a large retail chain must limit the number of different assortments that are used. This requires the problem to be solved simultaneously across all stores in the chain. We show how to model constraints on inventory productivity, display space, back room space, and inventory budgets. We present an algorithm based on column generation and provide extensive computational results to show how the algorithm performs.

References

1. Anderson, E.E. and H.N. Amato., A mathematical model for simultaneously determining the optimal brand-collection and display-area allocation, *Operations Research*, 22(1):13–21, 1974.
2. Bazaraa, M.S., J. Jarvis, and H. Sherali., *Linear Programming and Network Flows*, John Wiley and Sons, Inc., 2nd edition, 1990.
3. Borin, N., P. Farris, and J. Freeland., A model for determining retail product category and shelf space allocation, *Decision Sciences*, 25(3):359–384, 1994.
4. Brijs, T., G. Swinnen, K. Vanhoof, and G. Wets., Using association rules for product assortment decisions: A case study, *Proceedings of the Fifth ACM SIGKDD International Conference on Knowledge Discovery and Data Mining*, San Diego, pages 254–260, 1999.
5. Broniarcyzk, S.M., W.D. Hoyer, and L.M. McAlister., Consumers' perceptions of category assortment: The impact of item reduction, *Journal of Marketing Research*, 35:166–176, 1998.
6. Bultez, A. and P. Naert., S.H.A.R.P.: shelf allocation for retailer's profit, *Marketing Science*, 7:211–231, 1988.

7. Cairns, J.P., Suppliers, retailers, and shelf space, *Journal of Marketing*, 26 (3):34–36, 1962.
8. Chernev, A., Product assortment and individual decision processes, *Journal of Personality and Social Psychology*, 85(1):151–162, 2003.
9. Chong, J., T. Ho, and C.S. Tang., A modelling framework for category assortment planning, *Manufacturing and Service Operations Management*, 3(3):191–210, 2001.
10. Corstjens, M. and P. Doyle., A model for optimizing retail space allocations, *Management Science*, 27(7):822–833, 1981.
11. Cox, K.K., The effect of shelf space upon sales of branded products, *Journal of Marketing Research*, 7:55–58, 1970.
12. Curhan, R.C., The relationship between shelf space and unit sales in supermarkets, *Journal of Marketing Research*, 9(4):406–412, 1972.
13. Dhruv, G., M. Levy, A. Mehrotra, and A. Sharma., Planning merchandising decisions to account for regional and product assortment differences, *Journal of Retailing*, 75(3):405–424, 1995.
14. Fox, L.J., An integrated view of assortment management process: the next frontier for leading retailers, *Chain Store Age Executive*, 71(11), 62–68, November 1995.
15. Frank, R.E. and W.F. Massy., Shelf position and space effect on sales, *Journal of Marketing Research*, 7:59–66, 1970.
16. Green, P. and A. Krieger., Models and heuristics for product line selection, *Marketing Science*, 4:1–19, 1985.
17. Guadagni, P. and J. Little., A logit model of brand choice calibrated on scanner data, *Marketing Science*, 2(3):203–238, 1983.
18. Hoch, S.J., E.T. Bradlow, and B. Wansink., The variety of an assortment, *Marketing Science*, 18(4):527–546, 1999.
19. Jayaraman, V., R. Srivastava, and W.C. Benton., A joint optimization of product variety and ordering approach, *Computers and Operations Research*, 25(7/8):557–566, 1998.
20. Kahn, B. E. and D.R. Lehmann., Modelling choice among assortments, *Journal of Retailing*, 67(3):274–299, 1991.
21. Kahn, B. E. and L. McAlister., *Grocery Revolution: The New Focus on the Consumer*, Addison-Wesley Pub. Co, 1997.
22. Koelemeijer, K. and H. Oppewal., Assessing the effects of assortment and ambience: a choice experimental approach, *Journal of Retailing*, 75 (3):319–345, 1995.
23. Lancaster, K., The economics of product variety: a survey, *Marketing Science*, 9:189–210, 1990.
24. Larson, P.D. and R.A. DeMarais., Psychic stock: An independent variable category of inventory, *International Journal of Physical Distribution and Logistics Management*, 20(7):28–34, 1990.
25. Stassen, R.E., Mittelstaedt, J.D., and Mittelstaedt, R.A., Assortment overlap: its effect on shopping patterns in a retail market when the

distributions of prices and goods are known, *Journal of Retailing*, 75: 371–396, 1995.

26. Levin, R.I., C.P. McLaughlin, R.P. Lamone, and J.F. Kottas., *Production/ Operations Management: Contemporary Policy for Managing Operating Systems*, McGraw-Hill, N.Y., 1972.

27. Lilien, G.L., P. Kotler, and K.S.P. Moorthy., *Marketing Models*, Prentice-Hall, N.J., 1992.

28. Lim, A., B. Rodriques, and X. Zhang., Metaheuristics with local search techniques for retail shelf-space optimization, *Management Science*, 50: 117–131, 2004.

29. Mahajan, S. and G. Van Ryzin., A theory of retail assortments, *Working Paper no. 96-07-02 , The Department of Operations and Information Management, The Wharton School*, 1996.

30. Mahajan, S. and G. Van Ryzin., Stocking retail assortments under dynamic consumer substitution, *Operations Research*, 49(3):334–351, 2001.

31. Pashigan, P.B., Demand uncertainty and sales: A study of fashion and markdown pricing, *American Economic Review*, 78:936–953, 1988.

32. Rajaram, K., Assortment planning in fashion retailing: methodology, application and analysis, *European Journal of Operational Research*, 129: 186–208, 2001.

33. Schary, P.B. and B.W. Becker., Distribution and final demand: The influence of availability, *Mississippi Valley Journal of Business and Economics Industrial Management Review*, 8(1):17–26, 1972.

34. Shelby, H. M. and C.M. Miller., The selection and pricing of retail assortments: an empirical approach, *Journal of Retailing*, 75(3):295–318, 1995.

35. Shocker, A.D. and V. Srinivasan., Multi attribute approaches for product concept evaluation and generation: a critical review, *Journal of Marketing Research*, 16:159–180, 1979.

36. Silver, E.A. and R. Peterson., *Decision Systems for Inventory Management and Production Planning*, John Wiley and Sons, Inc., 2nd edition, 1985.

37. Simonson, I., The effect of product assortment on buyer preferences, *Journal of Retailing*, 75(3):347–370, 1995.

38. Smith, S.A. and N. Agrawal., Management of multi-item retail inventory systems with demand substitution, *Operations Research*, 48(1):50–64, 2000.

39. Urban, G.L., A mathematical modelling approach to product line decisions, *Journal of Marketing Research*, 6:40–47, 1969.

40. Urban, G.L., PERCEPTOR: A model for product positioning, *Management Science*, 21:858–871, 1975.

41. Urban, T.L., A mathematical modelling approach to product line decisions, *Journal of Retailing*, 74(1):15–35, 1998.

42. U.S.Census., *Annual Benchmark Report for Retail Trade and Food Services - January 1992 through February 2002*, accessed on September 7, 2004. http://www.census.gov/prod/2004pubs/br03-a.pdf, 2004.

43. Whitin, T.M., *The Theory of Inventory Management*, Princeton University Press, 1957.

44. Williard Bishop Consulting Ltd. and Information Resources Inc., Variety or duplication: A process to know where you stand, *Report Prepared for the Food Marketing Institute,* 1993.
45. Wolfe, H.B., A model for control of style merchandise, *Industrial Management Review,* 9(2):69–82, 1968.
46. Yang, M., An efficient algorithm to allocate shelf space, *European Journal of Operational Research,* 131:107–118, 2001.
47. Yang, M. and W. Chen., A study of shelf space allocation and management, *International Journal of Production Economics,* 60-61:309–317, 1999.
48. Zufryden, F.S., A dynamic programming approach for product selection and supermarket shelf-space allocation, *Journal of the Operational Research Society,* 37(4):413–422,1986.

10

Noncommercial Software for Mixed-Integer Linear Programming

Jeffrey T. Linderoth and Ted K. Ralphs

CONTENTS

10.1   Introduction

A mixed-integer linear program (MILP) is a mathematical program with linear constraints in which a specified subset of the variables are required to take on integer values. Although MILPs are difficult to solve in general, the past ten years have seen a dramatic increase in the quantity and quality of software — both commercial and noncommercial — designed to solve MILPs. Generally speaking, noncommercial MILP software tools cannot match the speed or robustness of their commercial counterparts, but they can provide a viable alternative for users who cannot afford the sometimes costly commercial offerings. For certain applications, open-source software tools can also be more extensible and easier to customize than their commercial counterparts, whose flexibility may be limited by the interface that is exposed to the user. Because of the large number of open-source and noncommercial packages available, it might be difficult for the casual user to determine which of these tools is the best fit for a given task. In this chapter, we provide an overview of the features of the available noncommercial and open source codes, compare selected alternatives, and illustrate the use of various tools. For an excellent overview of the major algorithmic components of commercial solvers, especially CPLEX, LINDO, and XPRESS, we refer to reader to the paper of Atamtürk and Savelsbergh [6].

To formally specify a MILP, let a polyhedron

$$\mathcal{P} = \{x \in \mathbb{R}^n \mid Ax = b, x \geq 0\} \tag{10.1}$$

be represented in standard form by a constraint matrix $A \in \mathbb{Q}^{m \times n}$ and a right-hand side vector $b \in \mathbb{Q}^m$. Without loss of generality, we assume that the variables indexed 1 through $p \leq n$ are the integer-constrained variables (the *integer variables*), so that the feasible region of the MILP is $\mathcal{P}^I = \mathcal{P} \cap \mathbb{Z}^p \times \mathbb{R}^{n-p}$. In contrast, the variables indexed $p + 1$ through n are called the *continuous variables*. A subset of the integer variables, called *binary variables*, may additionally be constrained to take on only values in the set $\{0, 1\}$. We will denote the set of indices of binary variables by $B \subseteq \{1, 2, \ldots, p\}$. The mixed-integer linear programming problem is then to compute the optimal value

$$z_{IP} = \min_{x \in \mathcal{P}^I} c^\top x, \tag{10.2}$$

where $c \in \mathbb{Q}^n$ is a vector that defines the objective function. The case in which all variables are continuous ($p = 0$) is called a *linear program* (LP). Associated with each MILP is an LP called the *LP relaxation*, obtained by relaxing the integer restrictions on the variables. For the remainder of the chapter, we use this standard notation to refer to the data associated with a given MILP and its LP relaxation.

In what follows, we review the relevant notions from the theory and practice of integer programming and refer to other sources when necessary for the full details of the techniques described. This chapter is largely self-contained, though we do assume that the reader is familiar with concepts from the theory of linear programming (see [18]). We also assume that the reader has at least a high-level knowledge of both object-oriented and functional programming interfaces. For an in-depth treatment of the theory of integer programming, we direct the reader to the works of Schrijver [76], Nemhauser and Wolsey [62], and Wolsey [85].

The chapter is organized as follows. In Section 10.2, we sketch the branch-and-cut algorithm, which is the basic method implemented by the solvers we highlight herein, and we describe in some detail the advanced bound improvement techniques employed by these solvers. In Section 10.3, we discuss the various categories of MILP software systems and describe how they are typically used. In Section 10.4, we describe the use and algorithmic features of eight different noncommercial MILP software systems: ABACUS, BCP, BonsaiG, CBC, GLPK, lp_solve, MINTO, and SYMPHONY. Section 10.5 illustrates the use of two solver frameworks to develop specialized algorithms for solving specific MILP problems. In Section 10.6, the six noncommercial solvers that can be used as "black-box" solvers are benchmarked on a suite of over 100 MILP instances. We conclude by assessing the current state of the art and trends for the future.

10.2 Branch and Bound

Branch and bound is a broad class of algorithms that is the basis for virtually all modern software for solving MILPs. Here, we focus specifically on *LP-based branch and bound*, in which LP relaxations of the original problem are solved to obtain bounds on the objective function value of an optimal solution. Roughly speaking, branch and bound is a divide and conquer approach that reduces the original problem to a series of smaller *subproblems* and then recursively solves each subproblem. More formally, recall that $\mathcal{P}^I$ is the set of feasible solutions to a given MILP. Our goal is to determine a least cost member of $\mathcal{P}^I$ (or prove $\mathcal{P}^I = \emptyset$). To do so, we first attempt to find a "good" solution $\bar{x} \in \mathcal{P}^I$ (called the *incumbent*) by a heuristic procedure or otherwise. If we succeed, then $\beta = c^\top \bar{x}$ serves as an initial upper bound on z_{IP}. If no such solution is found, then we set $\beta = \infty$. We initially consider the entire feasible region $\mathcal{P}^I$. In the *processing* or *bounding* phase, we solve the LP relaxation $\min_{x \in \mathcal{P}} c^\top x$ of the original problem in order to obtain a *fractional solution* $\hat{x} \in \mathbb{R}^n$ and a lower bound $c^\top \hat{x}$ on the optimal value z_{IP}. We assume this LP relaxation is bounded, or else the original MILP is itself unbounded.

After solving the LP relaxation, we consider $\hat{x}$. If $\hat{x} \in \mathcal{P}^I$, then $\hat{x}$ is an optimal solution to the MILP. Otherwise, we identify k disjoint polyhedral subsets of $\mathcal{P}$, $\mathcal{P}_1, \dots, \mathcal{P}_k$, such that $\cup_{i=1}^k \mathcal{P}_k \cap \mathbb{Z}^p \times \mathbb{R}^{n-p} = \mathcal{P}^I$. Each of these subsets defines a new MILP with the same objective function as the original, called a *subproblem*. Based on this partitioning of $\mathcal{P}^I$, we have

$$\min_{x \in \mathcal{P}^I} c^\top x = \min_{i \in 1 \dots k} \left(\min_{x \in \mathcal{P}_i \cap \mathbb{Z}^p \times \mathbb{R}^{n-p}} c^\top x \right), \qquad (10.3)$$

so we have reduced the original MILP to a family of smaller MILPs. The subproblems associated with $\mathcal{P}_1, \dots, \mathcal{P}_k$ are called the *children* of the original MILP, which is itself called the *root subproblem*. Similarly, a MILP is called the *parent* of each of its children. The set of subproblems is commonly associated with a tree, called the *search tree*, in which each node corresponds to a subproblem and is connected to both its children and its parent. We therefore use the term *search tree node* or simply *node* interchangeably with the term subproblem and refer to the original MILP as the *root node* or *root* of this tree.

After partitioning, we add the children of the root subproblem to the list of *candidate subproblems* (those that await processing) and associate with each candidate a lower bound either inherited from the parent or computed during the partitioning procedure. This process is called *branching*. To continue the algorithm, we select one of the candidate subproblems and process it, i.e., solve the associated LP relaxation to obtain a fractional solution $\hat{x} \in \mathbb{R}^n$, if one exists. Let the feasible region of the subproblem be $\mathcal{S} \subseteq \mathcal{P} \cap \mathbb{Z}^p \times \mathbb{R}^{n-p}$. There are four possible results:

1. If the subproblem has no solutions, then we discard, or *fathom* it.
2. If $c^\top \hat{x} \geq \beta$, then $\mathcal{S}$ cannot contain a solution strictly better than $\bar{x}$ and we may again fathom the subproblem.
3. If $\hat{x} \in \mathcal{S}$ and $c^\top \hat{x} < \beta$, then $\hat{x} \in \mathcal{P}^I$ and is the best solution found so far. We set $\bar{x} \leftarrow \hat{x}$ and $\beta \leftarrow c^\top \bar{x}$, and again fathom the subproblem.
4. If none of the above three conditions hold, we are forced to branch and add the children of this subproblem to the list of candidate subproblems.

We continue selecting subproblems in a prescribed order (called the *search order*) and processing them until the list of candidate subproblems is empty, at which point the current incumbent must be the optimal solution. If no incumbent exists, then $\mathcal{P}^I = \emptyset$.

This procedure can be seen as an iterative scheme for improvement of the difference between the current upper bound, which is the objective function value of the current incumbent, and the current lower bound, which is the minimum of the lower bounds of the candidate subproblems. The difference between these two bounds is called the *optimality gap*. We will see later that there is a tradeoff between improving the upper bound and improving the lower bound during the course of the algorithm.

The above description highlights the four essential elements of a branch-and-bound algorithm:

- *Lower bounding method*: A method for determining a lower bound on the objective function value of an optimal solution to a given subproblem.
- *Upper bounding method*: A method for determining an upper bound on the optimal solution value z_{IP}.
- *Branching method*: A procedure for partitioning a subproblem to obtain two or more children.
- *Search strategy*: A procedure for determining the search order.

With specific implementations of these elements, many different versions of the basic algorithm can be obtained. So far, we have described only the most straightforward implementation. In the sections that follow, we discuss a number of the most common improvements to these basic techniques. Further improvements may be possible through exploitation of the structure of a particular problem class.

10.2.1   Lower Bounding Methods

The effectiveness of the branch and bound algorithm depends critically on
the ability to compute bounds that are close to the optimal solution value.
In an LP-based branch-and-bound algorithm, the lower bound is obtained
by solving an LP relaxation, as we have indicated. There are a number of
ways in which this lower bound can be potentially improved using advanced
techniques. In the remainder of this section, we describe those that are implemented in the software packages reviewed in Section 10.4.

10.2.1.1   Logical Preprocessing

One method for improving the lower bound that can be applied even before
the solution algorithm is invoked is *logical preprocessing*. By application of
simple logical rules, preprocessing methods attempt to tighten the initial formulation and thereby improve the bound that will be produced when the LP
relaxation is solved. Formally, preprocessing techniques attempt to determine
a polyhedron $\mathcal{R}$ such that $\mathcal{P}^I \subseteq \mathcal{R} \subset \mathcal{P}$. The bound obtained by minimizing
over $\mathcal{R}$ is still valid but may be better than that obtained by minimizing over $\mathcal{P}$.

Preprocessing techniques are generally limited to incremental improvements of the existing constraint system. Although they are frequently
designed to be applied to the original formulation before the solution algorithm is invoked, they can also be applied to individual subproblems during
the search process if desired. Preprocessing methods include procedures for
identification of obviously infeasible instances, removal of redundant constraints, tightening of bounds on variables by analysis of the constraints,
improvement of matrix coefficients, and improvement of the right-hand side
value of constraints. For example, a constraint

$$\sum_{j=1}^{n} a_{ij}x_j \le b_i$$

can be replaced by the improved constraint

$$\sum_{j \in \{1,\dots,n\}\setminus\{k\}} a_{ij}x_j + (a_k - \delta)x_k \le b_i - \delta$$

for $k \in B$ such that $\sum_{j\in\{1,\dots,n\}\setminus\{k\}} a_{ij}x_j \le b_i - \delta \forall x \in \mathcal{P}^I$. This technique is called
coefficient reduction. Another technique, called *probing*, can be used to deter-
mine the logical implications of constraints involving binary variables. These
implications are used in a variety of ways, including the generation of valid
inequalities, as described next in Section 10.2.1.2. An extended discussion of
advanced preprocessing and probing techniques, can be found in the paper
of Savelsbergh [74].

10.2.1.2   Valid Inequalities

The preprocessing concept can be extended by dynamically generating entirely new constraints that can be added to the original formulation without

excluding members of $\mathcal{P}^I$. To introduce this concept formally, we define an *inequality* as a pair (a, a_0) consisting of a coefficient vector $a \in \mathbb{R}^n$ and a right-hand side $a_0 \in \mathbb{R}$. Any member of the half-space $\{x \in \mathbb{R}^n \mid a^\top x \le a_0\}$ is said to *satisfy* the inequality and all other points are said to *violate* it. An inequality is *valid for* $\mathcal{P}^I$ if all members of $\mathcal{P}^I$ satisfy it.

A valid inequality (a, a_0) is called *improving* for a given MILP if

$$\min\{c^\top x \mid x \in \mathcal{P}, a^\top x \le a_0\} > \min\{c^\top x \mid x \in \mathcal{P}\}.$$

A necessary and sufficient condition for an inequality to be improving is that it be violated by all optimal solutions to the LP relaxation, so violation of the fractional solution $\hat{x} \in \mathbb{R}^n$ generated by solution of the LP relaxation of a MILP is necessary for a valid inequality to be improving. If a given valid inequality violated by $\hat{x}$ is not improving, adding it to the current LP relaxation will still result in the generation of a new fractional solution, however, and might, in turn, result in the generation of additional candidate inequalities. By repeatedly searching for violated valid inequalities and using them to augment the LP relaxation, the bound may be improved significantly. If such an iterative scheme for improving the bound is utilized during the processing of each search tree node, the overall method is called *branch and cut*. Generation of valid inequalities in the root node only is called *cut and branch*. Branch and cut is the method implemented by the vast majority of solvers today.

Valid inequalities that are necessary to the description of $\text{conv}(\mathcal{P}^I)$ are called *facet-defining inequalities*. Because they provide the closest possible approximation of $\text{conv}(\mathcal{P}^I)$, facet-defining inequalities are typically very effective at improving the lower bound. They are, however, difficult to generate in general. For an arbitrary vector $\hat{x} \in \mathbb{R}^n$ and polyhedron $\mathcal{R} \subseteq \mathbb{R}^n$, the problem of either finding a facet-defining inequality (a, a_0) violated by $\hat{x}$ or proving that $\hat{x} \in \mathcal{R}$ is called the *facet identification problem*. The facet identification problem for a given polyhedron is polynomially equivalent to optimizing over the same polyhedron [37], so finding a facet-defining inequality violated by an arbitrary vector is in general as hard as solving the MILP itself. The problem of generating a valid inequality violated by a given fractional solution, whether facet-defining or not, is called the *separation problem*.

To deal with the difficulty of the facet identification problem, a common approach is to generate valid inequalities (possibly facet-defining) for the convex hull of solutions to a relaxation of the instance. In the following paragraphs, we describe commonly arising relaxations of general MILPs and classes of valid inequalities that can be derived from them. In most cases, the relaxations come from substructures that are not present in all MILPs, which means that the associated classes of valid inequalities cannot always be generated. Two exceptions are the Gomory and MIR inequalities, which can be generated for all MILP instances. Most solvers will try to determine automatically which substructures exist and activate the appropriate subroutines for cut generation. However, this is difficult at best. Therefore, if certain substructures are known *not* to exist a priori, then it is worthwhile for the user to disable subroutines that generate the corresponding classes of inequalities. On the other hand, if a particular substructure *is* known to exist, it is worthwhile to seek out a solver that can take advantage of this fact. The classes of valid inequalities covered here are those employed by at least one of the noncommercial solvers that we describe in Section 10.4.

Although subroutines for generating valid inequalities are generally integrated into the solver itself, free-standing subroutines for generating inequalities valid for generic MILPs can be obtained from the open-source Cut Generation Library (CGL), available for download as part of the Computational Infrastructure for Operations Research (COIN-OR) software suite [19]. The library is provided by the COIN-OR Foundation and includes separation routines for most of the classes of valid inequalities reviewed here. Three of the MILP solvers discussed in Section 10.4 make use of generic separation routines that are part of the CGL.

Knapsack Cover Inequalities. Often, a MILP has a row i of the form

$$\sum_{j \in B} a_{ij} x_j \leq b_i. \tag{10.4}$$

We assume without loss of generality that $a_{ij} > 0$ for all $j \in B$ (if not, we can complement the variables for which $a_{ij} < 0$). Considering only (10.4), we have a relaxation of the MILP, called the $0 - 1$ *knapsack problem*, with the feasible region

$$\mathcal{P}^{\text{knap}} = \left\{ x \in \{0, 1\}^B \mid \sum_{j \in B} a_j x_j \leq b \right\}.$$

Note that in all of the relaxations we consider, the feasible region is implicitly assumed to be contained in the space $\mathbb{Z}^p \times \mathbb{R}^{n-p}$, and the variables not explicitly present in the relaxation are treated as free. Many researchers have studied the structure of the knapsack problem and have derived classes of facet-defining inequalities for it [7, 42, 83].

A set $C \subseteq B$ is called a *cover* if $\Sigma_{j \in C} a_j > b$. A cover C is *minimal* if there does not exist a $k \in C$ such that $C \setminus \{k\}$ is also a cover. For any cover C, we must have

$$\sum_{j \in C} x_j \leq |C| - 1$$

for all $x \in \mathcal{P}^{\text{knap}}$. This class of valid inequalities are called *cover inequalities*. In general, these inequalities are not facet-defining for $\mathcal{P}^{\text{knap}}$, but they can be strengthened through a procedure called *lifting*. The interested reader is referred to the paper by Gu, Nemhauser, and Savelsbergh [40].

GUB Cover Inequalities. A *generalized upper bound* (GUB) inequality is an inequality of the form

$$\sum_{j \in Q} x_j \leq 1,$$

where $Q \subseteq B$. When MILP contains a knapsack row i of the form (10.4) and a set of GUB inequalities defined by disjoint sets $Q_k \subset B$ for $k \in K$, we obtain a relaxation of MILP with the feasible region

$$\mathcal{P}^{\text{GUB}} = \left\{ x \in \{0,1\}^B \mid \sum_{j \in B} a_{ij} x_j \leq b_i, \sum_{j \in Q_k} x_j \leq 1 \quad \forall k \in K \right\}.$$

This class of inequalities models a situation in which a subset of items not violating the knapsack inequality (10.4) must be selected, and within that subset of items, at most one element from each of the subsets Q_i can be selected. A *GUB cover* C_G is a cover that obeys the GUB constraints, (i.e., no two elements of the cover belong to the same Q_i). For any GUB cover C_G, the inequality

$$\sum_{j \in C_G} x_j \leq |C_G| - 1$$

is valid for $\mathcal{P}^{\text{GUB}}$. Again, a lifting procedure, in this case one that takes into account the GUB constraints, can lead to significant strengthening of these inequalities. For more details of the inequalities and lifting procedure, see the paper of Wolsey [84] and the paper of Gu, Nemhauser, and Savelsbergh [38].

Flow Cover Inequalities. Another important type of inequality commonly found in MILP problems is a *variable upper bound*. A variable upper bound is an inequality of the form

$$x_j \leq U_j x_k,$$

where x_j is a continuous variable ($j > p$), x_k is a binary variable ($k \in B$), and U_j is an upper bound on variable x_j. Such inequalities model the implication $x_k = 0 \Rightarrow x_j = 0$. Variable upper bound inequalities are often used to model a fixed charge associated with assignment of a positive value to variable x_j, and they are particularly prevalent in network flow models. In such models, we can often identify a relaxation with the following feasible region:

$$\mathcal{P}^{\text{flow}} = \left\{ (x^1, x^2) \in \mathbb{R}_+^{n'} \times \{0,1\}^{n'} \mid \sum_{j \in N^+} x_j^1 \right.$$
$$\left. - \sum_{j \in N^-} x_j^1 \leq d, \quad x_j^1 \leq U_j x_j^2, \quad j \in N \right\},$$

where N^+ and N^- are appropriate sets of indices, $N = N^+ \cup N^-$, and $n' = |N|$. A set $C = C^+ \cup C^-$ is called a *flow cover* if $C^+ \subseteq N^+$, $C^- \subseteq N^-$ and $\Sigma_{j \in C^+} m_j - \Sigma_{j \in C^-} m_j > d$. For any flow cover C, the inequality

$$\sum_{j \in C^+} x_j + \sum_{j \in C^{++}} (m_j - \lambda)(1 - y_j) \leq d + \sum_{j \in C^-} m_j + \sum_{j \in L^-} \lambda y_j + \sum_{j \in L^{--}} x_j,$$

where $\lambda = \Sigma_{j \in C^+} m_j - \Sigma_{j \in C^-} m_j - d$, $C^{++} = \{j \in C^+ \mid m_j > \lambda\}$, $L^- \subseteq (N^- \backslash C^-)$ and $m_j > \lambda$ for $j \in L^-$, and $L^{--} = N^- \backslash (L^- \cup C^-)$, is called a *simple generalized flow cover inequality* and is valid for $\mathcal{P}^{\text{flow}}$. Just as with knapsack cover inequalities, these inequalities can be strengthened through lifting to obtain an inequality called the *lifted simple generalized flow cover inequality*. The full details of obtaining such inequalities are given by Gu, Nemhauser, and Savelsbergh [39].

Clique Inequalities. Many MILPs contain logical restrictions on pairs of binary variables such as $x_i = 1 \Rightarrow x_j = 0$. In such cases, an auxiliary data structure, called a *conflict graph*, can be used to capture these logical conditions and further exploit them [4]. The conflict graph is a graph with vertex set B, and an edge between each pair of nodes that correspond to a pair of variables that cannot simultaneously have the value one in any optimal solution. The logical restrictions from which the conflict graph is derived may be present explicitly in the original model (for example, GUB inequalities lead directly to edges in the conflict graph), or may be discovered during preprocessing (see [5, 74]).

Because any feasible solution $x \in \mathcal{P}^I$ must induce a vertex packing in the conflict graph, inequalities valid for the vertex packing polytope of the conflict graph are also valid for the MILP instance from which the conflict graph was derived. Classes of inequalities valid for the vertex packing polytope have been studied by a number of authors [17, 43, 63, 65]. As an example, if C is the set of indices of nodes that form a clique in a conflict graph for a MILP instance, then the *clique inequality*

$$\sum_{j \in C} x_j \le 1$$

is satisfied by all $x \in \mathcal{P}^I$. Alternatively, if O is a cycle in a conflict graph for a MILP instance, and $|O|$ is odd, then the *odd-hole inequality*

$$\sum_{j \in O} x_j \le \frac{|O| - 1}{2}.$$

is also satisfied by all $x \in \mathcal{P}^I$. Again, these inequalities can be strengthened by lifting [63, 64].

Implication Inequalities. In some cases, the logical implications discovered during preprocessing are not between pairs of binary variables (in which case clique and odd-hole inequalities can be derived), but between a binary variable and a continuous variable. These logical implications can be enforced using inequalities known as *implication inequalities*. If x_i is a binary variable and x_j is a continuous variable with upper bound U, the implication

$$x_i = 0 \Rightarrow x_j \le \alpha$$

yields the implication inequality

$$x_j \le \alpha + (U - \alpha)x_i.$$

Other implication inequalities can also be derived. For more details, see the paper of Savelsbergh [74].

Gomory Inequalities. In contrast to the classes of inequalities we have reviewed so far, Gomory inequalities are generic, in the sense that they do not require the presence of any particular substructure other than integrality and nonnegativity. This means they can be derived for any MILP. Gomory inequalities are easy to generate in LP-based branch and bound. After solving the current LP relaxation, we obtain an optimal basis matrix $A_B \in \mathbb{R}^{m \times m}$. The vector $A_B^{-1}b$ yields the values of the basic variables in the current fractional solution $\hat{x}$. Assuming $\hat{x} \notin \mathcal{P}^I$, we must have $(A_B^{-1}b)_i \notin \mathbb{Z}$ for some i between 1 and m. Taking u above to be the i th row of A_B^{-1},

$$x_l + \sum_{j \in NB^I} u A_j x_j + \sum_{k \in NB^C} u A_k x_k = ub, \tag{10.5}$$

for all $x \in \mathcal{P}^I$, where NB^I is the set of nonbasic integer variables, NB^C is the set of nonbasic continuous variables, and A_l is the l th column of A. Let $f_j = uA_j - \lfloor uA_j \rfloor$ for $j \in NB^I \cup NB^C$, and let $f_0 = ub - \lfloor ub \rfloor$; then the inequality

$$\sum_{\substack{j \in NB^I: \\ f_j \le f_0}} f_j x_j + \sum_{\substack{j \in NB^I: \\ f_j > f_0}} \frac{f_0(1 - f_j)}{1 - f_0} x_j + \sum_{\substack{j \in NB^C: \\ ua_j > 0}} ua_j x_j - \sum_{\substack{j \in NB^C: \\ ua_j < 0}} \frac{f_0}{1 - f_0} ua_j x_j \ge f_0, \tag{10.6}$$

is called the *Gomory mixed-integer inequality* and is satisfied by all $x \in \mathcal{P}^I$, but not satisfied by the current fractional solution $\hat{x}$. It was first derived by Gomory [36], but also can be derived by a simple disjunctive argument, as in Balas et al. [8].

Mixed-Integer Rounding Inequalities. Gomory mixed-integer inequalities can be viewed as a special case of a more general class of inequalities known as *mixed-integer rounding inequalities*. Mixed-integer rounding inequalities are obtained as valid inequalities for the relaxed feasible region

$$\mathcal{P}^{MIR} = \left\{ (x^1, x^2, x^3) \in \mathbb{R}_+^1 \times \mathbb{R}_+^1 \times \mathbb{Z}_+^{n'} \mid \sum_{j=1}^{n'} a_j x_j^3 + x^1 \le b + x^2 \right\}. \tag{10.7}$$

The mixed-integer rounding inequality

$$\sum_{j=1}^{n'} \left(\lfloor a_j \rfloor + \frac{\max\{f_j - f, 0\}}{1 - f} \right) x_j^3 \le \lfloor b \rfloor + \frac{x^2}{1 - f},$$

where $f = b - \lfloor b \rfloor$, $f_j = a_j - \lfloor a_j \rfloor$ for $j = 1, 2, \ldots, n'$, is valid for $\mathcal{P}^{MIR}$ [55, 61]. Marchand [54] established that many classes of valid inequalities for structured problem instances are special cases of mixed-integer rounding inequalities, including certain subclasses of the lifted flow cover inequalities described above.

The process of generating a mixed-integer rounding inequality is a three-step procedure. First, rows of the constraint matrix are aggregated. Second, bound substitution of the simple or variable upper and lower bounds is performed, and variables are complemented to produce a relaxation of the form (10.7). Third, a heuristic separation procedure is used to find mixed-integer rounding inequalities valid for some such relaxation and violated by the current fractional solution. Marchand and Wolsey [55] discuss the application of mixed-integer rounding inequalities in detail.

10.2.1.3 *Reduced Cost Tightening*

After the LP relaxation of MILP is solved, the reduced costs of nonbasic integer variables can be used to tighten bounds on integer variables for the subtree rooted at that node. Although we have assumed a problem given in standard form, upper and lower bounds on variable values are typically present and are handled implicitly. Such bound constraints take the form $l \leq x \leq u$ for $l, u \in \mathbb{R}^n$ for all $x \in \mathcal{P}^I$. Even if no such bound constraints are initially present, they may be introduced during branching. Let $\bar{c}_j$ be the reduced cost of nonbasic integer variable j, obtained after solution of the LP relaxation of a given subproblem and let $\hat{x} \in \mathbb{R}^n$ be an optimal fractional solution. If $\hat{x}_j = l_j \in \mathbb{Z}$ and $\gamma \in \mathbb{R}_+$ is such that $c^\top \hat{x} + \gamma \bar{c}_j = \beta$, where β is the objective function value of the current incumbent, then $x_j \leq l_j + \lfloor \gamma \rfloor$ in any optimal solution. Hence, we can replace the previous upper bound u_j with $\min(u_j, l_j + \lfloor \gamma \rfloor)$. The same procedure can be used to potentially improve the lower bounds. This is an elementary form of preprocessing, but can be very effective when combined with other forms of logical preprocessing, especially when the optimality gap is small. Note that if this tightening takes place in the root node, it is valid everywhere and can be considered an improvement of the original model. Some MILP solvers store the reduced costs from the root LP relaxation, and use them to perform this preprocessing whenever a new incumbent is found.

10.2.1.4 *Column Generation*

A technique for improving the lower bound that can be seen as "dual" to the dynamic generation of valid inequalities is that of *column generation*. Most column generation algorithms can be viewed as a form of Dantzig-Wolfe decomposition, so we concentrate here on that technique. Consider a relaxation of the original MILP with feasible set $\mathcal{F} \supset \mathcal{P}^I$. We assume that $\mathcal{F}$ is finite and that it is possible to effectively optimize over $\mathcal{F}$, but that a minimal description of the convex hull of $\mathcal{F}$ is of exponential size. Let $\mathcal{Q} = \{x \in \mathbb{R}^n \mid Dx = d, x \geq 0\} \supset \mathcal{P}^I$ be a polyhedron that represents the feasible region of a second relaxation whose description is "small" and such that $\mathcal{F} \cap \mathcal{Q} \cap \mathbb{Z}^p \cap \mathbb{R}^{n-p} = \mathcal{P}^I$. We can then reformulate the original integer program as

$$\min\left\{ \sum_{s \in \mathcal{F}} c^\top s \lambda_s \mid \sum_{s \in \mathcal{F}} (Ds)\lambda_s = d, \lambda_s^\top \mathbf{1} = 1, \lambda \geq 0, \lambda \in \mathbb{Z}^{\mathcal{F}} \right\}, \quad (10.8)$$

where $\mathbf{1}$ is a vector of all ones of conformable dimension. Relaxing the integrality restriction on λ, we obtain a linear program whose optimal solution yields the (possibly) improved bound

$$\min_{x \in \mathcal{F} \cap Q} c^\top x \geq \min_{x \in \mathcal{P}} c^\top x. \tag{10.9}$$

Of course, this linear program generally has an exponential number of columns. Therefore, we must generate them dynamically in much the same fashion as valid inequalities are generated in branch and cut. If we solve the above linear program with a subset of the full set of columns to obtain a dual solution u, then the problem of finding the column with the smallest reduced cost among those not already present is equivalent to solving

$$\min_{s \in \mathcal{F}} (c^\top - uD)s, \tag{10.10}$$

which is referred to as the *column generation subproblem*. Because this is an optimization problem over the set $\mathcal{F}$, it can be solved effectively, and hence so can the linear program itself. When employed at each search tree node during branch and bound, the overall technique is called *branch and price*.

Because of the often problem-specific nature of the column generation subproblem, branch and price is frequently implemented using a solver framework. Recently two different groups have undertaken efforts to develop generic frameworks for performing column generation. Vanderbeck [82] is developing a framework for branch and price that will take care of many of the generic algorithmic details and allow the solver to behave essentially as a black box. Ralphs and Galati (see Chapter 4) have undertaken a similar effort in developing DECOMP, a general framework for computing bounds using decomposition within a branch-and-bound procedure. Currently, however, implementing a branch-and-price algorithm is a rather involved procedure that requires a certain degree of technical expertise. Section 10.5.2 also describes the implementation of a branch-and-price algorithm using the BCP framework.

10.2.2 Upper Bounding Methods

In branch and bound, upper bounds are obtained by discovering feasible solutions to the original MILP. Feasible solutions arise naturally if the branch-and-bound algorithm is allowed to run its full course. However, acceleration of the process of finding feasible solutions has three potential benefits. First, the solution process may be terminated prematurely and in such a case, we would like to come away with a solution as close to optimal as possible. Second, an improved upper bound β may lead to generation of fewer subproblems because of earlier fathoming (depending on the search strategy being employed). Third, a good upper bound allows the bounds on integer variables to be tightened on the basis of their reduced cost in the current relaxation (see Section 10.2.1.3). Such tightening can in turn enable additional logical preprocessing and may result in significant improvement to the lower bound as a result.

The process of finding feasible solutions during the search procedure can be accelerated in two ways. The first is to influence the search order, choosing to evaluate and partition nodes that are close to being integer feasible. This technique is further discussed in Section 10.2.4. The second is to use a heuristic procedure, called a *primal heuristic*, to construct a solution. Primal heuristics are applied during the search process and generally take an infeasible fractional solution as input. A very simple heuristic procedure is to round the fractional components of the infeasible solution in an attempt to produce a feasible solution. The current solution may be rounded in many ways and determination of a rounding that maintains feasibility with respect to the constraints $Ax = b$ may be difficult for certain problem classes. A more sophisticated class of primal heuristics, called *pivot and complement*, involves pivoting fractional variables out of the current linear programming basis in order to achieve integrality [9, 59, 11]. Still other classes of primal heuristics use the solution of auxiliary linear programs to construct a solution. One simple, yet effective example of such a heuristic is known as the *diving* heuristic. In the diving heuristic, some integer variables are fixed and the linear program resolved. The fixing and resolving is iterated until either an integral solution is found or the linear program becomes infeasible. Recent successful primal heuristics, such as local branching [29] and RINS [22], combine solution of auxiliary linear programs with methods that control the neighborhood of feasible solutions to be searched.

10.2.3   Branching

Branching is the method by which a MILP is divided into subproblems. In LP-based branch and bound, there are three requirements for the branching method. First, the feasible region must be partitioned in such a way that the resulting subproblems are also MILPs. This means that the subproblems are usually created by imposing additional linear inequalities. Second, the union of the feasible regions of the subproblems must contain at least one optimal solution. Third, because the primary goal of branching is to improve the overall lower bound, it is desirable that the current fractional solution not be contained in any of the members of the partition. Otherwise, the overall lower bound will not be improved.

Given a fractional solution to the LP relaxation $\hat{x} \in \mathbb{R}^n$, an obvious way to fulfill the above requirements is to choose an index $j < p$ such that $\hat{x}_j \notin \mathbb{Z}$ and to create two subproblems, one by imposing an upper bound of $\lfloor \hat{x}_j \rfloor$ on variable j and a second by imposing a lower bound of $\lceil \hat{x}_j \rceil$. This partitioning is valid because any feasible solution must satisfy one of these two linear constraints. Furthermore, $\hat{x}$ is not feasible for either of the resulting subproblems. This partitioning procedure is known as *branching on a variable*.

Typically, there are many integer variables with fractional values, so we must have a method for deciding which one to choose. A primary goal of branching is to improve the lower bound of the resulting relaxations. The most straightforward branching methods are those that choose the branching variable based solely on the current fractional solution and do not use any auxiliary information. Branching on the variable with the largest fractional part, the first variable (by index) that is fractional, or the last variable (by index) that is fractional are examples of such procedures. These rules tend to be too myopic to be effective, so many solvers use more sophisticated approaches. Such approaches fall into two general categories: *forward-looking methods* and *backward-looking methods*. Both types of methods try to choose the best partitioning by predicting, for a given candidate partitioning, how much the lower bound will actually be improved. Forward-looking methods generate this prediction based solely on locally generated information obtained by "presolving" candidate subproblems. Backward-looking methods take into account the results of previous partitionings to predict the effect of future ones. Of course, as one might expect, hybrids that combine these two basic approaches also exist [2].

A simple forward-looking method is the *penalty method* of Driebeek [26], which implicitly performs one dual simplex pivot to generate a lower bound on the bound improvement that could be obtained by branching on a given variable. Tomlin [79] improved on this idea by considering the integrality of the variables. *Strong branching* is an extension of this basic concept in which the solver explicitly performs a fixed and limited number of dual simplex pivots on the LP relaxations of each of the children that result from branching on a given variable. This is called *presolving* and again provides a bound on the improvement one might see as a result of a given choice. The effectiveness of strong branching was first demonstrated by Applegate et al. in their work on the traveling salesman problem [3] and has since become a mainstay for solving difficult combinatorial problems. An important aspect of strong branching is that presolving a given candidate variable is a relatively expensive operation, so it is typically not possible to presolve all candidates. The procedure is therefore usually accomplished in two phases. In the first phase, a small set of candidates is chosen (usually based on one of the simple methods described earlier). In the second phase, each of these candidates is presolved, and the final choice is made using one of the selection rules to be described below.

Backward-looking methods generally depend on the computation of *pseudocosts* [14] to maintain a history of the effect of branching on a given variable. Such procedures are based on the notion that each variable may be branched on multiple times during the search and the effect will be similar each time. Pseudocosts are defined as follows. With each integer variable j, we associate two quantities, P_j^- and P_j^+, that estimate the per unit increase in objective function value if we fix variable j to its floor and ceiling, respectively. Suppose that $f_j = \hat{x}_j - \lfloor \hat{x}_j \rfloor > 0$. Then by branching on variable j, we will estimate an increase of $D_j^- = P_j^- f_j$ on the "down branch" and an increase of $D_j^+ = P_j^+ (1 - f_j)$ on the "up branch".

The most important aspect of using pseudocosts is the method of obtaining the values P_j^- and P_j^+ for variable j. A popular way to obtain these values is to simply observe and record the true increase in objective function value whenever variable j is chosen as the branching variable. For example, if a given subproblem had lower bound z_{LP} and its children had lower bounds z_{LP}^- and z_{LP}^+ after branching on variable j, then the pseudocosts would be computed as

$$P_j^- = \frac{z_{LP}^- - z_{LP}}{f_j} \qquad P_j^+ = \frac{z_{LP}^+ - z_{LP}}{1 - f_j}, \tag{10.11}$$

where f_j is the fractional part of the value of variable j in the solution to the LP relaxation of the parent. The pseudocosts may be updated using the first observation, the last observation, or by averaging all observations. In contrast to the estimates generated in strong branching, generation of pseudocost estimates is inexpensive, so they are typically calculated for all variables.

Whether using a forward-looking or a backward-looking method, the final step is to select the branching variable. The goal is to maximize the improvement in the lower bound from the parent to each of its children. Because each parent has two (or more) children, however, no unique metric exists for this change. Suggestions in the literature have included maximization of the sum of the changes on both branches [35], maximization of the smaller of the two changes [12], or a combination of the two [27].

More general methods of branching can be obtained by branching on other disjunctions. For any vector $a \in \mathbb{Z}^n$ whose last $n - p$ entries are zero, we must have $a^\top x \in \mathbb{Z}$ for all $x \in \mathcal{P}^I$. Thus, if $a\hat{x} \notin \mathbb{Z}$, a can be used to produce a disjunction by imposing the constraint $a^\top x \leq \lfloor a^\top \hat{x} \rfloor$ in one subproblem and the constraint $a^\top x \geq \lceil a^\top \hat{x} \rceil$ in the other subproblem. This is known as *branching on a hyperplane*. Typically, branching on hyperplanes is a problem-specific method that exploits special structure, but it can be made generic by maintenance of a pool of inequalities that are slack in the current relaxation as branching candidates.

An example of branching on hyperplanes using special structure is *GUB branching*. If the MILP contains rows of the form

$$\sum_{j \in G} x_j = 1,$$

where $G \subseteq B$, then a valid partitioning is obtained by selection of a nonempty subset G^0 of G, and enforcement of the linear constraint $\sum_{j \in G^0} x_j = 0$ in one subproblem and the constraint $\sum_{j \in G \setminus G^0} x_j = 0$ in the other subproblem. These are linear constraints that partition the set of feasible solutions, and the current LP solution $\hat{x}$ is excluded from both resulting subproblems if G^0 is chosen so that $0 \leq \sum_{j \in G^0} \hat{x}_j < 1$. GUB branching is a special case of branching on *special ordered sets* (SOS)[1] [13]. Special ordered sets of variables can also be used in the minimization of separable piecewise-linear nonconvex functions.

Because of their open nature, noncommercial software packages are often more flexible and extensible than their commercial counterparts. This flexibility is perhaps most evident in the array of advanced branching mechanisms that can be implemented using the open source and noncommercial frameworks we describe in Section 10.4. Using a solver framework with advanced customized branching options, it is possible, for instance, to branch directly on

---

[1] Some authors refer to GUBs as special ordered sets of type 1.

a disjunction, rather than introducing auxiliary integer variables. An important example of this is in the handling of semicontinuous variables, in which a variable is constrained to take either the value 0 or a value larger than a parameter K. Additionally, frameworks can make it easy for the user to specify prioritization schemes for branching on integer variables or to implement complex partitioning schemes based on multiple disjunctions.

10.2.4 Search Strategy

As mentioned in Section 10.2.3, branching decisions are made with the goal of improving the lower bound. In selecting the order in which the candidate subproblems should be processed, however, our focus may be on improving either the upper bound, the lower bound, or both. Search strategies, or *node selection methods*, can be categorized as either *static* methods, *estimate-based* methods, *two-phase* methods, or *hybrid* methods.

Static node selection methods employ a fixed rule for selecting the next subproblem to process. A popular static method is *best-first search*, which chooses the candidate node with the smallest lower bound. Because of the fathoming rule employed in branch and bound, a best-first search strategy ensures that no subproblem with a lower bound above the optimal solution value can ever be processed. Therefore, the best-first strategy minimizes the number of subproblems processed and improves the lower bound quickly. However, this comes at the price of sacrificing improvements to the upper bound. In fact, the upper bound will only change when an optimal solution is located. At the other extreme, *depth-first search* chooses the next candidate to be a node at maximum depth in the tree. In contrast to best-first search, which will produce no suboptimal solutions, depth-first search tends to produce many suboptimal solutions, typically early in the search process, because such solutions tend to occur deep in the tree. This allows the upper bound to be improved quickly. Depth-first search also has the advantage that the change in the relaxation being solved from subproblem to subproblem is very slight, so the relaxations tend to solve more quickly when compared to best-first search. Some solvers also allow the search tree to be explored in a breadth-first fashion, but this method offers little advantage over best-first search.

Neither best-first search nor depth-first search make any intelligent attempt to select nodes that may lead to improved feasible solutions. Estimate-based methods such as the *best-projection method* [31, 58] and the *best-estimate method* [14] are improvements in this regard. The best-projection method measures the overall "quality" of a node by combining its lower bound with the degree of integer infeasibility of the current solution. Alternatively, the best-estimate method combines a node's lower bound, integer infeasibility, and pseudocost information to rank the desirability of exploring a node.

Because we have two goals in node selection — finding "good" feasible solutions (i.e., improving the upper bound) and proving that the current incumbent is in fact a "good" solution (i.e., improving the lower bound) — it is natural to develop node selection strategies that switch from one goal to the other during the course of the algorithm. This results in a two-phase search strategy. In the first phase, we try to determine "good" feasible solutions, while in the second phase, we are interested in proving this goodness. Perhaps the simplest "two-phase" algorithm is to perform depth-first search until a feasible solution is found, and then switch to best-first search. A variant of this two-phase algorithm is used by many of the noncommercial solvers that we describe in Section 10.4.

Hybrid methods also combine two or more node selection methods, but in a different manner. In a typical hybrid method, the search tree is explored in a depth-first manner until the lower bound of the child subproblem being explored rises above a prescribed level in comparison to the overall lower or upper bounds. After this, a new subproblem is selected by a different criterion (e.g., best-first or best-estimate), and the depth-first process is repeated. For an in-depth discussion of search strategies for mixed-integer programming, see the paper of Linderoth and Savelsbergh [49].

10.3 User Interfaces

An important aspect of the design of software for solving MILPs is the user interface, which determines both the way in which the user interacts with the solver and the form in which the MILP instance must be specified. The range of purposes for noncommercial MILP software is quite large, so it stands to reason that the number of user interface types is also large. In this section, we broadly categorize the software packages available. The categorization provided here is certainly not perfect — some tools fall between categories or into multiple categories. However, it does represent the most typical ways in which software packages for MILP are employed in practice.

10.3.1 Black-Box Solvers

Many users simply want a "black box" that takes a given MILP as input and returns a solution as output. For such black-box applications, the user typically interacts with the solver through a command-line interface or an interactive shell, invoking the solver by passing the name of a file containing a description of the instance to be solved. One of the main differences between various black-box solvers from the user's perspective is the format in which the model can be specified to the solver. In the two sections below, we describe two of the most common input mechanisms — raw (uninterpreted) file formats and modeling language (interpreted) file formats. Table 10.1 in Section 10.4 lists the packages covered in this chapter that function as black-box solvers, along with the file formats they accept and modeling languages they support. In Section 10.6, we provide computational results that compare all of these solvers over a wide range of instances.

10.3.1.1 Raw File Formats

One of the first interfaces used for black-box solvers was a standard file format for specifying a single instance of a mathematical program. Such a file format provides a structured way of constructing a file containing a description of the constraint matrix and rim vectors (objective function vector, right hand side vector, and variable lower and upper bound vectors) in a form that can be easily read by the solver. The oldest and most pervasive file format is the longstanding Mathematical Programming System (MPS) format, developed by IBM in the 1970s. In MPS format, the file is divided into sections, each of which specifies one of the elements of the input, such as the constraint matrix, the right-hand side, the objective function, and upper and lower bounds on the variables. MPS is a column-oriented format, which means that the constraint matrix is specified column-by-column in the MPS file. Another popular format, LP format, is similar to MPS in that the format consists of a text file divided into different sections, each of which specifies one of the elements of the input. However, LP format is a row-oriented format, so the constraint matrix is specified one row at a time in the file. This format tends to be slightly more readable by humans.

Because MPS was adopted as the *de facto* standard several decades ago, there has not been much deviation from this basic approach. MPS, however, is not an easily extensible standard, and is only well-suited for specifying integer and linear models. Several replacements based on the extensible markup language (XML), a wide-ranging standard for portable data interchange, have been proposed. One of the most well-developed of these is an open standard called LPFML [32]. The biggest advantage of formats based on XML is that they are far more extensible and are based on an established standard with broad support and well-developed tools.

10.3.1.2 Modeling Languages

Despite their wide and persistent use, raw file formats for specifying instances have many disadvantages. The files can be tedious to generate, cumbersome to work with, extremely difficult to debug, and are not easily readable by humans. For these reasons, most users prefer to work with a *modeling language*. Modeling languages allow the user to specify a model in a more intuitive (e.g., algebraic) format. An interface layer then interprets the model file and translates it into a raw format that the underlying solver can interpret directly. Another powerful feature of modeling languages is that they allow for the separation of the model specification from the instance data.

Full-featured modeling languages are similar to generic programming languages, such as C and C++, in that they have constructs such as loops and conditional expressions. They also contain features designed specifically to allow the user to specify mathematical models in a more natural, human-readable form. Two modeling language systems that are freely available are ZIMPL [87] and GNU Mathprog (GMPL) [53]. ZIMPL is a stand-alone parser that reads in a file format similar to the popular commercial modeling language AMPL [33] and outputs the specified math program in either MPS or LP format. GMPL is the GNU Math Programming Language, which is again similar to AMPL. The parser for GMPL is included as part of the GLPK package described in Section 10.4.5, but it can also be used as a free-standing parser.

10.3.2 Callable Libraries

A more flexible mechanism for invoking a MILP solver is through a callable library interface. Most often, callable libraries are still treated essentially as a "black box," but they can be invoked directly from user code, allowing the development of custom applications capable of generating a model, invoking the solver directly without user intervention, parsing the output and interpreting the results. Callable libraries also make possible the solution of more than one instance within a single invocation of an application or the use of the solver as a subroutine within a larger implementation. The interfaces to the callable libraries discussed in Section 10.4 are implemented in either C or C++, with each solver generally having its own Application Programming Interface (API). The column labeled *Callable API* in Table 10.1 of Section 10.4 indicates which of the software packages discussed in this chapter have callable library interfaces and the type of interface available.

The fact that each solver has its own API makes development of portable code difficult, as this fosters an inherent dependence on one particular API. Recently, however, two open standards for calling solvers have been developed that remove the dependence on a particular solver's API. These are discussed below.

10.3.2.1 Open Solver Interface

The Open Solver Interface (OSI), part of the COIN-OR software suite mentioned earlier, is a standard C++ interface for invoking solvers for LPs and MILPs [51]. The OSI consists of a C++ base class with containers for storing instance data, as well as a standard set of problem import, export, modification, solution, and query routines. Each supported solver has a corresponding derived class that implements the methods of the base class and translates the standard calls into native calls to the API of the solver in question. Thus, a code that uses only calls from the OSI base class could be easily interfaced with any supported solver without change. At the time of this writing, eleven commercial and noncommercial solvers with OSI implementations are available, including several of the solvers reviewed in Section 10.4.

10.3.2.2 Object-Oriented Interfaces

In an object-oriented interface, there is a mapping between the mathematical modeling objects that comprise a MILP instance (variables, constraints, etc.) and programming language objects. Using this mapping, MILP models can be easily built in a natural way directly within C++ code. The commercial package called Concert Technology [45] by ILOG was perhaps the first example of such an object-oriented interface. FLOPC++ [44] is an open source C++ object-oriented interface for algebraic modeling of LPs and MILPs that provides functionality similar to Concert. FLOPC++ allows linear models to be specified in a declarative style, similar to algebraic modeling languages such as GAMS and AMPL, within a C++ program. As a result, the traditional strengths of algebraic modeling languages, such as the ability to declare a model in a human-readable format, are preserved, while still allowing the user to embed model generation and solution procedures within a larger applications. To achieve solver independence, FLOPC++ uses the OSI to access the underlying solver, and may therefore be linked to any solver with an OSI implementation. Another interesting interface that allows users to model LP instances in the python language is PuLP [73].

10.3.3 Solver Frameworks

A solver framework is an implementation of a branch-and-bound, branch-and-cut, or branch-and-price algorithm with hooks that allow the user to provide custom implementations of certain aspects of the algorithm. For instance, the user may wish to provide a custom branching rule or problem-specific valid inequalities. The customization is generally accomplished either through the use of C language callback functions, or through a C++ interface in which the user must derive certain base classes and override default implementations for the desired functions. Some frameworks have the ability to function as black-box solvers, but others, such as BCP and ABACUS, do not include default implementations of certain algorithmic components. Table 10.1 in Section 10.4 indicates the frameworks available and their style of customization interface.

10.4 MILP Software

In this section, we summarize the features of the noncommercial software packages available for solving MILPs. Table 10.1 to Table 10.3 summarize the packages reviewed here. In Table 10.1, the columns have the following meanings:

- *Version Number*: The version of the software reviewed for this chapter. Note that BCP does not assign version numbers, so we have listed the date obtained instead.
- *LP Solver*: The LP software used to solve the relaxations arising during the algorithm. The MILP solvers listed as OSI-compliant can use any LP solver with an OSI interface. ABACUS can use either CPLEX, SOPLEX, or XPRESS-MP.
- *File Format*: The file formats accepted by packages that include a black-box solver. File formats are discussed in Section 10.3.1.1.
- *Callable API*: The language in which the callable library interface is implemented (if the package in question has one). Some packages

support more than one interface. Two of the solvers, SYMPHONY and GLPK, can also be called through their own OSI implementation.

- *Framework API*: For those packages that are considered frameworks, this indicates how the callback functions must be implemented — through a C or a C^{++} interface.
- *User's Manual*: Indicates whether the package has a user's manual.

TABLE 10.1

List of Solvers and Main Features

	Version Number	LP Solver	File Format	Callable API	Framework API	User's Manual
ABACUS	2.3	C/S/X	no	none	$C++$	yes
BCP	11/1/04	OSI	no	none	$C++$	yes
bonsaiG	2.8	DYLP	MPS	none	none	yes
CBC	0.70	OSI	MPS	C++/C	$C++$	no
GLPK	4.2	GLPK	MPS/GMPL	OSI/C	none	yes
lp_solve	5.1	lp_solve	MPS/LP/GMPL	C/VB/Java	none	yes
MINTO	3.1	OSI	MPS/AMPL	none	C	yes
SYMPHONY	5.0	OSI	MPS/GMPL	OSI/C	C	yes

Table 10.2 indicates the algorithmic features of each solver, including whether the solver has a preprocessor, whether it can dynamically generate valid inequalities, whether it can perform column generation, whether it includes primal heuristics, and what branching and search strategies are available. For the column that indicates available branching methods, the letters stand for the following methods:

- e: pseudocost branching
- f: branching on the variables with the largest fractional part
- h: branching on hyperplanes
- g: GUB branching

TABLE 10.2

Algorithmic Features of Solvers

	Preproc	Built-in Cut Generation	Column Generation	Primal Heuristic	Branching Rules	Search Strategy
ABACUS	no	no	yes	no	f,h,s	b,r,d,2(d,b)
BCP	no	no	yes	no	f,h,s	h(d,b)
bonsaiG	no	no	no	no	p	h(d,b)
CBC	yes	yes	no	yes	e,f,g,h,s,x	2(d,p)
GLPK	no	no	no	no	i,p	b,d,p
lp_solve	no	no	no	no	e,f,i,x	d,r,e,2(d,r)
MINTO	yes	yes	yes	yes	e,f,g,p,s	b,d,e,h(d,e)
SYMPHONY	no	yes	yes	no	e,f,h,p,s	b,r,d,h(d,b)

- i: branching on first or last fractional variable (by index)
- p: penalty method
- s: strong branching
- x: SOS(2) branching and branching on semicontinuous variables

For the column that indicates available search strategies, the codes stand for the following:

- b: best-first
- d: depth-first
- e: best-estimate
- p: best-projection
- r: breadth-first
- h(x,z): a hybrid method switching from strategy 'x' to strategy 'z'
- 2(x,z): a two-phase method switching from strategy 'x' to strategy 'z'

Finally, Table 10.3 indicates the classes of valid inequalities generated by those solvers that generate valid inequalities.

In the following sections, we provide an overview of each solver, then describe the user interface, and finally describe the features of the underlying algorithm in terms of the four categories listed in Section 10.2. In what follows, the reader should keep in mind that solver performance can vary significantly with different parameter settings, and it is unlikely that one set of parameters will work best for all classes of MILP instances. When deciding on a MILP package to use, users are well-advised to consider the ability of a packages to meet their performance requirements through customization and parameter tuning. An additional caveat about performance is that MILP solver performance can be affected significantly by the speed with which the LP relaxations are solved, so users may need to pay special attention to the parameter tuning of the underlying LP solver as well.

TABLE 10.3

Classes of Valid Inequalities Generated by Black-Box Solvers

Name	Knapsack	GUB	Flow	Clique	Implication	Gomory	MIR
CBC	yes	no	yes	yes	yes	yes	yes
MINTO	yes	yes	yes	yes	yes	no	no
SYMPHONY	yes	no	yes	yes	yes	yes	yes

10.4.1 ABACUS

10.4.1.1 Overview

ABACUS [46] is a pure solver framework written in C++. It has a flexible, object-oriented design that supports the implementation of a wide variety of sophisticated and powerful variants of branch and bound. The object-oriented design of the library is similar in concept to BCP, described below. From the user's perspective, the framework is centered around $C++$ objects that represent the basic building blocks of a mathematical model — constraints and variables. The user can dynamically generate variables and valid inequalities by defining classes derived from the library's base classes. ABACUS supports the simultaneous generation of variables and valid inequalities for users requiring this level of sophistication. Another notable feature of ABACUS is its very general implementation of *object pools* for storing previously generated constraints and variables for later use.

ABACUS was a commercial code for some time, but has recently been released under the open source [78] GNU Library General Public License (LGPL). Because of the generality of its treatment of dynamically generated classes of constraints and variables, it is one of the most full-featured solver frameworks available. ABACUS does not, however, have a callable library interface and it cannot be used as a black-box solver, though it can be called recursively. It comes with complete documentation and a tutorial that shows how to use the code. Compared to the similar MILP framework BCP, ABACUS has a somewhat cleaner interface, with fewer classes and a more straightforward object-oriented structure. The target audience for ABACUS consists of sophisticated users who need a powerful framework for implementing advanced versions of branch and bound, but who do not need a callable library interface.

10.4.1.2 User Interface

There are four main $C++$ base classes from which the user may derive problem-specific implementations in order to develop a custom solver. The base classes are the following:

- `ABA_VARIABLE`: The base class for defining problem-specific classes of variables.
- `ABA_CONSTRAINT`: The base class for defining problem-specific classes of constraints.
- `ABA_MASTER`: The base class for storing problem data and initializing the root node.
- `ABA_SUB`: The base class for methods related to the processing of a search tree node.

In addition to defining new template classes of constraints and variables, the latter two $C++$ classes are used to implement various user callback routines to further customize the algorithm. The methods that can be implemented in these classes are similar to those in other solver frameworks and can be used to customize most aspects of the underlying algorithm.

10.4.1.3 Algorithm Control

ABACUS does not contain built-in routines for generating valid inequalities, but the user can implement any desirable separation or column generation procedures to improve the **lower bound** in each search tree node. ABACUS does not have a default primal heuristic for improving the **upper bound** either, but again, the user can implement one easily. ABACUS has a general notion of **branching** in which one may branch on either a variable or a constraint (hyperplane). Several strategies for selecting a branching variable are provided. In addition, a strong branching facility is provided, in which a number of variables or constraints are selected and presolved before the final branching is performed. To select the candidates, a sophisticated mechanism for selection and ranking of candidates using multiple user-defined branching rules is employed. The **search strategies** include depth-first, breadth-first, and best-first search, as well as a strategy that switches from depth-first to best-first after the first feasible solution is found.

10.4.2 BCP

10.4.2.1 Overview

BCP is a pure solver framework developed by Ladányi [48]. It is a close relative of SYMPHONY, described below. Both frameworks are descended from the earlier COMPSys framework of Ralphs and Ladányi [47, 68] like ABACUS. BCP is implemented in C++ and is design centered around problem-specific template classes of cuts and variables, but it takes a more "function-oriented" approach that is similar to SYMPHONY. The design is very flexible and supports the implementation of the same range of sophisticated variants of branch and bound that ABACUS supports, including simultaneous generation of columns and valid inequalities. BCP is a pure solver framework and does not have a callable library interface. The BCP library provides its own main function, which means that it cannot easily be called recursively or as a subroutine from another code. Nonetheless, it is still one of the most full-featured solver frameworks available, because of the generality with which it handles constraint and variable generation, as well as branching.

Although BCP is not itself a black-box solver, two different black-box codes [56] have been built using BCP and are available along with BCP itself as part of the COIN-OR software suite [48]. BCP is open source software licensed under the Common Public License (CPL). It has a user's manual, though it is slightly out of date. The code itself contains documentation that can be parsed and formatted using the Doxygen automatic documentation system [81]. A number of applications built using BCP are available for download, including some simple examples that illustrate its use. Tutorials developed by Galati that describe the implementation of two problem-specific solvers — one implementing branch and cut and one implementing branch and price — are also available [34]. The target audience for BCP is similar to that of ABACUS — sophisticated users who need a powerful framework for implementing advanced versions of branch and bound without a callable library interface. BCP is also capable of solving MILPs in parallel and is targeted at users who need such a capability.

10.4.2.2 User Interface

To use BCP, the user must implement application-specific C++ classes derived from the virtual base classes provided as part of the BCP library. The classes that must be implemented fall broadly into two categories: *modeling object classes*, which describe the variables and constraints associated with the user's application, and *user callback classes*, which control the execution of various specific parts of the algorithm.

From the user's point of view, a subproblem in BCP consists primarily of a *core relaxation* present in every subproblem and a set of *modeling objects* — the extra constraints and variables that augment the core relaxation. To define new template classes of valid inequalities and variables, the user must derive from the classes BCP_cut and BCP_var. The derivation involves defining an abstract data structure for describing a member of the class and providing methods for *expanding* each object (i.e., adding the object to a given LP relaxation).

To enable parallel execution, the internal library and the set of user callback functions are divided along functional lines into five separate computational modules. The modular implementation facilitates code maintenance and allows easy, configurable parallelization. The five modules are *master, tree manager, node processor, cut generator*, and *variable generator*. The master module includes functions that perform problem initialization, input/output, and overall execution control. The tree manager is responsible for maintaining the search tree and managing the search process. The node processor is responsible for processing a single search tree node (i.e., producing a bound on the solution value of the corresponding subproblem by solving a dynamically generated LP relaxation). Finally, the cut and variable generators are responsible for generating new modeling objects for inclusion in the current LP relaxation.

Associated with each module "xx" is a class named BCP_xx_user that contains the user callbacks for the module. For each module, the user must implement a derived class that overrides those methods for which the user wishes to provide a customized implementation. Most, but not all, methods have default implementations. The user callback classes can also be used to store the data needed to execute the methods in the class. Such data could include the original input data, problem parameters, and instance-specific auxiliary information such as graph data structures.

10.4.2.3 Algorithm Control

Like ABACUS, BCP does not contain built-in routines for generating valid inequalities, but the user can implement any separation or column generation procedure that is desired in order to improve the **lower bound**. BCP tightens variable bounds on the basis of reduced cost and allows the user to tighten bounds on the basis of logical implications that arise from the model. BCP does not yet have a built-in integer preprocessor and also has no built-in primal heuristic to improve the **upper bound**. The user can, however, pass an initial upper bound if desired. The default **search strategy** is a hybrid depth-first/best-first approach in which one of the children of the current node is retained for processing as long as the lower bound is not more than a specified percentage higher than the best available. The user can also specify a customized search strategy by implementing a new comparison function for sorting the list of candidate nodes in the BCP_tm_user class.

BCP has a generalized **branching** mechanism in which the user can specify *branching sets* that consist of any number of hyperplanes and variables. The hyperplanes and variables in these branching sets do not have to be present in the current subproblem. In other words, one can branch on any arbitrary hyperplane (see Section 10.2.3), whether or not it corresponds to a known valid inequality. After the desired candidate branching sets have been chosen, each one is presolved as usual by performance of a specified number of simplex pivots to estimate the bound improvement that would result from the branching. The final branching candidate can then be chosen by a number of standard built-in rules. The default rule is to select a candidate for which the smallest lower bound among its children is maximized.

10.4.3 BonsaiG

10.4.3.1 Overview

BonsaiG is a pure black-box solver developed by Lou Hafer that comes with complete documentation and descriptions of its algorithms and is available as open source software under the GNU General Public License (GPL) [41]. BonsaiG does not have a documented callable library interface or the customization options associated with a solver framework. It does, however, have two unique features worthy of mention. The first is the use of a partial arc consistency algorithm proposed in [77] to help enforce integrality constraints and dynamically tighten bounds on variables. Although the approach is similar to that taken by today's integer programming preprocessors, the arc consistency algorithm can be seen as a constraint programming technique and is applied aggressively during the processing of each subproblem. From this perspective, bonsaiG is perhaps one of the earliest examples of integrating constraint programming techniques into an LP-based branch-and-bound algorithm. The integration of constraint programming and traditional mathematical programming techniques has recently become a topic of increased interest among researchers. Achterberg is currently developing a solver called SCIP that will also integrate these two approaches [1].

The second feature worthy of mention is the use of DYLP, an implementation of the dynamic LP algorithm of Padberg [66], as the underlying LP solver. DYLP was designed specifically to be used for solving the LP relaxations arising in LP-based branch and bound. As such, DYLP automatically selects the subsets of the constraints and variables that should be active in a relaxation and manages the process of dynamically updating the active constraints and variables as the problem is solved. This management must be performed in some fashion by all MILP solvers, but it can be handled more efficiently if kept internal to the LP solver. The **target audience** for bonsaiG consists of users who need a lightweight black-box solver capable of solving relatively small

MILPs without incurring the overhead associated with advanced bounding techniques and those who do not need a callable library interface.

10.4.3.2 *User Interface and Algorithm Control*

BonsaiG can only be called from the command-line, with instances specified in MPS format. Algorithm control in bonsaiG is accomplished through the setting of parameters that are specified in a separate file. To improve the **lower bound** for generic MILPs, bonsaiG aggressively applies the arc consistency algorithm discussed earlier in combination with reduced-cost tightening of bounds in an iterative loop called the *integrated variable forcing loop*. No generation of valid inequalities or columns is supported. BonsaiG does not have any facility for improving the **upper bound**. The default **search strategy** is a hybrid of depth-first and best-first, but with a slight modification. When a subproblem is partitioned, all children are fully processed and among those that are not fathomed, the best one, according to an evaluation function that takes into account both the lower bound and the integer infeasibility, is retained for further partitioning. The others are added to the list of candidates, so that the list is actually one of candidates for branching, rather than for processing. When all children of the current node can be fathomed, then the candidate with the best bound is retrieved from the list and another depth-first search is initiated.

For **branching**, BonsaiG uses a penalty method strengthened by integrality considerations. Only branching on variables or groups of variables is supported. The user can influence branching decisions for a particular instance or develop custom branching strategies through two different mechanisms. First, bonsaiG makes the specification of relative branching priorities for groups of variables easy. This tells the solver which variables the user thinks will have the most effect if branched upon. The solver then attempts to branch on the highest-priority fractional variables first. The second mechanism is for specifying *tours* of variables. Tours are groups of variables that should be branched on as a whole. The group of child subproblems (called a *tour group*) is generated by adjustment of the bounds of each variable in the tour so that the feasible region of the parent is contained in the union of the feasible regions of the children, as usual.

10.4.4 CBC

10.4.4.1 *Overview*

CBC is a black-box solver distributed as part of the COIN-OR software suite [30]. CBC was originally developed by John Forrest as a lightweight branch-and-cut code to test CLP, the COIN-OR LP solver. However, CBC has since evolved significantly and is now quite sophisticated, even sporting customization features that allow it to be considered a solver framework. CBC has a native C++ callable library API similar to the Open Solver Interface, as well as a C interface built on top of that native interface. The CBC solver framework consists of a collection of C++ classes whose methods can be overridden to customize the algorithm. CBC does not have a user's manual, but it does come with some well-commented examples and the source code is also well-commented. It is distributed as part of the COIN-OR software suite and is licensed as open source software under the Common Public License (CPL). The target audience for CBC consists of users who need a full-featured black-box solver with a callable library API and very flexible, yet relatively lightweight, customization options.

10.4.4.2 User Interface

The user interacts with CBC as a black-box solver either by invoking the solver on the command line or through a command-based interactive shell. In either case, instances must be specified in MPS format. The callable library API is a hybrid of the API of the underlying LP solver, which is accessed through the Open Solver Interface, and the methods in the CbcModel class. To load a model into CBC, the user creates an OSI object, loads the model into the OSI object and then passes a pointer to that object to the constructor for the Cbc-Model object. CBC uses the OSI object as its LP solver during the algorithm.

To use CBC as a solver framework, a number of classes can be reimplemented in order to arrive at a problem-specific version of the basic algorithm. The main classes in the CBC library are:

- CbcObject, CbcBranchingObject, and CbcBranchDecision: The classes used to specify new branching rules. CBC has a very general notion of branching that is similar to that of BCP. CBC's branching mechanism is described in more detail in Section 10.4.4.3.

- CbcCompare and CbcCompareBase: The classes used to specify new search strategies by specifying the method for sorting the list of candidate search tree nodes.

- CbcHeuristic: The class used to specify new primal heuristics.

- CbcCutGenerator: The class that interfaces to the Cut Generation Library.

As seen from the above list, custom branching rules, custom search strategies, custom primal heuristics, and custom generators for valid inequalities can all be introduced. Cut generator objects must be derived from the Cut Generation Library base class and are created by the user before they are passed to CBC. Primal heuristic objects are derived from the CbcHeuristic class and are also created before being passed to CBC.

10.4.4.3 Algorithm Control

CBC is one of the most full-featured black-box solver of those reviewed here in terms of available techniques for improving bounds. The **lower bound** can be improved through the generation of valid inequalities using all of the separation algorithms implemented in the CGL. Problem-specific methods for generation of valid inequalities can be implemented, but column generation is not supported. CBC has a logical preprocessor to improve the initial model and tightens variable bounds using reduced cost information. Several primal heuristics to improve the **upper bound** are implemented and provided as part of the distribution, including a rounding heuristic and two different local search heuristics. The default **search strategy** in CBC is to perform depth-first search until the first feasible solution is found and then to select nodes for evaluation on the basis of a combination of bound and number of unsatisfied integer variables. Specification of new search strategies can also be accomplished easily.

CBC has a strong **branching** mechanism similar to that of other solvers, but the type of branching that can be done is more general. An abstract CBC branching object can be anything that (1) has a feasible region whose degree of infeasibility with respect to the current solution can be quantified, (2) has an associated action that can be taken to improve the degree of infeasibility in the child nodes, and (3) supports some comparison of the effect of branching. Specification of a CBC branching object involves implementation of three methods: `infeasibility()`, `feasibleRegion()`, and `createBranch()`. These methods allow CBC to perform strong branching on any sort of branching objects. Default implementations are provided for branching on integer variables, branching on cliques, and branching on special ordered sets.

10.4.5 GLPK

10.4.5.1 Overview

GLPK is the GNU Linear Programming Kit, a set of subroutines that comprise a callable library and black-box solver for solving linear programming and MILP instances [53]. GLPK also comes equipped with GNU MathProg (GMPL), an algebraic modeling language similar to AMPL. GLPK was developed by Andrew Makhorin and is distributed as part of the GNU Project, under the GNU General Public License (GPL). Because GLPK is distributed through the Free Software Foundation (FSF), it closely follows the guidelines of the FSF with respect to documentation and automatic build tools. The build system relies on `autoconf`, which ensures that users can easily build and execute the library on a wide variety of platforms. The documentation consists of a reference manual and a description of the GNU MathProg language. The distribution includes examples of the use of the callable library, and models demonstrate the MathProg language.

GLPK is a completely self-contained package that does not rely on external components to perform any part of the branch-and-bound algorithm. In particular, GLPK includes the following main components:

- revised primal and dual simplex methods for linear programming
- a primal-dual interior point method for linear programming
- a branch-and-bound method
- a parser for GNU MathProg
- an application program interface (API)
- a black-box LP/MILP solver

The target audience for GLPK consists of users who want a lightweight, self-contained MILP solver with both a callable library and modeling language interface.

10.4.5.2   User Interface

The default build of GLPK yields the callable library and a black-box solver. The callable library consists of nearly 100 routines for loading and modifying a problem instance, solving the loaded instance, querying the solver, and getting and setting algorithm parameters. There are also utility routines to read and write files in MPS format, LP format, and GMPL format. The subroutines operate on a data structure for storing the problem instance that is passed explicitly, so the code is thread safe and it is possible to work on multiple models simultaneously.

10.4.5.3   Algorithm Control

Because GLPK was first and foremost developed as a solver of linear programs, it does not yet contain advanced techniques for improving the **lower bound**, such as those for preprocessing and generating valid inequalities. It also does not include a primal heuristic for improving the **upper bound**. The user can set a parameter (either through the callable library or in the black-box solver) to choose from one of three methods for selecting a **branching** variable — the index of the first fractional variable, the index of the last fractional variable, or the penalty method discussed in Section 10.2.3. The user can change the **search strategy** to either depth-first-search, breadth-first-search, or the best-projection method described in Section 10.2.4.

10.4.6   lp_solve

10.4.6.1   Overview

Lp_solve is a black-box solver and callable library for linear and mixed-integer programming. The original solver was developed by Michel Berkelaar at Eindhoven University of Technology, and the work continued with Jeroen Dirks, who contributed a procedural interface, a built-in MPS reader, and fixes and enhancements to the code. Kjell Eikland and Peter Notebaert took over development starting with version 4. There is currently an active group of users. The most recent version bears little resemblance to earlier versions and includes a number of unique features, such as a modular LP basis factorization engine and a large number of language interfaces. Lp_solve is distributed as open source under the GNU Library General Public License (LGPL). The main repository for lp_solve information, including a FAQ, examples, the full source, precompiled executables, and a message board, is available at the YAHOO lp_solve group [15]. The target audience for lp_solve is similar to that of GLPK — users who want a lightweight, self-contained solver with a callable library API implemented in a number of popular programming languages, including C, VB, and Java, as well as an AMPL interface.

10.4.6.2   User Interface

Lp_solve can be used as a black-box solver or as a callable library through its native C API. The lp_solve API consists of over 200 functions for reading and writing problem instances, building or querying problem instances, setting algorithm parameters, invoking solution algorithms, and querying the results. The lp_solve API has methods that can read and write MPS files, LP format files, and an XLI (External Language Interface) that allows users to implement their own readers and writers. At present, XLI interfaces are in place for GNU MathProg, LPFML, and for the commercial LP formats of CPLEX and LINDO. An interface to ZIMPL will be available in lp_solve v5.2.

Of the noncommercial MILP software reviewed here, lp_solve has interfaces to the largest number of different programming languages. With lp_solve, there are examples that illustrate how to call its API from within a VB.NET or C# program. Also, a Delphi library and a Java Native Interface (JNI) to lp_solve, are available, so lp_solve can be called directly from Java programs. AMPL, MATLAB, R, S-Plus, and Excel driver programs are also available. Lp_solve supports several types of user callbacks and an object-like facility for revealing functionality to external programs.

10.4.6.3   Algorithm Control

Lp_solve does not have any special procedures for improving the **upper** or **lower bounds**. Users can set parameters either from the command line or through the API to control the branch-and-bound procedure. The **search strategy** is one of depth-first search, breadth-first search, or a two-phase method that initially proceeds depth-first, followed by breadth-first. Lp_solve contains a large number of built-in **branching** procedures and can select the branching variable on the basis of the lowest indexed noninteger column (default), the distance from the current bounds, the largest current bound, the most fractional value, the simple, unweighted pseudocost of a variable, or pseudocosts combined with the number of integer infeasibilities.

The algorithm also allows for GUB branching and for branching on semi-continuous variables (variables that have to take a value of zero or a positive value above some given lower bound). The branching rule and search strategy used by lp_solve are set through a call to set_bb_rule(), and the branching rule can even be modified using parameter values beyond those listed here. As with many solvers, MILP performance can be expected to vary significantly based on parameter settings and model class. The default settings in lp_solve are inherited from v3.2, and tuning is therefore necessary to achieve desirable results. The developers have indicated that MILP performance improvement and more robust default settings will be a focus in lp_solve v5.3.

10.4.7   MINTO

10.4.7.1   Overview

MINTO (Mixed INTeger Optimizer) is a black-box solver and solver framework whose chief architects were George Nemhauser and Martin Savelsbergh, with a majority of the software development done by Savelsbergh. MINTO was developed at the Georgia Institute of Technology and is available in binary form under terms of an agreement created by the Georgia Tech Research Institute. The current maintainer of MINTO is Jeff Linderoth of Lehigh University. It is available as a precompiled library for a number of common platforms [60]. MINTO relies on external software to solve the linear programming relaxations that arise during the algorithm. Since version 3.1, MINTO has been equipped to use the OSI, so any of the linear programming solvers for which there is an OSI interface can be used with MINTO. MINTO can also be built to directly use the commercial LP solvers CPLEX, XPRESS-MP, and OSL. MINTO comes with a user's manual that contains instructions for building an executable, and a description of the API for user callbacks that allow it to be used as a solver framework, and examples of the use of each routine. The target audience for MINTO are users who require the power of a sophisticated solver framework for implementing advanced versions of branch and bound, but with a relatively simple C-style interface, or who need a full-featured black-box solver without a callable API.

10.4.7.2 User Interface

MINTO can be accessed as a black-box solver from the command line with parameters set through a number of command-line switches. The most common way of using MINTO is to pass problem instances in MPS file format. However, beginning with version 3.1, MINTO can also be used directly with the AMPL modeling language. MINTO can be customized through the use of "user application" functions (callback functions) that allow MINTO to operate as a solver framework. At well-defined points of the branch-and-cut algorithm, MINTO will call the user application functions, and the user must return a value that signals to MINTO whether or not the algorithm is to be modified from the default. For example, consider the MINTO user application function `appl_constraints()`, which is used for generating user-defined classes of valid inequalities. The input to `appl_constraints()` is the solution to the current LP relaxation. The outputs are arrays describing any valid inequalities that the user wishes to have added to the formulation. The return value from `appl_constraint()` should be SUCCESS or FAILURE, depending on whether or not the user-supplied routine was able to find inequalities violated by the input solution. If so, MINTO will add these inequalities and pass the new formulation to the linear programming solver.

10.4.7.3 Algorithm Control

To strengthen the **lower bound** during the course of the algorithm, MINTO relies on advanced preprocessing and probing techniques, as detailed in the paper of Atamtürk, Nemhauser, and Savelsbergh [5], and also tightening of bounds on the basis of reduced cost. MINTO has separation routines for a number of classes of valid inequalities, including clique inequalities, implication inequalities, knapsack cover inequalities, GUB cover inequalities, and flow cover inequalities. The user can perform dynamic column generation by implementing the callback function `appl_variables()`. For improving the **upper bound**, MINTO has a primal diving heuristic. A number of built-in **branching** methods are included, such as branching on the most fractional variable, penalty method strengthened with integrality considerations, strong branching, a pseudocost-based method, a dynamic method that combines both the penalty method and pseudocost-based branching, and a method that favors branching on GUB constraints. For **search strategies**, the user can choose best-bound, depth-first, best-projection, best-estimate, or an adaptive mode that combines depth-first with the best-estimate mode. Of course, by using the solver framework, the user can override any of the branching or node selection methods.

10.4.8 SYMPHONY

10.4.8.1 *Overview*

SYMPHONY is a black-box solver, callable library, and solver framework for MILPs that evolved from the COMPSys framework of Ralphs and Ladányi [47, 68]. The source code for packaged releases, with full documentation and examples, is available for download and is licensed under the Common Public License (CPL) [69]. The latest source is also available from the CVS repository of the COIN-OR Foundation [19]. SYMPHONY is fully documented and seven different specialized solvers built with SYMPHONY are available as examples of how to use the code. A step-by-step example that illustrates the building of a simple branch-and-cut solver for the matching problem [80] is available and summarized in Section 10.5. The core solution methodology of SYMPHONY is a highly customizable branch-and-bound algorithm that can be executed sequentially or in parallel [67]. SYMPHONY calls on several other open source libraries for specific functionality, including COIN-OR's Cut Generation Library, Open Solver Interface, and MPS file parser components, GLPK's GMPL file parser, and a third-party solver for linear-programming problems (LPs), such as COIN-OR's LP Solver (CLP).

Several unique features of SYMPHONY are worthy of mention. First, SYMPHONY contains a generic implementation of the WCN algorithm described in [72] for solving bicriteria MILPs, and methods for approximating the set of Pareto outcomes. The bicriteria solver can be used to examine tradeoffs between competing objectives, and for solving parametric MILPS, a form of global sensitivity analysis. SYMPHONY also contains functions for local sensitivity analysis based on ideas suggested by Schrage and Wolsey [75]. Second, SYMPHONY has the capability to warm start the branch-and-bound process from a previously calculated branch-and-bound tree, even after modifying the problem data. These capabilities are described in more detail in the paper of Ralphs and Guzelsoy [71]. The target audience for SYMPHONY is similar to that of MINTO — users who require the power of a sophisticated solver framework, primarily for implementing custom branch and cut algorithms, with a relatively simple C-style interface, or users who require other advanced features such as parallelism, the ability to solve multi-criteria instances, or the ability to warm start solution procedures.

10.4.8.2 User Interface

As a black-box solver, SYMPHONY can read GMPL files using an interface to GLPK's file parser or MPS files using COIN-OR's MPS file reader class. It can also be used as a callable library through the API described below. SYMPHONY's callable library consists of a complete set of subroutines for loading and modifying problem data, setting parameters, and invoking solution algorithms. The user invokes these subroutines through the native C API, which is exactly the same whether SYMPHONY is invoked sequentially or in parallel. The choice between sequential and parallel execution modes is made at compile-time. SYMPHONY has an OSI implementation that allows solvers built with SYMPHONY to be accessed through the OSI.

The user's main avenues for customization are the tuning of parameters and the implementation of SYMPHONY's callback functions. SYMPHONY contains over 50 callback functions that allow the user to override SYMPHONY's default behavior for branching, generation of valid inequalities, management of the cut pool, management of the LP relaxation, search and diving strategies, program output, and others. Each callback function is called from a SYMPHONY *wrapper function* that interprets the user's return value and determines what action should be taken. If the user performs the required function, the wrapper function exits without further action. If the user requests that SYMPHONY perform a certain default action, then this is done. Files that contain default function stubs for the callbacks are provided along with the SYMPHONY source code. These can then be modified by the user as desired. Makefiles and Microsoft Visual C++ project files are provided for automatic compilation. A full list of callback functions is contained in the user's manual [70]. For an example of the use of callbacks, see the SYMPHONY case study in Section 10.5.1.

10.4.8.3 Algorithm Control

To improve the **lower bound** for generic MILPs, SYMPHONY generates valid inequalities using COIN-OR's Cut Generation Library (CGL) described in Section 10.2.1.2. The user can easily insert custom separation routines and can perform column generation, though the implementation is not yet fully general and requires that the set of variables be indexed *a priori*. This limitation makes the column generation in SYMPHONY most appropriate for situations in which the set of columns has a known combinatorial structure and is of relatively small cardinality. In each iteration, SYMPHONY tightens bounds by reduced cost and also allows the user to tighten bounds on the basis of logical implications arising from the model. SYMPHONY does not yet have its own logical preprocessor or primal heuristics to improve the **upper bound**, although it is capable of using CBC's primal heuristic if desired. The user can also pass an initial upper bound.

SYMPHONY uses a strong **branching** approach by default. Branching candidates can be either constraints or variables and are chosen by any one of a number of built-in rules, such as most fractional, or by a customized rule. After the candidates are chosen, each one is presolved to determine an estimate of the bound improvement that would result from the branching. The final branching candidate can then be chosen by a number of standard built-in rules. A naive version of pseudocost branching is also available.

The default **search strategy** is a hybrid depth-first/best-first approach in which one of the children of the current node is retained for processing as long as the lower bound is not more than a specified percentage higher than the best available. Another option is to stop diving when the current node is more than a specified percentage of the gap higher than the best available. By tuning various parameters, one can obtain a number of different search strategies that run the gamut between depth-first and best-first.

10.5 Case Studies

In this section, we describe two examples that illustrate the power of solver frameworks for developing custom optimization codes. In Section 10.5.1, we describe a custom branch-and-cut algorithm for solving the matching problem developed using SYMPHONY. In Section 10.5.2, we describe a custom branch-and-price algorithm for the axial assignment problem developed using BCP. Full source code and more detailed descriptions of both solvers are available [34, 80].

10.5.1 Branch and Cut

The Matching Problem. Given a complete, undirected graph $G = (V, E)$, the *Matching Problem* is that of selecting a set of pairwise disjoint edges of minimum weight. The problem can be formulated as follows:

$$\min \sum_{e \in E} c_e x_e$$

$$\sum_{e \in \{(i,j) \mid j \in V, (i,j) \in E\}} x_e = 1 \quad \forall i \in V, \tag{10.12}$$

$$x_e \geq 0 \quad \forall e \in E, \tag{10.13}$$

$$x_e \in \mathbb{Z} \quad \forall e \in E,$$

where x_e is a binary variable that takes value 1 if edge e is selected and 0 otherwise.

Implementing the Solver. The first thing needed is a data structure in which to store the description of the problem instance and any other auxiliary information required to execute the algorithm. Such a data structure is shown

```
typedef struct MATCH_DATA{
    int                 numnodes;
    int                 cost [MAXNODES] [MAXNODES] ;
    int                 endpoint1 [MAXNODES* (MAXNODES-1) /2] ;
    int                 endpoint2 [MAXNODES* (MAXNODES-1) /2] ;
    int                 index [MAXNODES] [MAXNODES] ;
}match_data;
```

FIGURE 10.1
Data structure for matching solver.

in Figure 10.1. We assume a complete graph, so a problem instance can be described simply by the objective function coefficients, stored in the two-dimensional array cost. Each primal variable is identifiable by an index, so we must have a way of mapping an edge $\{i, j\}$ to the index that identifies the corresponding variable and vice versa. Such mappings between problem instance objects and variable indices are a common construct when using solver frameworks. Thusly, recent commercial modeling frameworks such as Concert [45] and Mosel [20] and the noncommercial modeling system FLOPC^{++} [44] have an interface that allows for a more explicit coupling of problem objects and instance variables. In the data structure shown, endpoint1 [k] returns the first endpoint of the edge with index k and endpoint2 [k] returns the second endpoint. On the other hand index [i] [j] returns the index of edge $\{i, j\}$.

Next, functions for reading in the problem data and creating the instance are needed. The function match_read_data () (not shown) reads the problem instance data (a list of objective function coefficients) in from a file. The function match_load_problem (), shown in Figure 10.2, constructs the instance in column-ordered format. In the first part of this routine, a description of the MILP is built, while in the second part, this representation is loaded into the solver through the subroutine sym_explicit_load _problem ().

The main () routine for the solver is shown in Figure 10.3. In this routine, a SYMPHONY environment and a user data structure are created, the data are read in, the MILP is created and loaded into SYMPHONY and then one instance is solved. Results are automatically printed, but one could also implement a custom subroutine for displaying these if desired.

We next show how to add the ability to generate some simple problem-specific valid inequalities. The odd-set inequalities

$$\sum_{e \in E(O)} x_e \leq \frac{|O| - 1}{2} \qquad O \subseteq V, |O| \text{ odd,} \tag{10.14}$$

with $E(O) = \{e = \{i, j\} \in E \mid i \in O, j \in O\}$ are satisfied by all matchings. Indeed, Edmonds [28] showed that the inequalities (10.12) to (10.14) describe the convex hull of matchings, so the matching problem can, in

```
int match_load_problem(sym_environment *env, match_data
  *prob){int i, j, index, n, m, nz, *column_starts,
  *matrix_indices;
  double *matrix_values, *lb, *ub, *obj, *rhs, *rngval;
  char *sense, *is_int;

n = prob->numnodes*(prob->numnodes-1)/2;
                                  /* Number of columns */
m = 2 * prob->numnodes;           /* Number of rows */
nz = 2 * n;                       /* Number of nonzeros */
/* Normally, allocate memory for the arrays here
(left out to save space) */
for (index = 0, i = 0; i < prob->numnodes; i++) {
for (j = i+1; j < prob->numnodes; j++) {
prob->match1[index] = i; /*The 1st component of
assignment 'index'*/
prob->match2[index] = j; /*The 2nd component of
assignment 'index'*/
prob->index[i][j] = prob->index[j][i] = index; /*To
recover later*/
obj[index] = prob->cost[i][j]; /* Cost of assignment
(i, j) */
is_int[index] = TRUE; /* Indicates the variable is
integer */
column_starts[index] = 2*index;
matrix_indices[2*index] = i;
matrix_indices[2*index+1] = j;
matrix_values[2*index] = 1;
matrix_values[2*index+1] = 1;
ub[index] = 1.0;
index++;
}
}
column_starts[n] = 2 * n; /* We have to set the ending
position */
for (i = 0; i < m; i++) { /* Set the right-hand side */
    rhs[i] = 1;
    sense[i] = 'E';
  }
  sym_explicit_load_problem(env, n, m, column_starts,
     matrix_indices, matrix_values, lb, ub, is_int,
     obj, 0, sense, rhs, rngval, true);
  return (FUNCTION_TERMINATED_NORMALLY);
}
```

FIGURE 10.2
Function to load the problem for matching solver.

```
int main(int argc, char **argv)
{
    int termcode;
    char * infile;

    /* Create a SYMPHONY environment */
    sym_environment *env = sym_open_environment();

    /* Create the data structure for storing the
    instance.*/
    user_problem *prob =
    (user_problem *)calloc(1, sizeof (user_problem));

    sym_set_user_data(env, (void *)prob);
    sym_parse_command_line(env, argc, argv);
    sym_get_str_param(env, "infile_name", &infile);
    match_read_data(prob, infile);
    match_load_problem(env, prob);
    sym_solve(env);
    sym_close_environment(env);
    return(0);
}
```

FIGURE 10.3
Main function for the matching solver.

theory, be solved as a linear program, albeit with an exponential number of constraints.

The textbook of Grötschel, Lovász, and Schrijver [37] describes how to efficiently separate the odd-set inequalities in full generality, but for simplicity, we implement separation only for odd-set inequalities for sets of cardinality three. This is done by brute force enumeration of triples, as shown in Figure 10.4. The function user_find_cuts() is the SYMPHONY callback for generating valid inequalities. The user is provided with the current fractional solution in a sparse vector format and asked to generate violated valid inequalities. The function cg_add_explicit_cut() is used to report any inequalities found. Even this simple separation routine can significantly reduce the number of nodes in the branch-and-cut tree.

10.5.2 Branch and Price

The Three-Index Assignment Problem. The *Three-Index Assignment Problem* (3AP) is that of finding a minimum-weight clique cover of the complete tri-partite graph $K_{n,n,n}$, where n is redefined here to indicate the size of the underlying graph. Let I, J, and K be three disjoint sets with $|I| = |J| = |K| = n$

```
int user_find_cuts(void *user, int varnum, int iter_num,
                int level, int index, double objval,
                int *indices, double *values, double
                ub, double etol, int *num_cuts, int
                *alloc_cuts, cut_data ***cuts)
{
    user_problem *prob = (user_problem *) user;
    double cut_val[3], edge_val[200][200];
    /* Matrix of edge values */
    int i, j, k, cutind[3], cutnum = 0;

    /* Allocate the edge_val matrix to zero (we could
    also just calloc it) */ memset((char *)edge_val,
    0, 200*200*ISIZE);

    for (i = 0; i < varnum; i++) {
        edge_val[prob->node1[indices[i]]][prob->node2
        [indices[i]]] = values[i];
    }
    for (i = 0; i < prob->nnodes; i++){
        for (j = i+1; j < prob->nnodes; j++){
        for (k = j+1; k < prob->nnodes; k++) {
            if (edge_val[i][j]+edge_val[j][k]+edge_val[i]
            [k] > 1.0 + etol) {
            /* Found violated triangle cut */
            /* Form the cut as a sparse vector */
            cutind[0] = prob->index[i][j];
            cutind[1] = prob->index[j][k];
            cutind[2] = prob->index[i][k];
            cutval[0] = cutval[1] = cutval[2] = 1.0;
            cg_add_explicit_cut(3, cutind,
            cutval, 1.0, 0, 'L',
                    TRUE, num_cuts, alloc_cuts, cuts);
            cutnum++;
            }
        }
        }
    }
    return(USER_SUCCESS);
}
```

FIGURE 10.4
Cut generator for matching solver.

and set $H = I \times J \times K$. Then, 3AP can be formulated as follows:

$$\min \sum_{(i,j,k)\in H} c_{ijk} x_{ijk},$$

$$\sum_{(j,k)\in J \times K} x_{ijk} = 1 \quad \forall i \in I, \tag{10.15}$$

$$\sum_{(i,k)\in I \times K} x_{ijk} = 1 \quad \forall j \in J,$$

$$\sum_{(i,j)\in I \times J} x_{ijk} = 1 \quad \forall k \in K, \tag{10.16}$$

$$x_{ijk} \in \{0, 1\} \quad \forall (i, j, k), \in H.$$

In [10], Balas and Saltzman consider the use of the classical Assignment Problem (AP) as a relaxation of 3AP in the context of Lagrangian relaxation. We use the same relaxation to reformulate the 3AP using a Dantzig-Wolfe decomposition (see Section 10.2.1.4 for a discussion of this technique). The AP is a relaxation of the 3AP obtained by relaxing constraint (10.15). Let $\mathcal{F}$ be the set of feasible solutions to the AP. The Dantzig Wolfe (DW) reformulation of 3AP is then:

$$\min \sum_{s\in\mathcal{F}} c_s \lambda_s,$$

$$\sum_{s\in\mathcal{F}} \left(\sum_{(j,k)\in J \times K} s_{ijk} \lambda_s \right) = 1 \quad \forall i \in I,$$

$$\sum_{s\in\mathcal{F}} \lambda_s = 1,$$

$$\lambda_s \in \mathbb{Z}_+ \forall s \in \mathcal{F},$$

where $c_s = \Sigma_{(i,j,k)\in H} c_{ijk} s_{ijk}$ for each $s \in \mathcal{F}$. Relaxing the integrality constraints of this reformulation, we obtain a relaxation of 3AP suitable for use in an LP-based branch-and-bound algorithm. Because of the exponential number of columns in this linear program, we use a standard column generation approach to solve it.

Implementing the Solver. The main classes to be implemented are the BCP_xx_user classes mentioned earlier and a few problem-specific classes. We describe the problem-specific classes first:

- AAP: This class is a container for holding an instance of 3AP. Data members include the dimension of the problem n and the objective function vector.

- AAP_user_data: This class is derived from BCP_user_data and is used to store problem-specific information in the individual nodes of the search tree. Because we branch on the original variables x_{ijk} and not the master problem variables λ_s, we must keep track of which variables have been fixed to 0 or 1 at a particular node, so that we can enforce these conditions in our column generation subproblem.

- `AAP_var`: Each variable in the Dantzig-Wolfe reformulation represents an assignment between members of sets J and K. In each instance of the class, the corresponding assignment is stored using a vector that contains the indices of the assignment along with its cost.

Because `BCP` is a parallel code, every data structure must be accompanied by a subroutine that can both pack it into and unpack it from a character buffer for the purposes of communicating the contents to other parallel processes. For most built-in types, default pack and unpack routines are predefined. For user-defined data structures, however, they must be provided. Typically, such routines consist simply of a collection of calls to either `BCP_buffer::pack()` or `BCP_buffer::unpack()`, as appropriate, packing or unpacking each data member of the class in turn. The user callback classes that must be modified are as follows:

- `AAP_tm`: This class is derived from `BCP_tm_user` and contains the callbacks associated with initialization and tree management. The main callbacks implemented are `initialize_core()` and `create_ root()`. These methods define the core relaxation and the initial LP relaxation in the root node.

- `AAP_lp`: This class is derived from `BCP_lp_user` and contains the callbacks associated with the solution of the LP relaxations in each search tree node. The main methods implemented are:

  `generate_vars_in_lp()`: the subroutine that generates new variables,

  `compute_lower_bound()`: returns a true lower bound in each iteration (the LP relaxation does not yield a true lower bound unless no variables with negative reduced cost exist),

  `restore_feasibility()`: a subroutine that tries to generate columns that can be used to restore the feasibility of a relaxation that is currently infeasible,

  `vars_to_cols()`: a subroutine that generates the columns corresponding to a set of variables, so that they can be added to the current relaxation,

  `select_branching_candidates()`: a subroutine that selects candidates for strong branching. We branch on the original variables x_{ijk}. Candidates are selected by the usual "most fractional" rule using the helper function `branch_close_to_half()`. A second function, `append_branching_vars()`, is called to create the branching objects, and

set_user_data_for_children(): Stores the appropriate data regarding what original variables were branched on in each child node.

In addition to defining these classes, a few important parameters must be set. We have glossed over some details here, but the source code and a more thorough description of this example are available for download [34].

10.6 Benchmarking

In this section, we present computational results that show the relative performance of the black-box solvers reviewed in Section 10.4. Each solver was built with gcc 3.3 using the default build parameters. The experiments were run under Linux RedHat v7.3 on an Intel Pentium III with an 1133MHz clock speed and 512 MB of memory. The maximum CPU time allowed for each solver and each instance was two hours. For each of the codes, the default parameter settings were used on the instances, except for lp_solve, in which the default branching and node selection rules were changed to ones based on pseudocosts.[2]

There were 122 problem instances included in the test: the instances of miplib3[3][16], miplib2003 [57], and 45 instances collected by the authors from various sources. The instances collected by the authors are available from the Computational Optimization Research at Lehigh (COR@L) Web site [50]. Table 10.4 shows the number of rows m, number of variables n, number of continuous variables $n - p$, number of binary variables $|B|$, and number of general integer variables $p - |B|$ for each instance in the test suite that is not already available through MIPLIB.

In order to succinctly present the results of this extensive computational study, we use performance profiles, as introduced by Dolan and Moré [25]. A performance profile is a relative measure of the effectiveness of a solver s when compared to a group of solvers S on a set of problem instances P. To completely specify a performance profile, we need the following definitions:

- γ_{ps} is a quality measure of solver s when solving problem p
- $r_{ps} = \gamma_{ps}/(\min_{s \in S} \gamma_{ps})$
- $\rho_s(\tau) = |\{p \in P \mid r_{ps} \leq \tau\}|/|P|$

---

[2] The lp_solve command line was: lp_solve -mps name.mps -bfp ./bfp_LUSOL -timeout 7200 -time -presolve -presolvel -piva -pivla -piv2 -ca -B5 -Bd -Bg -si -s5 -se -degen -S1 -v4.
[3] With the exception of the instances air03, blend2, egout, enigma, flugpl, gen, khb05250 lseu misc03 misc06, mod008, mod010, p0033, p0201, p0282, rgn, stein27, vpm1, which at least five of the six solvers were able to solve in less than 2 minutes.

TABLE 10.4

Characteristics of (non MIPLIB) Problem Instances

Name	m	n	$n-p$	$\|B\|$	$p-\|B\|$	Name	m	n	$n-p$	$\|B\|$	$p-\|B\|$
22433	198	429	198	231	0	23588	137	368	137	231	0
aligninq	340	1831	1	1830	0	bc1	1913	1751	1499	252	0
bienst1	576	505	477	28	0	bienst2	576	505	470	35	0
dano3_3	3202	13873	13804	69	0	dano3_4	3202	13873	13781	92	0
dano3_5	3202	13873	13758	115	0	fiball	3707	34219	1	33960	258
mcsched	2107	1747	2	1731	14	mkc1	3411	5325	2238	3087	0
neos10	46793	23489	0	23484	5	neos11	2706	1220	320	900	0
neos12	8317	3983	847	3136	0	neos13	20852	1827	12	1815	0
neos14	552	792	656	136	0	neos15	552	792	632	160	0
neos16	1018	377	0	336	41	neos17	486	535	235	300	0
neos18	11402	3312	0	3312	0	neos1	5020	2112	0	2112	0
neos20	2446	1165	198	937	30	neos2	1103	2101	1061	1040	0
neos3	1442	2747	1387	1360	0	neos4	38577	22884	5712	17172	0
neos5	63	63	10	53	0	neos6	1036	8786	446	8340	0
neos7	1994	1556	1102	434	20	neos8	46324	23228	0	23224	4
neos9	31600	81408	79309	2099	0	npmv07	76342	220686	218806	1880	0
nsa	1297	388	352	36	0	nug08	912	1632	0	1632	0
pg5_34	225	2600	2500	100	0	pg	125	2700	2600	100	0
qap10	1820	4150	0	4150	0	ramos3	2187	2187	0	2187	0
ran14x18_1	284	504	252	252	0	roy	162	149	99	50	0
sp97ic	2086	1662	0	718	944	sp98ar	4680	5478	0	2796	2682
sp98ic	2311	2508	0	1139	1369	sp98ir	1531	1680	0	871	809
Test3	50680	72215	39072	7174	25969						

Hence, $\rho_s(\tau)$ is the fraction of instances for which the performance of solver s was within a factor of τ of the best. A performance profile for solver s is the graph of $\rho_s(\tau)$. In general, the "higher" the graph of a solver, the better the relative performance.

Comparison of MILP solvers directly on the basis of performance is problematic in a number of ways. By compiling these codes with the same compiler on the same platform and running them under identical conditions, we have eliminated some of the usual confounding variables, but some remain. An important consideration is the feasibility, optimality, and integrality tolerances used by the solver. Dolan, Moré, and Munson [24] performed a careful study of the tolerances used in nonlinear programming software and concluded that trends of the performance profile tend to remain the same when tolerances are varied. The differences in solver tolerances for these tests were relatively minor, but these minor differences could lead to large differences in runtime performance. Another potential difficulty is the verification of solvers' claims with respect to optimality and feasibility of solutions. The authors made little attempt to verify *a posteriori* that the solutions claimed as optimal or feasible were indeed optimal or feasible. The conclusions drawn here about the relative effectiveness of the MILP solvers must be considered with these caveats in mind.

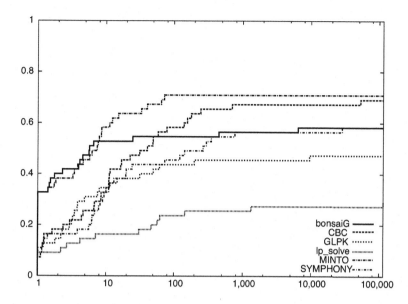

FIGURE 10.5
Performance profile of MILP solvers on solved instances.

For instances that were solved to provable optimality by one of the six solvers, the solution time was used as the quality measure γ_{ps}. Under this measure, $\rho_s(1)$ is the fraction of instances for which solver s was the fastest solver, and $\rho_s(\infty)$ is the fraction of instances for which solver s found a provably optimal solution. Figure 10.5 shows a performance profile of the instances that were solved in two CPU hours by at least one of the solvers. The graph shows that bonsaiG and MINTO were able to solve the largest fraction of the instances the fastest. The solvers MINTO and CBC were able to find a provably optimal solution within the time limit for the largest largest fraction of instances, most likely because these two solvers contain the largest array of specialized MILP solution techniques.

For instances that were not solved to optimality by any of the six solvers in the study, we used the value of the best solution found as the quality measure. Under this measure, $\rho_s(1)$ is the fraction of instances for which solver s found the best solution among all the solvers, and $\rho_s(\infty)$ is the fraction of instances for which solver s found at least one feasible solution. In Figure 10.6, we give the performance profile of the six MILP solvers on the instances for which no solver was able to prove the optimality of the solution. SYMPHONY was able to obtain the largest percentage of good feasible solutions, and the performance of GLPK was also laudable in this regard. This conclusion is somewhat surprising since neither SYMPHONY nor GLPK contain a specialized primal heuristic designed for finding feasible solutions. This seems to indicate that the primal heuristics existing in these noncommercial codes are relatively ineffective. Implementation of a state-of-the-art primal heuristic in a noncommercial code would be a significant contribution.

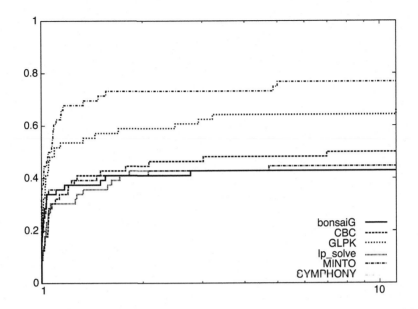

FIGURE 10.6
Performance profile of MILP solver on unsolved instances.

10.7   Future Trends and Conclusions

In closing, we want to mention the work of two groups that have been strong supporters and developers of open code and are well-positioned to support such development for many years into the future. The first is the COIN-OR Foundation, a nonprofit foundation mentioned several times in this chapter, part of whose mission is to promote the development of open source software for operations research [52]. This foundation, originally a loose consortium of researchers from industry and academia founded by IBM, has become a major provider of open source optimization software and is poised to have a large impact on the field over the coming decade. The second is the NEOS (Network Enabled Optimization System) project [21, 23]. NEOS provides users with the ability to solve optimization problems on a remote server using any of a wide range of available solvers. At current count, 55 different solvers for a variety of different optimization problem types are available for use. Interfaces to the noncommercial MILP solvers CBC, GLPK, and MINTO are available on NEOS. To use NEOS, the user submits a problem represented in a specified input format (e.g., MPS, AMPL, GMPL, or LP) through either an e-mail interface, a web interface, or a specialized client program running on the user's local machine. The instance is sent to the NEOS server, which locates resources to run the instance and sends the results back to the user.

Because the job runs remotely, this provides the user with the ability to test multiple solvers without downloading and installing each of them individually. The NEOS project has been solving optimization problems on-line since 1991, and currently handles over 10,000 optimization instances each month.

As for the future, we see no slowing of the current trend toward the development of competitive noncommercial software for solving MILPs. A number of exciting new open source projects are currently under development and poised to further expand the offerings to users of optimization software. Among these is the Abstract Library for Parallel Search (ALPS), a C++ class library for implementing parallel search algorithms planned that is the planned successor to BCP [86]. ALPS will further generalize many of the concepts present in BCP, providing the ability to implement branch-and-bound algorithms for which the bounding procedure is not necessarily LP-based. A second framework, called DECOMP, will provide the ability to automate the solution of decomposition-based bounding problems, i.e., those based on Lagrangian relaxation or Dantzig-Wolfe decomposition. Both of these frameworks will be available as part of the COIN-OR software suite. A new generation of the Open Solver Interface supporting a much wider range of problem types and with better model-building features is under development by COIN-OR, along with a new open standard based on LPFML for describing mathematical programming instances [32]. Finally, a MILP solver that integrates techniques from constraint programming with those described here is also under development and due out soon [1].

On the whole, we were impressed by the vast array of packages and auxiliary tools available, as well as the wide variety of features exhibited by these codes. The most significant features still missing in open codes are effective logical preprocessors and primal heuristics. More effort is needed in developing tools to fill this gap. Although noncommercial codes will most likely continue to lag behind commercial codes in terms of raw speed in solving generic MILPs out of the box, they generally exhibit a much greater degree of flexibility and extensibility. This is especially true of the solver frameworks, which are designed specifically to allow the development of customized solvers. A number of features that appear in noncommercial codes, such as parallelism, the ability to support column generation, and the ability to solve multi-criteria MILPs, simply do not exist in most commercial codes. Although the noncommercial codes are in general slower than the best commercial codes, we believe that many users will be genuinely satisfied with the features and performance of the codes reviewed here and we look forward to future developments in this fast-growing area of software development.

Acknowledgment

The authors would like to thank Kjell Eikland, John Forrest, Matthew Galati, Lou Hafer, and Matthew Saltzman for their insightful comments that helped to improve the chapter.

References

1. Achterberg, T., SCIP — a framework to integrate constraint and mixed integer programming, Technical Report ZIB-Report 04-19, Konrad-Zuse-Zentrum für Informationstechnik Berlin, Takustr. 7, Berlin, 2005.
2. Achterberg, T., Koch, T., and Martin, A., Branching rules revisited, *Operations Research Letters*, 33, 42, 2004.
3. Applegate, D., Bixby, R., Cook, W., and Chvátal, V., 1996, Personal communication.
4. Atamtürk, A., Nemhauser, G., and Savelsbergh, M. W. P., Conflict graphs in integer programming, Technical Report LEC-98-03, Georgia Institute of Technology, 1998.
5. Atamtürk, A., Nemhauser, G., and Savelsbergh, M. W. P., Conflict graphs in solving integer programming problems, *European J. Operational Research*, 121, 40, 2000.
6. Atamtürk, A. and Savelsbergh, M., Integer-programming software systems, Annals of Operations Research, 2005, To appear, Available from http://www.isye.gatech.edu/faculty/Martin_Savelsbergh/publications/ipsoftware-final.pdf.
7. Balas, E., Facets of the knapsack polytope, *Mathematical Programming*, 8, 146, 1975.
8. Balas, E., Ceria, S., Cornuejols, G., and Natraj, N., Gomory cuts revisited, *Operations Research Letters*, 19, 1, 1999.
9. Balas, E. and Martin, R., Pivot and complement: a heuristic for 0-1 programming, *Management Science*, 26, 86, 1980.
10. Balas, E. and Saltzman, M., An algorithm for the three-index assignment problem, *Operations Research*, 39, 150, 1991.
11. Balas, E., Schmieta, S., and Wallace, C., Pivot and shift — A mixed integer programming heuristic, *Discrete Optimization*, 1, 3, 2004.
12. Beale, E. M. L., Branch and bound methods for mathematical programming systems, in *Discrete Optimization II*, edited by Hammer, P. L., Johnson, E. L., and Korte, B. H., pages 201–219, North Holland Publishing Co., 1979.
13. Beale, E. W. L. and Tomlin, J. A., Special facilities in a general mathematical programming system for nonconvex problems using ordered sets of variables, in *Proceedings of the 5th International Conference on Operations Research*, edited by Lawrence, J., pages 447–454, 1969.
14. Bénichou, M. et al., Experiments in mixed-integer linear programming, *Mathematical Programming*, 1, 76, 1971.
15. Berkelaar, M., lp_solve 5.1, 2004, Available from http://groups.yahoo.com/group/lp_solve/.
16. Bixby, R. E., Ceria, S., McZeal, C. M., and Savelsbergh, M. W. P., An updated mixed integer programming library: MIPLIB 3.0, *Optima*, 58, 12, 1998.
17. Borndörfer, R. and Weismantel, R., Set packing relaxations of some integer programs, *Mathematical Programming*, 88, 425, 2000.
18. Chvátal, V., *Linear Programming*, W. H. Freeman and Co., New York, 1983.
19. COIN-OR: Computational Infrastructure for Operations Research, 2004, http://www.coin-or.org.
20. Colombani, Y. and Heipcke, S., Mosel: An extensible environment for modeling and programming solutions, in *Proceedings of the Fourth International Workshop*

on Integration of AI and OR Techniques in Constraint Programming for Combinatorial Optimisation Problems (CP-AI-OR'02), edited by Jussien, N. and Laburthe, F., pages 277–290, 2002.

21. Czyzyk, J., Mesnier, M., and Moré, J., The NEOS server, *IEEE Journal on Computational Science and Engineering*, 5, 68, 1998.

22. Danna, E., Rothberg, E., and LePape, C., Exploring relaxation induced neighborhoods to improve MIP solutions, *Mathematical Programming*, 2004, To appear.

23. Dolan, E., Fourer, R., Moré, J., and Munson, T., Optimization on the NEOS server, *SIAM News*, 35, 8, 2002.

24. Dolan, E., Moré, J., and Munson, T., Optimality measures for performance profiles, Preprint ANL/MCS-P1155-0504, Mathematics and Computer Science Division, Argonne National Lab, 2004.

25. Dolan, E. and Moré, J., Benchmarking optimization software with performance profiles, *Mathematical Programming*, 91, 201, 2002.

26. Driebeek, N. J., An algorithm for the solution of mixed integer programming problems, *Management Science*, 12, 576, 1966.

27. Eckstein, J., Parallel branch-and-bound methods for mixed integer programming, *SIAM News*, 27, 12, 1994.

28. Edmonds, J., Maximum matching and a polyhedron with 0-1 vertices, *Journal of Research of the National Bureau of Standards*, 69B, 125, 1965.

29. Fischetti, M. and Lodi, A., Local branching, *Mathematical Programming*, 98, 23, 2002.

30. Forrest, J., CBC, 2004, Available from http://www.coin-or.org/.

31. Forrest, J. J. H., Hirst, J. P. H., and Tomlin, J. A., Practical solution of large scale mixed integer programming problems with UMPIRE, *Management Science*, 20, 736, 1974.

32. Fourer, R., Lopes, L., and Martin, K., LPFML: A W3C XML schema for linear programming, 2004, Available from http://www.optimization-online.org/DB_HTML/2004/02/817.html.

33. Fourer, R., Gay, D. M., and Kernighan, B. W., *AMPL: A Modeling Language for Mathematical Programming*, The Scientific Press, 1993.

34. Galati, M., COIN-OR tutorials, 2004, Available from http://coral.ie.lehigh.edu/~coin/.

35. Gauthier, J. M. and Ribière, G., Experiments in mixed-integer linear programming using pseudocosts, *Mathematical Programming*, 12, 26, 1977.

36. Gomory, R. E., An algorithm for the mixed integer problem, Technical Report RM-2597, The RAND Corporation, 1960.

37. Grötschel, M., Lovász, L., and Schrijver, A., *Geometric Algorithms and Combinatorial Optimization*, Springer-Verlag, New York, 1988.

38. Gu, Z., Nemhauser, G. L., and Savelsbergh, M. W. P., Cover inequalities for 0-1 linear programs: Computation, *INFORMS Journal on Computing*, 10, 427, 1998.

39. Gu, Z., Nemhauser, G. L., and Savelsbergh, M. W. P., Lifted flow covers for mixed 0-1 integer programs, *Mathematical Programming*, 85, 439, 1999.

40. Gu, Z., Nemhauser, G. L., and Savelsbergh, M. W. P., Sequence independent lifting, *Journal of Combinatorial Optimization*, 4, 109, 2000.

41. Hafer, L., bonsaiG 2.8, 2004, Available from http://www.cs.sfu.ca/~lou/BonsaiG/dwnldreq.html.

42. Hammer, P. L., Johnson, E. L., and Peled, U. N., Facets of regular 0-1 polytopes, *Mathematical Programming*, 8, 179, 1975.

43. Hoffman, K. and Padberg, M., Solving airline crew-scheduling problems by branch-and-cut, *Management Science*, 39, 667, 1993.
44. Hultberg, T., FlopC++, 2004, Available from `http://www.mat.ua.pt/thh/flopc/`.
45. ILOG concert technology, `http://www.ilog.com/products/optimization/tech/concert.cfm`.
46. Jünger, M. and Thienel, S., The ABACUS system for branch and cut and price algorithms in integer programming and combinatorial optimization, *Software Practice and Experience*, 30, 1325, 2001.
47. Ladányi, L., *Parallel Branch and Cut and Its Application to the Traveling Salesman Problem*, PhD thesis, Cornell University, 1996.
48. Ladányi, L., BCP, 2004, Available from `http://www.coin-or.org/`.
49. Linderoth, J. T. and Savelsbergh, M. W. P., A computational study of search strategies in mixed integer programming, *INFORMS Journal on Computing*, 11, 173, 1999.
50. Linderoth, J., MIP instances, 2004, Available from `http://coral.ie.lehigh.edu/mip-instances`.
51. Lougee-Heimer, R., The Common Optimization INterface for Operations Research, *IBM Journal of Research and Development*, 47, 57, 2003.
52. Lougee-Heimer, R., Saltzman, M., and Ralphs, T., 'COIN' of the OR Realm, *OR/MS Today*, October, 2004.
53. Makhorin, A., GLPK 4.2, 2004, Available from `http://www.gnu.org/software/glpk/glpk.html`.
54. Marchand, H., *A Study of the Mixed Knapsack Set and Its Use to Solve Mixed Integer Programs*, PhD thesis, Facult'e des SciencesAppliquées, Université Catholique de Louvain, 1998.
55. Marchand, H. and Wolsey, L., Aggregation and mixed integer rounding to solve MIPs, *Operations Research*, 49, 363, 2001.
56. Margot, F., BAC: A BCP based branch-and-cut example, Report RC22799, *IBM Research*, 2004.
57. Martin, A., Achterberg, T., and Koch, T., MIPLIB 2003, Avaiable from `http://miplib.zib.de`.
58. Mitra, G., Investigation of some branch and bound strategies for the solution of mixed integer linear programs, *Mathematical Programming*, 4, 155, 1973.
59. Nediak, M. and Eckstein, J., Pivot, cut, and dive: A heuristic for mixed 0-1 integer programming, Technical Report RUTCOR Research Report RRR 53-2001, Rutgers University, 2001.
60. Nemhauser, G. and Savelsbergh, M., MINTO 3.1, 2004, Available from `http://coral.ie.lehigh.edu/minto/`.
61. Nemhauser, G. and Wolsey, L., A recursive procedure for generating all cuts for 0-1 mixed integer programs, *Mathematical Programming*, 46, 379, 1990.
62. Nemhauser, G. and Wolsey, L. A., *Integer and Combinatorial Optimization*, John Wiley and Sons, New York, 1988.
63. Nemhauser, G. L. and Sigismondi, G., A strong cutting plane/branch-and-bound algorithm for node packing, *Journal of the Operational Research Society*, 43, 443, 1992.
64. Nemhauser, G. L. and Trotter Jr., L. E., Properties of vertex packing and independence system polyhedra, *Mathematical Programming*, 6, 48, 1974.
65. Padberg, M., On the facial structure of set packing polyhedra, *Mathematical Programming*, 5, 199, 1973.

66. Padberg, M., *Linear Optimization and Extensions*, Springer-Verlag, New York, 1995.
67. Ralphs, T. K., Ladányi, L., and Saltzman, M. J., Parallel branch, cut, and price for large-scale discrete optimization, *Mathematical Programming*, 98, 253, 2003.
68. Ralphs, T., *Parallel Branch and Cut for Vehicle Routing*, PhD thesis, Cornell University, 1995.
69. Ralphs, T., SYMPHONY 5.0, 2004, Available from http://www.branchand-cut.org/SYMPHONY/.
70. Ralphs, T., SYMPHONY Version 5.0 User's Manual, Technical Report 04T-020, Lehigh University Industrial and Systems Engineering, 2004, Available from http://www.lehigh.edu/~tkr2/research/pubs.html.
71. Ralphs, T. and Guzelsoy, M., The SYMPHONY callable library for mixed integer programming, in *The Proceedings of the Ninth Conference of the INFORMS Computing Society*, 2005, To appear, Available from http://www.lehigh.edu/~tkr2/research/pubs.html.
72. Ralphs, T., Saltzman, M., and Wiecek, M., An improved algorithm for biobjective integer programming, *Annals of Operations Research*, 2005, To appear, Available from http://www.lehigh.edu/~tkr2/research/pubs.html.
73. Roy, J.-S., PuLP : A linear programming modeler in Python, Available from http://www.jeannot.org/~js/code/index.en.html#PuLP.
74. Savelsbergh, M. W. P., Preprocessing and probing techniques for mixed integer programming problems, *ORSA Journal on Computing*, 6, 445, 1994.
75. Schrage, L. and Wolsey, L. A., Sensitivity analysis for branch and bound linear programming, *Operations Research*, 33, 1008, 1985.
76. Schrijver, A., *Theory of Linear and Integer Programming*, Wiley, Chichester, 1986.
77. Sidebottom, G., Satisfaction of constraints on nonnegative integer arithmetic expressions, Open File Report 1990-15, Alberta Research Council, 6815 8th Street, Calgary, Alberta, CA T2E 7H7, 1990.
78. Thienel, S., ABACUS 2.3, 2004, Available from http://www.informatik.uni-koeln.de/abacus/.
79. Tomlin, J. A., An improved branch-and-bound method for integer programming, *Operations Research*, 19, 1070, 1971.
80. Trick, M. and Guzelsoy, M., Simple walkthrough of building a solver with symphony, 2004, Available from ftp://branchandcut.org/pub/reference/SYMPHONY-Walkthrough.pdf.
81. van Heesch, D., Doxygen documentation system, 2004, Available from http://www.doxygen.org/.
82. Vanderbeck, F., A generic view of Dantzig-Wolfe decomposition in mixed integer programming, Working paper, 2003, Available from http://www.math.u-bordeaux.fr/~fv/papers/dwcPap.pdf.
83. Wolsey, L. A., Faces for a linear inequality in 0-1 variables, Mathematical Programming, 8, 165, 1975.
84. Wolsey, L. A., Valid inequalities for mixed integer programs with generalized and variable upper bound constraints, *Discrete Applied Mathematics*, 25, 251, 1990.
85. Wolsey, L. A., *Integer Programming*, John Wiley and Sons, New York, 1998.
86. Xu, Y., Ralphs, T., Ladányi, L., and Saltzman, M., ALPS: A framework for implementing parallel search algorithms, in *The Proceedings of the Ninth Conference of the INFORMS Computing Society*, 2005, To appear, Available from http://www.lehigh.edu/~tkr2/research/pubs.html.
87. ZIMPL, Available from http://www.zib.de/koch/zimpl/.

Index

Simple generalized flow cover
 inequality, 262
Simplex methods, 96
Size, grouping by, 235, 240, 241,
 242
SKU (stock keeping units), *see* Retail
 assortment planning
Sliding flights, airline, 113
Slot constraints, set packing integer
 programming model, data
 cycle map optimization,
 210–212
Software, 254–299
 benchmarking, 294–298
 branch-and-bound algorithm,
  256–270
  branching, 266–269
  lower bounding methods,
   258–265
  search strategy, 269–270
  upper bounding methods,
   265–266
 case studies, 288–294
  branch and cut, 288–291
  branch and price, 291–294
 decomposition, 97–98
 future trends, 298–299
 generalized assignment problem,
  52
 MILP programs
  ABACUS, 275–277
  BCP, 277–279
  BonsaiG, 279–280
  CBC, 280–282
  GLPK, 282–283
  lp_solve, 283–284
  MINTO, 284–286
  SYMPHONY, 286–288
 user interfaces, 270–273
  black-box solvers, 270–272
  callable libraries, 272–273
  solver frameworks, 273
Solver frameworks, user interfaces,
 273
Solvers
 black-box, 270–272
 user interfaces, 272
SOPLEX, 274
SOS (special ordered sets), 268
Special ordered sets (SOS), 268

Stabilization methods, master
 problem solution, 96–97
Starting solution, search basics, 3
Static search strategies, 269
Steiner tree problem, decomposition
 applications, 105–107
Stock keeping units (SKU), *see* Retail
 assortment planning
Stopping criterion, search basics, 3
Structured separation, 92, 93–94
Subcommutation, data cycle map
 frames, 198, 201, 202–203, 204
Subgradient methods, 97
Subproblem
 branch-and-bound algorithm, 256
 decomposition
  cutting, 92–96
  defined, 58
  synergy between, 59
 formulation for retail assortment
  planning, 236–237
Subtour elimination constraints
 (SECs)
 separating solutions with known
  structure, 93–94
 Traveling Salesman Problem, 64
Successive shortest path (SSP)
 method, airline flights, 127,
 128, 131
Supercommutation, data cycle map
 frames, 198, 201, 202, 203
SYMPHONY, 274, 277
 benchmarking, 296, 297, 298
 branch and cut case study, 288–291
 features of, 255, 286–288

T

Tabu search, 15–16
Tabu tenure, 3
Telemetry frames, *see* Data cycle map
 optimization
Template paradigm, 92, 93
Three-index assignment problem
 branch and price case study,
  291–293
 decomposition applications,
  103–105
Time-space networks, airline
 management, 113, 114–115